Challenges in Estuarine and Coastal Science

Challenges in Estuarine and Coastal Science

Estuarine and Coastal Sciences Association
50th Anniversary Volume

edited by
John Humphreys and Sally Little

Estuarine & Coastal
Sciences Association

Pelagic Publishing | www.pelagicpublishing.com

First published in 2022 by
Pelagic Publishing
PO Box 874
Exeter, EX3 9BR
UK

www.pelagicpublishing.com

Challenges in Estuarine and Coastal Science:
Estuarine and Coastal Sciences Association 50th Anniversary Volume

British Library Cataloguing in Publication Data
A catalogue record for this book is available
from the British Library

ISBN 978-1-78427-285-2 Pbk
ISBN 978-1-78427-286-9 ePub
ISBN 978-1-78427-287-6 PDF

https://doi.org/10.53061/BDIX4458

Cover image: Lismore Lighthouse on Eilean Musdile in the Inner Hebrides off the west
coast of mainland Scotland, taken from the Sound of Mull. © Sally Little

Typeset in Palatino and Myriad by BBR Design, Sheffield

Printed in Wales by Cambrian Printers, The Pensord Group

Contents

Contributors

Amir AghaKouchak is a Professor of Civil and Environmental Engineering at the University of California, Irvine. His research crosses the boundaries between hydrology, climatology, statistics and remote sensing to address global water resource issues. He utilises satellite data and ground-based observations to develop integrated drought, flood and landslide modelling systems.

William Austin is a Professor at the Scottish Oceans Institute, University of St Andrews, and Chairs the Scottish Government's Blue Carbon Forum. His research spans marine environments, with a particular focus in recent times on blue carbon as a nature-based solution for climate change, people and biodiversity.

Natasha Barlow is Associate Professor of Quaternary Environmental Change at the University of Leeds. Her research focuses on how landscapes respond to changing climate and sea level. She utilises core material and proxy-based methods, alongside geophysical techniques, to better understand past and future ice-sheet melt, sea-level rise, sediment mobility and carbon storage.

Richard Barnes was a zoology undergraduate at University College London (1962–65) and a postgraduate at the University of Queensland (1965–67), before becoming a SERC Postdoctoral Research Fellow at the University of Bristol. After a spell at an industrial laboratory on Southampton Water (looking at the effect of hot water discharges), he became what is now termed an Assistant Professor in the Cambridge Zoology Department in 1972 and in one way or another has been in Cambridge ever since. He currently researches the macroecology of seagrass faunas, mainly via honorary positions at Rhodes University (South Africa) and back at the University of Queensland.

Paula Birocchi holds an MSc in Physical Oceanography and is a PhD student at the University of São Paulo, Brazil. She studies physical oceanographic processes, including the hydrodynamic conditions of coastal areas and the continental shelf, with a focus on the dispersion of pollutants and biophysical interactions.

Simon Blott is a geomorphologist and sedimentologist with over 20 years' experience in research and consultancy relating to coastal, estuarine and shallow marine environments. He is a Principal Consultant at Kenneth Pye Associates Ltd, having previously studied at Reading University and Royal Holloway University of London.

Oliver Brooks is an aquatic ecological consultant with a background in marine biology, conservation and marine microplastic research.

Christina Buelow is an ecologist whose research focuses on monitoring and evaluating coastal wetland health at local, regional and global scales. She endeavours to provide managers and policy-makers with information to effectively conserve these ecosystems and the valuable services that they provide.

Annette Burden is a wetland biogeochemist at the UK Centre for Ecology & Hydrology, researching effects of land management and land-use change on carbon storage and pathways in saltmarsh and peatlands. She has a particular focus on restoration of habitat and greenhouse gas emissions, using field experimentation and long-term national-scale monitoring to quantify relationships between ecosystem function and potential climate change mitigation.

Helene Burningham is Professor of Physical Geography at University College London. She is a coastal scientist specialising in the geomorphology and dynamics of shorelines and coastal landforms, and the assessment of forcing of coastal change over different timescales within the Anthropocene.

Henrique Cabral is a Senior Researcher in marine and estuarine ecology at INRAE, in France. He is the director of a research unit dedicated to the study of global change and aquatic ecosystems. Previously, he was a professor at the Universidade de Lisboa, Portugal (1997–2018), and the director of two research centres: MARE (2015–17) and the Centre of Oceanography (2012–14). He has taught courses at several universities worldwide, and supervised more than 80 MSc and 30 PhD students. He has authored or co-authored more than 320 articles in international scientific journals.

Guillaume Corbeau began his biogeography PhD at the University of Nantes (France) in 2020. He works with a geohistorical approach to the connection between human activities and benthic biodiversity in the western English Channel. Using a unique benthic fauna dataset from the Normanno-Breton Gulf, he studies the changes of community composition over the last two centuries and particularly their link with shellfish farming.

Nicolas Desroy is a marine ecologist at Ifremer, Laboratoire Environnement et Ressources Bretagne nord in France.

Marcelo Dottori graduated in physics (Bachelor's degree from IF-USP), holds a Master's degree in physical oceanography from the University of Sao Paulo and a PhD in oceanography from Florida State University. His research has an emphasis on fluid dynamics, working mainly on physical–biological interactions, low-frequency variability, Rossby waves and continental shelf dynamics.

Awa Bousso Dramé is a research student in Geography at University College London, where she is undertaking a PhD on the relative roles of estuarine dynamics, anthropogenic intervention and geopolitical context on the dynamics of cross-border coastal systems in West Africa.

Rachel Dunk is a Principal Lecturer at Manchester Metropolitan University. She has over 20 years' experience researching the chemistry of greenhouse gases and carbon storage and management, with a particular interest in marine carbon stocks and fluxes.

Mike Elliott is Director of International Estuarine & Coastal Specialists (IECS) Ltd and Professor of Estuarine and Coastal Sciences at the University of Hull. He is a marine biologist with wide experience and interests and his teaching, research, advisory and consultancy includes estuarine and marine ecology, policy, governance and management. Mike is a past-President of ECSA and is a Co-Editor-in-Chief of the international journal *Estuarine, Coastal & Shelf Science*; he currently is or has had Adjunct Professor and research positions at several universities and research institutes worldwide. He was awarded Laureate of the Honorary Winberg Medal 2014 of the Russian Hydrobiological Academic Society.

Juliana Correa Neiva Ferreira holds a BSc in biomedicine and MSc in biological oceanography from the University of São Paulo (Brazil). She has studied the microbial ecology of Antarctic environments with a focus on archaeal micro-organisms and bioinformatics.

Louise Firth is a marine ecologist who works in both natural and artificial coastal environments. She is interested in the relationship between humans and coastal ecosystems and how this has changed over time. She is particularly keen to develop novel ways of making space for nature in human-dominated environments.

Anthony Gallagher is a marine scientist and planner working extensively in marine plastics. He is Director of Evolved Ocean, Chair of the Clyde Marine Planning Partnership and a Visiting Professor with the University of Southampton.

Angus Garbutt is a Senior Ecologist at the UK Centre for Ecology & Hydrology. His research has taken him to all the major saltmarsh complexes in the UK, north-west Europe, the Baltic and the Mediterranean, giving him a unique insight into their flora and fauna. His work focuses on habitat management, restoration and long-term, large-scale monitoring.

Roland Gehrels is a Professor of Physical Geography at the University of York, UK. His research aims to unravel the history of late Holocene sea-level changes on regional and global scales, with particular emphasis on linking instrumental and geological sea-level records during the historical period. He utilises microfossils preserved in saltmarsh sediments, in combination with various dating techniques, to produce high-resolution proxy records of sea-level change.

William Glamore is an Associate Professor of Civil and Environmental Engineering at the Water Research Laboratory, UNSW Sydney, Australia. His research team investigates estuarine and tidal wetland dynamics in response to climate change and tidal restoration. He aims to integrate fundamental research within applied outcomes for better on-ground management.

Laurent Godet is a Senior Scientist at the CNRS (Centre national de la recherche scientifique) in France. He works in ecology and geography and focuses on the spatial impacts of anthropisation on biodiversity at large scale, both regional and local. He is mainly interested in coastal areas, birds and marine invertebrates.

John Goss-Custard retired in 2002 from the Natural Environmental Research Council where he had developed individual-based models to predict the effect of environmental change on shorebird populations. He is now a Visiting Professor in the Department of Life and Environmental Sciences at Bournemouth University. He continues with field research and modelling, particularly on resolving conflicts between conservationists and other users of the coast, such as shellfisheries.

Alice Hall is a postdoctoral marine ecologist at Plymouth University who specialises in ecological engineering. Her recent work focuses on using artificial reefs to enhance coastal infrastructure and restore temperate ecosystems.

Glenn Havelock is an Associate Lecturer and Research Associate in the Department of Environment and Geography at the University of York, UK.

Stephen Hawkins has researched the ecology of coastal ecosystems since 1975, using experimental, long-term and broad-scale descriptive approaches to understand structuring of rocky shore communities and the consequences for ecosystem functioning. He also charts long-term responses of marine ecosystems to climate change and other impacts in order to understand their recovery and restoration.

Roger Herbert is Professor of Marine and Coastal Biology at Bournemouth University and has worked on Channel ecosystems for 40 years. His research is broadly focused on the impact of environmental change and approaches to marine conservation. This has included the ecology and management of marine invasive species and the population dynamics of intertidal organisms and their responses to climate change.

Malcolm Hudson is a researcher and academic in environmental science. He directs the interdisciplinary plastic pollution group at the University of Southampton.

John Humphreys is a Visiting Professor at Bournemouth University and former Pro Vice-chancellor at the University of Greenwich, London, for whom his work in Africa won a Queen's Award for Higher Education. He has served as non-executive director of a commercial port and harbour authority, and as chair of a UK Inshore Fisheries and Conservation Authority. His books include *Marine Protected Areas: Science, Policy and Management* (2020).

Jordan Iles is an aquatic scientist interested in biogeochemical processes occurring in freshwater rivers, streams and wetlands. Jordan completed his PhD at the University of Western Australia on intermittent rivers and ephemeral streams, where his research investigated how nutrients and organic matter are utilised and conserved throughout these systems.

Laurence Jones is a group leader at the UK Centre for Ecology & Hydrology and Professor of Geography and Environmental Science at Liverpool Hope University. His research spans three areas: natural capital and ecosystem services, climate change, and pollution. His work focuses on ecology and biogeochemical cycling in coastal systems, and more widely on the interactions between people and the environment.

Danial Khojasteh is a Scientia PhD student at the UNSW Sydney Water Research Laboratory and has over 10 years' experience in fluid mechanics and hydrodynamic numerical modelling studies. Danial's research aims to improve evidence-based management plans and decision-making for tidal rivers and estuaries under climate-driven sea-level rise.

Cai Ladd is a Postdoctoral Research Associate in Biogeomorphology at the University of Glasgow, Scotland. His research examines how physical and biological processes interact across local and regional scales to shape coastal wetlands and the communities that rely on them. He utilises hydrological monitoring, plant-soil analyses and geographic information systems to assess the value and vulnerability of the world's saltmarshes and mangrove forests.

Jonathan P. Lewis is a Visiting Research Fellow in the Department of Geography and Environment at Loughborough University and a catchment manager for Trent Rivers Trust. Jonathan specialises in coastal environmental change and whole-catchment aquatic conservation.

Sally Little is President of the Estuarine and Coastal Sciences Association and a Senior Lecturer in Environmental Science at Nottingham Trent University. Sally is an estuarine ecologist, specialising in assessing the impact of environmental, climatic and other anthropogenic pressures on the benthic ecology of upper estuarine tidal freshwater zones.

Donald McLusky was formerly a Senior Lecturer in Marine Biology, and Head of the Department of Biological and Molecular Sciences at the University of Stirling. His estuarine studies began on the Ythan estuary, Aberdeenshire and continued at Stirling on the Forth estuary. He particularly studied the impact of waste discharges on the intertidal areas adjacent to the oil refinery at Grangemouth. He is the author of textbooks on estuaries, including *Ecology of Estuaries* and *The Estuarine Ecosystem*, now in its third edition. He served as Secretary and Bulletin Editor of ECSA, and was ECSA's editor of *Estuarine Coastal and Shelf Science*. In 2015 he received a Lifetime Achievement award from ECSA.

Lucy McMahon is a PhD student in the Department of Environment and Geography at the University of York, UK. Her current research explores the major environmental drivers that influence spatial and temporal variation of blue carbon in coastal wetlands. She utilises plant-soil analyses and social science techniques to inform the management and restoration of saltmarshes as a nature-based solution to climate change.

Krysia Mazik is a Lecturer in marine biology in the Department of Biological and Marine Sciences at the University of Hull. She specialises in the structural and functional ecology of marine and estuarine benthic communities.

Nova Mieszkowska is a marine physiological ecologist who applies an omics to ecosystems approach to study the impacts of multiple stressors and mechanistic responses of coastal marine species. Her biogeographical time-series studies have shown some of the fastest responses to climate change in any natural system.

Robert Mills is a Lecturer (Assistant Professor) of Environmental Science in the Department of Environment & Geography at the University of York, UK.

Steven Benjamin Mitchell obtained a BSc Hons. in Civil Engineering from the University of Durham, UK and then an MSc and PhD in Coastal Engineering at the University of Birmingham. He has undertaken fieldwork and modelling in the UK, New Zealand, France and Ghana, and has published more than 60 papers and book chapters. He currently serves as Editor-in-Chief for the *Estuarine Coastal and Shelf Science Journal* and is a Fellow of the Institution of Civil Engineers. Since 2010 he has been a Lecturer in the School of Civil Engineering and Surveying at the University of Portsmouth. He has a particular interest in hydraulics and sediment transport in macrotidal systems.

Hamed Moftakhari is an Assistant Professor at the Department of Civil, Construction, and Environmental Engineering at the University of Alabama. His research interests are in the area of coastal hydrology, and involve multi-hazard risk analysis and integrated coastal/estuarine hydrodynamic modelling.

Hannah Mossman is an Applied Ecologist at Manchester Metropolitan University. She has a particular interest in studying the development of saltmarshes restored by managed realignment. Her research seeks to understand factors constraining the development of restored marshes and to identify management actions to improve restoration success.

David M. Paterson holds a personal chair in Coastal Ecology at the University of St Andrews and is Executive Director of the Marine Alliance for Science and Technology for Scotland. He also Chairs the Sullom Voe Oil Terminal Environmental Advisory Group. He has 25 years of experience in research on marine systems and held a Royal Society University Research Fellowship at Bristol before moving to St Andrews. He established the Sediment Ecology Research Group and continues his work on the dynamics and ecology of coastal systems.

Helen Pietkiewicz is a PhD research student in the School of Animal, Rural and Environmental Science at Nottingham Trent University. Her research focuses on the impact of estuarine squeeze on the structure and functioning of tidal freshwater zones.

Silvia Palotti Polizel is a research student in Geography at University College London, where she is pursuing a PhD on changes in the morphosedimentary behaviour of Brazilian delta shorelines during the Anthropocene.

Nigel Pontee is a geomorphologist with over 28 years of experience in coastal management, including shoreline management plans and strategies and numerous coastal habitat-creation projects. He is the Global Technology Leader for Coastal Planning and Engineering at Jacobs and a visiting professor at the National Oceanography Centre in the UK.

Kenneth (Ken) Pye is a geomorphologist and sedimentologist with more than 40 years' research and consultancy experience relating to coastal, estuarine and shallow marine environments. He is the Director of Kenneth Pye Associates Ltd and previously held academic appointments at the Universities of Cambridge, Reading and Royal Holloway University of London.

Stuart Rae is a Postdoctoral Research Associate at Manchester Metropolitan University, specialising in environmental geochemistry. His research focuses on the analysis and measurement of greenhouse gas fluxes and carbon storage within wetland ecosystems.

Jamie Ruprecht is based at the UNSW Sydney Water Research Laboratory, and has more than 10 years of experience in field, laboratory and numerical modelling studies, focused on aquatic ecosystem restoration and improving waterway management. He is currently completing a PhD investigating the biophysical linkages in rivers and estuaries.

Paulina Ruranska is a PhD student at the School of Geography & Sustainable Development and the School of Chemistry at the University of St Andrews. Her research project addresses the problem of climate change in the context of carbon sequestration in British saltmarshes. She performs organic carbon and organic matter calculations on field-collected samples and uses solid-state NMR spectroscopy to derive information on the chemical form and structure of the carbon within soil profiles.

Martin W. Skov is a Reader in marine ecology at Bangor University, UK. His research on coastal wetlands focuses on landscape-scale biophysical processes underpinning ecosystem service delivery of flood protection, carbon storage, biodiversity functioning and human wellbeing. Martin helped initiate the world's first blue carbon-trading project, Mikoko Pamoja, in Kenya (www.aces-org.co.uk).

Craig Smeaton is a Research Fellow at the University of St Andrews. His research focuses on sedimentary carbon dynamics across the land–ocean interface. He utilises biogeochemical and sedimentological approaches alongside geophysical and geospatial techniques to better understand the drivers of carbon storage and burial in intertidal and subtidal sedimentary environments.

Robert Sparkes is an organic geochemist at Manchester Metropolitan University. Using an array of analytical tools, he studies the movement of organic carbon through the Earth system, and its potential for long-term storage in marine systems.

Jessica Stead is a PhD researcher working on understanding the fate and transport of microplastics in estuaries.

Richard Stillman is an applied ecologist with an interest in predicting how environmental change and management influence animal populations. His research aims to advise policy-makers, conservationists and industry on the best ways of reconciling the interests of wildlife with those of humans. His main study species have been wading birds; he conducts research on foraging behaviour and decision-making to search for general rules in the ways that individual animals interact with each other and their environment. His applied research incorporates these general rules into computer individual-based models to predict how environmental change influences animal populations.

Martin Sullivan is an Ecologist at Manchester Metropolitan University. His research investigates how species and ecosystems respond to environmental change, in study systems ranging from tropical forests to saltmarshes.

James Tempest is a geomorphologist with over eight years' experience in coastal research and consultancy roles, specialising in intertidal habitat creation. He is currently a consultant at Jacobs, having previously studied and held research roles at Cambridge University and Queen Mary University of London.

Reginald James Uncles is an Associate of the Royal College of Science and a Fellow of the Institute of Physics. He is a past-President of ECSA and has led research teams at the Plymouth Marine Laboratory and served on its senior management team. He is currently a Research Fellow at the Plymouth Marine Laboratory, where his research includes theoretical and experimental work on physical processes in estuaries. Special areas of interest include estuarine hydrodynamics and the mechanisms responsible for fine sediment accumulation and saltwater intrusion in the upper reaches of estuaries. He has worked overseas for extended periods of time in Australia, Africa, Argentina, Germany, Malaysia and the USA on problems concerned with estuaries, publishing more than 165 articles in international research journals, books, and bulletins, in addition to authoring more than 50 reports and book reviews.

Katrina Waddington has over 25 years' engineering experience, consulting in flood risk, land and water management plans, urban drainage, water and wastewater treatment, river rehabilitation and environmental impact assessment. She is currently undertaking a PhD at the UNSW Water Research Laboratory, Sydney, investigating the impact of sea-level rise in estuaries.

Sophie Walker is currently enrolled as a PhD student at James Cook University, Australia, where her project centres around addressing barriers to the uptake new technology for more effective management of coastal environments. She hopes to solve those barriers by investigating ways in which broad- and fine-scale habitat complexity in coastal systems can be mapped and quantified.

Nathan Waltham is an aquatic ecologist with a deep interest in coastal landscape ecology and urbanisation, having worked in local government for 13 years. Nathan is currently a Lecturer in coastal aquatic science and management with James Cook University Australia, where his research takes a transformative approach to solving coastal conservation and protection issues with the increasing challenges of urbanisation.

Martin Wilkinson is Emeritus Professor of Marine Biology at Heriot-Watt University, Edinburgh, and Treasurer of ECSA. Initially studying taxonomy and ecology of seaweeds, he has extended this to long-term studies of macroalgae in estuaries and the use of seaweeds in environmental assessment. After 50 years of studying seaweed assemblages on various British shores he is now examining long-term change both anthropogenic and natural.

Alan Whitfield is Emeritus Chief Scientist at the South African Institute for Aquatic Biodiversity and also an Honorary Professor at the Department of Ichthyology & Fisheries Science at Rhodes University in the Eastern Cape Province of South Africa. His speciality is estuarine ecology, with wide experience and interests in research, teaching and the promotion of wise management and conservation of fishes in estuaries. Alan has published widely and serves on a number of journal editorial boards, including that of Associate Editor on *Estuarine, Coastal & Shelf Science*. In 2008 he received the prestigious Gilchrist Memorial Medal for research contributions to South African marine and coastal science, and in 2015 was the recipient of the Southern African Society of Aquatic Scientists Gold Medal for contributions to aquatic research in Africa.

Preface

JOHN HUMPHREYS and SALLY LITTLE

In the early 1970s that part of the earth where land and sea interact most intensely was defined by Ketchum (1972) among others as the coastal zone, the aquatic component of which is made up of a heterogeneous band extending from the upper reaches of estuaries to the edge of the continental shelf. For the field of estuarine and coastal science, the 1970s marked an emergent appreciation that this 'coastal realm' (Pielou 1979) exhibited features and functions different from either land or sea alone, where production, consumption and exchange processes occur at higher rates of intensity. It has, for example, been estimated that estuaries and coastal seas, while making up 8% of ocean surface, are responsible for about 50% of global denitrification (Ray and McCormick-Ray 2014).

One manifestation of this growing awareness of the unique features of such environments was the establishment, in October 1971, of what is now the Estuarine and Coastal Sciences Association (ECSA). The inaugural symposium in London on *The Estuarine Environment* was followed in 1972 by a volume of proceedings with the same title (Barnes and Green 1972). Considering how best to celebrate the 50th anniversary of the Association, the current Council determined to follow the lead of their antecedents in terms of both a symposium and proceedings. Unfortunately, the Coronavirus pandemic, with its tragic consequences

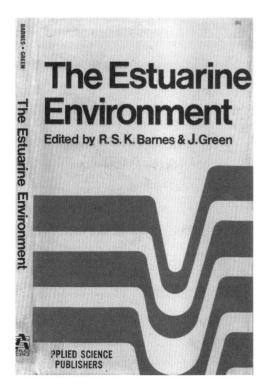

Figure 1 Cover of the 1972 proceedings of the inaugural symposium of the Association.

across the world, intervened to curtail our ambitions. Consequently, our celebrations have been postponed and will be integrated into one of our forthcoming international conferences, with a small celebratory event in London in 2022. However, it was decided to go ahead with an anniversary volume, not in the form of a proceedings, but instead on the basis of a call for contributions. This book is the result.

Our intention in this collection has been to cover a representative selection of current issues, illustrated by the activities and preoccupations of academic and practitioner scientists working in the field today. A glance at our contents pages shows that the book considers a wide range of the major current challenges in estuarine and coastal science, exemplified in various geographical contexts that illustrate the international scope of our members' interests.

The 1972 book contents (Table P1) reveal a number of pure science research programmes that helped to establish the parameters of the subject. At its most basic level the introductory chapter (Stewart) notes at least seven different definitions of estuary. In fact, most of the 1972 chapters grapple with this question and tend to emphasise the variable salinity and chemistry of waters of mixed origin. An early chapter reports classic work on categories of estuary on the basis of salinity profiles and mixing (Dyer). Several chapters consider the physiological implications of this directly (Beadle on animals and Jefferies on plants), and these are followed by a chapter reporting a first synthesis of early research on faunal productivity and trophic relations in the Ythan (Milne and Dunnet). These chapters are essentially reporting pure science investigations of the estuarine environment with little reference to human activity, in contrast to the two final chapters on fisheries (Walne) and water quality (Barrett).

In our own concluding chapter, we have taken the opportunity to contrast the content of the ECSA inaugural and anniversary volumes, along with some additional literature, to reflect on the development of the field over the last 50 years and the challenges ahead. In this respect, readers will not be surprised to know that whereas none of the chapters in the 1972 volume mentioned climate change, all of the chapters in this volume do. Reflection on the development of the subject over the half-century is also facilitated by a specially commissioned foreword, which outlines the history of the Association with reference *inter alia* to its name changes and its linked journals, along with the many diverse and changing themes of ECSA conferences since its inauguration.

As editors we must thank many people for their contribution in bringing this volume to fruition. We are delighted to have a prologue by Richard Barnes, a founder member who describes our 'pre-history', when the idea of the Association coalesced by a brackish lagoon. He also co-edited the original 1972 volume. Similarly, we are particularly grateful to Martin

Table P1 Chapter contents in the 1972 book: *The Estuarine Environment*

1.	Estuarine and Brackish Waters – an Introduction	W.D.P. Stewart
2.	Sedimentation in Estuaries	K.R. Dyer
3.	Chemical Processes in Estuaries	J. Phillips
4.	Physiological Problems for Animal Life in Estuaries	L.C. Beadle
5.	Aspects of Salt Marsh Ecology with Particular Reference to Inorganic Plant Nutrition	R.L. Jefferies
6.	Standing Crop, Productivity and Trophic Relations of the Fauna of the Ythan Estuary	H. Milne and G.M. Dunnet
7.	The Importance of Estuaries to Commercial Fisheries	P.R. Walne
8.	The Effects of Pollution on the Thames Estuary	M.J. Barrett
9.	Summarising Review	D.R. Arthur

Table P2 List of current ECSA council members and council positions held

Name	Position
Dr Sally Little	President
Prof. John Humphreys	President-Elect
Dr Gillian Glegg	Honorary Secretary
Prof. Martin Wilkinson	Honorary Treasurer
Dr Clare Scanlan	Membership Secretary
Dr Tim Jennerjahn	Conference Coordinator
Mr Andrew Wither	Focus Meetings Coordinator
Dr Jonathan Dale	Communications Officer
Dr Anita Franco	E-Newsletter Editor
Prof. Teresa Fernandes	Workshop Coordinator
Dr Steve Mitchell	Publications Secretary
Dr Ross Brown	Webmaster
Prof. Axel Miller	Sponsor Representative (Scottish Association for Marine Science)
Dr Mike Best	Sponsor Representative (Environment Agency)
Mr John Baugh	Sponsor Representative (HR Wallingford)
Prof. Patrick Meire	Council Member
Dr Alastair Lyndon	Council Member
Dr Lysann Schneider	Council Member
Dr Lois Calder	Council Member
Dr Dan Harries	Honorary Auditor
Prof. Geoff Millward	ECSA Trustee
Dr Donald McLusky	ECSA Trustee
Dr Reg Uncles	ECSA Trustee

Wilkinson and Donald McLusky who not only wrote our history chapter but also lived through it! We also thank all our current ECSA Council members (Table P2) who supported the project, not least Axel Miller under whose tenure as president this anniversary volume was conceived. Many thanks also to our publisher Nigel Massen and production editor David Hawkins both of Pelagic Publishing. Finally, and above all, we must thank our authors for contributing such a stimulating and wide-ranging collection of chapters.

References

Barnes, R.S.K. and Green, J. (1972) *The Estuarine Environment.* London: Applied Science Publishers Ltd.

Ketchum, B.H. (ed.) (1972) *The Water's Edge: Critical Problems of the Coastal Zone.* Cambridge, MA: MIT Press.

Pielou, E.C. (1979) *Biogeography.* New York: John Wiley & Sons.

Ray, G.C. and McCormick-Ray, J. (2014) *Marine Conservation: Science, Policy and Management.* Chichester: Wiley & Sons.

Prologue

RICHARD S.K. BARNES

Before embarking on celebratory or other significant activities, it is the custom among various Australian aboriginal peoples to pay their respects to those who brought their society into being. Perhaps then we might do likewise and place a metaphorical blue plaque on the 7 ha reed-fringed pool located at 50°08'34"N, 05°04'42"W on (what was then) the western fringe of the Cornish town of Falmouth, to acknowledge the seminal role that it played in the origin of our association.

In 1968, a junior lecturer and two postdocs from Bristol University's Department of Zoology set out in the departmental VW Dormobile to have another look at a peculiar body of water 'discovered' during an earlier field course. Lynyeyn Pryskelow as it was originally known (in the seventeenth century renamed Swanpool by the non Kernewek-speaking emmets) was dominated by a curious mixture of animals: insects (*Sigara*, *Corixa*, *Notonecta*, *Noterus*, etc.), non-freshwater crustaceans (*Neomysis*, *Palaemonetes* and an uncommon *Gammarus*, *G. chevreuxi*), the microgastropod then known as *Hydrobia jenkinsi*, a goby *Pomatoschistus* and the bryozoans *Victorella pavida* and *Plumatella repens*; indeed, it proved to be the UK's only known locality for living *Victorella*. Like many other coastal water bodies in south-west England, in origin Swanpool was a small arm of the sea cut off by the deposition of a shingle bar across its mouth during post-glacial sea-level rise, and thereafter a freshwater lake because of inflowing streams. In 1826, however, to reduce its size (to some 33% of the original), it was decided to lower the lake's level by constructing an outlet culvert (hitherto water had simply percolated through the shingle barrier). This was done, but at a height some 0.25 m below that of mean high water of spring tide in the adjacent sea, with the result that while freshwater did indeed flow out for most of the time, during spring tides seawater flowed in instead. A brackish lagoon was (presumably unwittingly) created, with a highly unusual oligohaline salinity regime including a distinct halocline separating the lower seawater mass from the surface freshwater.

We were none of us trained ecologists (a cell biologist and two physiologists, albeit with interests in osmotic and ionic regulation) but we found this situation fascinating. The pool was, however, apparently too salty to fall within the body of knowledge of freshwater biologists but too fresh for the likes of their marine counterparts. As we investigated this no-man's-water (and others like it) in greater detail (thereby conducting surely one of the very few studies to be published by both the *Journal of Animal Ecology* and *Studhyansow Kernewek*!), so we felt that if these brackish lagoons and equivalent water bodies fell outside the remit of other habitat-based scientific societies and associations then they merited their own parallel organisation and research outlet. So we set up a Committee for the Establishment of a Brackish-Water and Estuarine Biological Association (CEBEBA) with a few of our like-minded colleagues. Perhaps there was something in the air at the time (it was the late 1960s, after all!): monographs devoted to brackish systems were then beginning to appear, including such gems as Bent Muus's 1967 *The Fauna of Danish Estuaries and Lagoons* (Copenhagen: Høst), Jim Green's 1968 *The Biology of Estuarine Animals* (London: Sidgwick & Jackson), Kenneth Emery's 1969 account of Oyster Pond near the other town of Falmouth in Massachusetts, *A Coastal Pond Studied by Oceanographic Methods* (New York: Elsevier) and Adolf Remane and Carl Schlieper's 1971 English edition of their earlier *Die Biologie des Brackwassers* (Stuttgart: Schweizerbart'sche). In any event, on 13 October 1971 we did manage formally to establish such an association, followed 15 months later by a journal, *Estuarine and Coastal Marine Science* (also, of course, later to change its name), and during the association's juvenile and early teenage years one of those two original Swanpool postdocs was its secretary and the other its treasurer (although both by now no longer hand-to-mouth Research Fellows), while Jim Green (one of the CEBEBA) served a term as president.

And now 50 years later, the embryo conceived in Swanpool and hatched in the rooms of the Zoological Society of London has matured into a successful adult (perhaps via EBBA and EBSA larval stages?), and we who remain from the halcyon days of conception and birth – not many now, I fear (and perhaps that period is more halcyon in recollection than it was at the time) – have good reason to thank those who came later for the success that they have made of our initial dreams and hopes.

Meanwhile, life has changed almost beyond recognition. No more hours on the typewriter to produce a manuscript, or days addressing envelopes by hand to send out material to EBSA members. No more being stuck on the bench with a burette, pipette or Lovibond comparator. We can now do in a split second on the laptop what used to take hours (if not longer) on the university mainframe. And we have cell phones, GPS, digital photographs and books, Google Earth, search engines, multivariate statistical software and now a vast volume of estuarine and brackish-water literature. Then and now could be on different planets – apart, of course, from David Attenborough, Mick Jagger and Elizabeth Windsor; and it still takes a long time to drive to Falmouth where Swanpool survives as (from 1995) an SSSI and Local Nature Reserve, and, according to *Cornwall Today*, 'an oasis of calm and tranquillity'.

So, next it'll be the centenary I suppose; might have to miss out on that one, though. Hopefully, however, some remnants of our planet's natural estuarine and other brackish-water habitats will survive through to 2071, and hopefully there will still be an ECSA (or whatever new acronym will have replaced it) that has continued to go from strength to strength.

Cambridge (UK), 27 November 2021

The Origins and History of the Estuarine and Coastal Sciences Association

MARTIN WILKINSON and DONALD McLUSKY

Abstract

The Estuarine and Coastal Sciences Association (ECSA) was established in 1971. Following a large attendance from many disciplines at the inaugural symposium, the intended name was changed from Estuarine and Brackish-Water Biological Association to Estuarine and Brackish-Water Sciences Association (EBSA). EBSA changed its name to ECSA in 1989, reflecting the wider coastal breadth of our activities. The purpose of ECSA is to promote and support research and education in our estuaries and coastal waters, and the introductory meeting proposed that close contact with industry be maintained. These aims are achieved through the organisation of scientific meetings and the publication of the proceedings of those meetings, as well as production of practical handbooks. The meetings range in scope from international symposia through to meetings that focus on a single estuary; there has also been a wide range of workshops both on policy discussion and practical techniques. The Association has also developed the provision of research grants to enable estuarine research to be undertaken and to support attendance at scientific meetings. It has close links to the academic journals *Estuarine, Coastal and Shelf Science* and *Ocean & Coastal Management*. Initially, ECSA was focused on the estuarine environment of the British Isles, but it has expanded its remit so that it now encompasses the estuarine (or transitional waters) environment throughout Europe as well as Africa, Asia and Australasia. ECSA has always encouraged an interdisciplinary approach to the study of estuaries and coasts, ranging from the biological through to physical and chemical studies. It has also fostered links between the academic world of research and the world of applications to estuarine management such as pollution control.

Correspondence: martin.wilkinson8@btopenworld.com

Introduction: what triggered the development of EBSA/ECSA?

There were a number of factors at play in the world of aquatic science in the late 1960s that led to the idea that a scientific society was needed to enhance the study of estuaries and brackish-water areas.

The aquatic science world of Britain of the first half of the twentieth century and through to the 1970s was firmly divided into two camps. The freshwater group was centred on the Freshwater Biological Association with an established laboratory at Windermere. The marine group was centred on the Marine Biological Association and the Scottish Marine

Martin Wilkinson and Donald McLusky, 'The Origins and History of the Estuarine and Coastal Sciences Association' in: *Challenges in Estuarine and Coastal Science*. Pelagic Publishing (2022). © Martin Wilkinson and Donald McLusky. DOI: 10.53061/LRJC3469

Biological Association with laboratories at Plymouth and Millport (soon to move to Oban), as well as the large government fishery laboratories at Lowestoft and Aberdeen. In addition, there were several university-led marine laboratories at Cullercoats, Menai Bridge, Port Erin and elsewhere. None of these largely government funded institutions considered estuaries and brackish waters to be within its remit.

The polluted state of rivers (and estuaries) in Britain has been reported on for over two centuries, although for far too long little remedial action was undertaken. The Royal Commission on Environmental Pollution of the 1930s instigated studies of the estuaries of the River Tay and River Tees in an attempt to understand the causes and effects of aquatic pollution. The reports of these studies, comparing the 'clean' Tay estuary with the 'polluted' Tees estuary, formed the basis for legislation aimed at controlling the severe effects of aquatic organic and industrial pollution, which was finally introduced in 1951. The legislation created the River Purification Boards in Scotland and the River Authorities in England. In 1965 their powers were extended, but it was not until 1974 that the Control of Pollution Act established the basic provisions for defining pollution offences. Subsequent legislation, such as the Water Framework Directive of 2000 has further strengthened pollution control in estuaries and on coasts.

In order to understand the effects of pollution it was essential to grasp what was the "normal" situation in estuaries, so there was a variety of studies undertaken in the late 1950s and the 1960s around the developed world aimed at developing an understanding of the basic ecology of estuaries and brackish waters. These included Remane and Schlieper's *Die biologie des brackwassers* (1958), Muss's *The Fauna of Danish Estuaries and Lagoons* (1967), Lauff's *Estuaries* (1967) and Green's *The Biology of Estuarine Animals* (1968).

Seven of the ten largest cities in the world are situated adjacent to estuaries, and in Britain many of the main cities border an estuary. The estuarine environment is thus accessible for study to many students, and within an estuary it is possible to cover a wide range of scientific subjects including biology, chemistry and physics. The 1960s saw a major expansion of universities in Britain, including both new universities as well as the expansion of existing universities. As the number of students expanded there was a demand for suitable research projects for undergraduates, postgraduates and staff, so inevitably many looked to their local estuaries. Here they found willing partnerships with pollution control authorities who were trying to implement pollution control measures but who often lacked basic knowledge of the state of the estuarine environment.

The development triggers for EBSA were thus mainly concerned with the apparent neglect of the study of estuaries and brackish waters by the existing established marine and freshwater science laboratories. This was coupled with the demand by the relatively new pollution control authorities for better information on the areas that they had to administer and the willingness of the expanding university sector to undertake research in these areas. Thus, right from the start EBSA/ECSA has seen collaboration between fundamental science and environmental management authorities as being the basis of its existence, and its conferences and publications have always reflected this collaboration. Naylor (2002) reviewed the history of ECSA during its first 30 years and the present review seeks to look at the entire 50-year span of ECSA.

Origins of ECSA

The society that we now call ECSA was inaugurated as the Estuarine and Brackish-Water Biological Association (EBBA) during a special inaugural symposium to celebrate the occasion at the Zoological Society of London on 13 October 1971. This was the culmination of about a year of planning. Fortunately the first EBBA minute book includes minutes of the committee for the establishment of the association, which met at least five times before the inaugural symposium in 1971, and of the discussion with the amazing total of close to 300 people who attended the inaugural symposium, who discussed, during a break in the papers, what they expected from the new association.

Colin Little and Richard Barnes recounted the background to this inaugural meeting in the ECSA Bulletin in 2019. The group of founders was mainly made up of young marine zoologists who had begun a study of The Swanpool at Falmouth. Initially they were going to call the prospective society the Brackish-Water Biological Association, but this quickly changed to Estuarine and Brackish-Water Biological Association. Within six months of the inauguration, it was to change again at the first AGM held in 1972 to Estuarine and Brackish-Water Sciences Association (EBSA).

The establishment committee was chaired by an older, very experienced worker, Ronald Bassindale, from Bristol University Zoology Department. In the context of ECSA he was best known for the monumental work by Alexander, Southgate and Bassindale on the Tees estuary published in 1935 by UK Government's Department of Scientific and Industrial Research. He was then involved in further similar publications in a series on the Mersey, Thames and Tees estuaries. This was the start of the UK government looking to set up aquatic pollution management leading to the formation of the English and Welsh River Authorities and the Scottish River Purification Boards, but the Second World War delayed things. It was clear from the start that the society was being planned to bring together high standards of pure marine biology with practical environmental management.

The founding committee appointed three acting officials in May 1971 to run the fledgling association until it could have its first annual general meeting in April 1972. These were the chair Ronald Bassindale; secretary Richard Barnes; treasurer Colin Little. The minutes of the committee, which are in the ECSA archive, showed great attention to detail, and this gave the association a good start. Considerable publicity was given to the inaugural symposium, which was supported by the Fishmongers' Company, ICI Ltd, Unilever Ltd, Esso Petroleum Ltd and the Zoological Society of London. As a special deal to accumulate reserve funds for the initially penniless association, the close to 300 people at the inaugural meeting were offered life membership for only £15. This life membership option was only at this meeting and never repeated. The association protected this life membership income as a reserve and lived off the subscriptions of annual members.

The inaugural meeting had a short, general programme of eight talks using selected, invited speakers to introduce the estuarine environment, to consider its importance and how we abuse it. The papers from the inaugural symposium were published by Applied Science Publishers as a 140-page volume, *The Estuarine Environment*, edited by Barnes and Green (1972).

EBSA was renamed the Estuarine and Coastal Sciences Association (ECSA) in 1989 as a reflection of the growth of the subject area and the widening scope of the association, and also to match more closely its journal (*Estuarine Coastal and Shelf Science*). As Little and Barnes (2019) commented, it is one of life's little ironies that a body originally founded because of a particular interest in non-estuarine (i.e. lagoonal) brackish waters dropped the term 'brackish water' from its name, retaining only 'estuarine'.

Meetings

Over the past 50 years, EBSA/ECSA has organised 183 meetings, as shown in Table F1. From the beginning, most meetings were held in the UK. Meetings in the first years were either collaborative meetings, for example with the Geological Society or the Challenger Society, or were 'local' meetings that focused on a specific estuary or coastal area, looking at all aspects from hydrology, chemistry, geology through to biology, and often to consider pollution and its control.

Table F1 Conferences, meetings and workshops of EBSA and ECSA 1971–2021

Date	Topic	Place
13 Oct 1971	The Estuarine Environment – EBBA Inaugural meeting	London Zoo
14 Apr 1972	Productivity and Energetics of Shallow Marine and Estuarine Systems *with BES*	York Univ
21 Feb 1973	Characteristics of Estuarine Sedimentation *with Geological Society*	London
23 May 1973	Science and Strategy in Estuarine Research – talk by RJH Beverton of NERC	London
23 Aug 1973	Estuaries session *during British Association meeting*	Canterbury
4–6 Sept 1973	Estuaries *with Challenger Society*	UWIST Cardiff
15–19 Oct 1973	2nd International Conference on Estuarine Research *with ERF*	South Carolina
16 May 1974	Chemical Processes in Estuaries *with Marine Chemistry Discussion Group*	London
19–20 Sept 1974	Estuarine & Coastal Land Reclamation & Water Storage	PCL London
28–29 Sept 1974	Severn Estuary local meeting	Bristol Univ
12–13 Apr 1975	Mersey Estuary and Liverpool Bay local meeting	Liverpool Univ
21–23 Apr 1975	Marine Ecology & Oil Pollution *with FSC and Inst Petroleum*	Aviemore
24–25 Sept 1975	Role of Meiofauna in Estuaries	Durham Univ
10–11 Apr 1976	Thames Estuary local meeting	London
30 Sept–1 Oct 1976	Lagoons and Saline Enclosures	Port Erin, IoM
16–17 Apr 1977	Clyde Estuary local meeting	Stirling Univ
20–21 Sept 1977	Role of Models in Estuarine Studies	Plymouth
12 Nov 1977	Estuarine Molluscs in Science and Commerce *with Malacological Society*	London
15–16 Apr 1978	Milford Haven local meeting	FSC Orielton
13–16 Sept 1978	Environmental Impacts of Large Hydraulic Engineering Projects *with Dutch Oceanografenclub and Hydrobiologische Vereniging*	Middelburg, Netherlands
11–16 Sept 1978	Workshop on Estuarine Hydrography & the Measurement of Dilution & Dispersion	HWU Edinburgh
7–8 Apr 1979	The Firth of Forth and Forth Estuary local meeting	HWU Edinburgh
26–28 Sept 1979	Problems of an Industrialised Estuary: Swansea Bay	Swansea Univ
19–20 Apr 1980	Solent local meeting	Southampton
17–20 Sept 1980	Feeding & Survival Strategies of Estuarine Organisms *with Dutch Oceanografenclub and Hydrobiologische Vereniging*	Hull Univ
23–29 Mar 1981	Taxonomy & Identification of Marine & Estuarine Organisms workshop	HWU Edinburgh
11–12 Apr 1981	Morecambe Bay local meeting	Lancaster Univ
14–16 Sept 1981	First International Symposium on Cohesive Sediments – ECSA co-sponsor with FBA	FBA Windermere
16–17 Apr 1982	Tamar Estuary local meeting	Plymouth
7–10 Sept 1982	Echinoderms Taxonomic Workshop	Stirling Univ
21–22 Sept 1982	Estuarine Management and Quality Assessment	TCD Dublin
8–9 Apr 1983	Tees Estuary local meeting	Durham Univ
21–22 Sept 1983	Structure and Functioning of Brackish-Water and Inshore Communities	HWU Edinburgh
3 Nov 1983	Scottish Sea Lochs local meeting *with SMBA*	Stirling Univ

Date	Topic	Place
9–13 Apr 1984	Sedentary Polychaetes Taxonomic Workshop	Swansea Univ
13–14 Apr 1984	Humber Estuary local meeting	Hull Univ
11–14 Sept 1984	Nutrients in Estuaries. Eutrophication and Other Effects *with Dutch Oceanografenclub and Hydrobiologische Vereniging*	Texel, Netherlands
24–28 Sept 1984	Mixing and Dispersal of Effluents in Inshore Waters workshop	Brixham
20–22 Mar 1985	Numerical Models and Community Analysis workshop	HWU Edinburgh
25–29 Mar 1985	Errant Polychaetes Taxonomic Workshop	HWU Edinburgh
12–13 Apr 1985	Mersey Estuary local meeting	Liverpool Univ
23–27 Sept 1985	Aquatic Oligochaetes Taxonomic Workshop	Hull Univ
26–27 Sept 1985	Problems of Development in Tropical Estuaries EBSA15	Wallingford
11–12 Apr 1986	Poole Harbour local meeting	Swanage
28 Apr–2 May 1986	Mollusca Taxonomic Workshop	SMBA Oban
1–5 Sept 1986	Dynamics of Turbid Coastal Environments EBSA16	Plymouth
15–18 Sept 1986	Estuarine and Marine Coastal Pollution Microbiology workshop	Glasgow
8–10 Apr 1987	The Wash and its Estuaries local meeting *with NCC*	Horncastle
6–10 Jul 1987	Fringing Habitats *with NCC*	Hull Univ
14–18 Sept 1987	Marine and Estuarine Methodologies EBSA17 *with Challenger Society and Marine Chem Discussion Group*	Dundee Univ
8–9 Apr 1988	Conwy Estuary local meeting	Betws-y-Coed
11–15 Apr 1988	Infaunal and Small Epifaunal Crustacea Taxonomic Workshop	Bangor Univ
18–22 Jul 1988	Fish in Estuaries *with FSBI*	Southampton
29 Aug–2 Sep 1988	North Sea – Estuaries Interactions EBSA18 *with Dutch Oceanografenclub and Hydrobiologische Vereniging*	Newcastle
Oct 1988	Fate and Effects of Toxic Chemicals in Larger Rivers and their Estuaries *ECSA was co-sponsor*	Quebec, Canada
3–7 Apr 1989	Estuarine Algal Taxonomic Workshop	HWU Edinburgh
13–16 Apr 1989	Evolution and Change in the Bristol Channel and Severn Estuary *with Linnean Society, FSC and BES*	Taunton
4–8 Sept 1989	Spatial and Temporal Intercomparisons ECSA19 *with GEMEL*	Caen, France
19–21 Apr 1990	South-East Estuaries and Coasts local meeting	Wye, Kent
23–27 Apr 1990	Polychaetes Taxonomic Workshop *with FSC*	Fort Popton
10–15 Sept 1990	The Changing Coastline ECSA20	Hull Univ
12–14 Apr 1991	West Wales Estuaries local meeting	Aberystwyth Uni
15–19 Apr 1991	Phyto- and Zooplankton Taxonomic Workshop	Cullercoats
Apr 1991	First World Fisheries Congress *ECSA was co-sponsor*	Athens, Greece
9–13 Sept 1991	Estuarine and Marine Gradients ECSA21	Gent, Belgium
28 Mar 1992	Ecology and Behaviour of Aquatic Molluscs *joint with Malacological Society*	Bristol Univ
2–4 Apr 1992	Southampton Water the Solent and Hampshire Coast local meeting	Southampton
14–18 Sept 1992	Changing Fluxes to Estuaries: Implications from Science to Management ECSA22 *with ERF*	Plymouth
15–17 Apr 1993	Solway, Cumbria, Galloway local meeting	Penrith College

Date	Topic	Place
30 Aug–3 Sept 1993	Role of Particles in Estuaries and Coastal Waters ECSA23 *with Netherlands Society of Aquatic Ecology*	Groningen
21–25 Mar 1994	Estuarine & Coastal Mollusca Taxonomic Workshop	HWU Edinburgh
28–30 Mar 1994	Numerical Methods in Biology Workshop	HWU Edinburgh
7 to 9 Apr 1994	Estuaries and Coasts of North-east England local meeting	Durham Univ
5–9 Sept 1994	Northern and Southern European Estuaries ECSA24	Aveiro
14–18 Nov 1994	Oligochaetes Taxonomic Workshop *with UNICOMarine*	FSC Preston Montford
29–31 Mar 1995	Biological Sampling Methods and Strategies, workshop	Bangor Univ
6–8 Apr 1995	Estuaries of Central Scotland local meeting	HWU Edinburgh
26–30 June 1995	Estuarine Sediments workshop	St Andrews Univ
11–16 Sept 1995	Strategies and Methods in Coastal and Estuarine Management ECSA25	Dublin TCD
Feb 1996	Estuarine and Coastal Pollution workshop *with AWT-Ensight (Australia)*	NSW, Australia
26–29 Jan 1996	Estuarine Fishes workshop	Groningen
1–3 April 1996	Biology of Crustacea *with MBA and SEB*	Plymouth
11–13 Apr 1996	Humber Estuary and the Adjacent Yorkshire and Lincolnshire Coasts local meeting	Hull Univ
12–14 Sept 1996	Estuarine Habitat Restoration	Brussels
16–20 Sept 1996	Transport Retention and Transformation in Estuarine and Coastal Systems ECSA26 *with ERF*	Middelburg
4–8 Nov 1996	Taxonomy workshop and Discussion *with UNICOMarine and NMBAQC*	Millport
12–21 Mar 1997	Marine Benthic Field and Laboratory Methods Workshop *with IECS and NMBAQC*	Hull Univ
6–11 Apr 1997	Marine and Estuarine Algae Identification Workshop	HWU Edinburgh
9–13 June 1997	Comparison of Enclosed and Semi-enclosed Marine Systems ECSA27 *with Baltic Marine Biologists*	Mariehamn, Finland
25–27 Jun 1997	The Tagus Estuary and Adjoining Coastlines Local Meeting	Lisbon, Portugal
1–5 Sept 1997	Remote Sensing in Estuaries ECSA28 *with Remote Sensing Society*	St Andrews Univ
10–12 Sept 1997	Suffolk & Essex Estuaries Local Meeting	Colchester
17–18 Apr 1998	Devon Estuaries Local Meeting	Plymouth Univ
13–17 Jul 1998	Estuarine Research and Management in Developed and Developing Countries ECSA29	Port Elizabeth, South Africa
7–13 Aug 1998	Seagrass and Algal Mats Workshop	Sylt, Germany
11–13 Feb 1999	Infocoast 99 *with EUCC and Coastlink*	Leeuwenhorst, Netherlands
8–9 Apr 1999	Estuaries and Coasts of North-west England Local Meeting	Ormskirk
24–28 May 1999	Benthic Sampling Strategies: Design and Methods Workshop *with NMBAQC*	Millport
9–13 Aug 1999	Impact of Climate Change on the Coastal Zone ECSA30	Hamburg, Germany
25–29 Oct 1999	Beginners Coastal and Estuarine Taxonomy Workshop *with NMBAQC*	Millport
13–15 Apr 2000	Estuaries and Coasts of South Wales Local Meeting	Cardiff Univ
19–22 May 2000	Teaching & Fieldwork in Marine Science *UMBSM event with ECSA support*	Millport

Date	Topic	Place
3–7 Jul 2000	Eutrophication of Estuaries and Nearshore Waters: A Challenge for the New Millennium ECSA31	Bilbao, Spain
12–13 Jul 2000	Marine Environment, Science and Law Workshop	Scarborough
2–6 Aug 2000	Community Ecology of Mussel Beds Workshop	Sylt, Germany
29 Mar–1 Apr 2001	Estuaries and Coasts of North-East Scotland Local Meeting	Aberdeen Univ
25–28 Apr 2001	Marine Biodiversity in Ireland and Adjacent Waters	Ulster Museum, Belfast
15–17 Jun 2001	Estuaries and Coasts of Northern Portugal Local Meeting	Porto, Portugal
8–11 Jul 2001	Changing Coastal Margins: Chemical Processes and Dynamics *joint with IECS*	Scarborough
9–13 Jul 2001	Fish Diversity and Conservation *joint with FSBI*	Leicester
1–5 Oct 2001	Workshop on Difficult Taxa *with NMBAQC and IRTU(NI)*	Portaferry
3–8 Nov 2001	An Estuarine Odyssey ECSA32 *with ERF*	Florida
2001	EU-Life Algae Project Local Meeting	Gothenburg, Sweden
11–13 Apr 2002	Coast and Estuaries of the Isle of Man and Adjacent Waters Local Meeting	Port Erin, IoM
8–12 Jul 2002	Estuarine and Lagoon Fish and Fisheries ECSA33 *with FSBI*	Hull Univ
15–20 Sept 2002	Estuaries and Other Brackish Areas – Pollution Barriers or Sources to the Sea ECSA34	Gdansk, Poland
2002	Climate Change Workshop *with ERF*	
16–20 2002	Seaweed Identification Workshop	HWU Edinburgh
7–9 Oct 2002	Ecological Structures and Functions in the Scheldt Estuary: from Past to Future	Antwerp, Belgium
7–8 Oct 2002	Interpreting Change in Coastal Systems Workshop	Antwerp, Belgium
28 Apr–2 May 2003	Scientific Research as a Support Strategy to Estuarine and Coastal Management ECSA35 *with ERF*	Sonora, Mexico
17–22 Aug 2003	Impacts of Diffuse Pollution on Estuarine and Marine Environments ECSA36 *joint with 7th International Diffuse Pollution Conference of International Water Association*	Dublin TCD
27–28 Aug 2003	Recent Developments in Estuarine Ecology and Management	Scarborough
20–25 Jun 2004	Recent Developments in Estuarine Ecology ECSA37 *with ERF*	Ballina, NSW
11–13 Sept 2004	Aquatic Field Survey Design Workshop	Slapton, Devon
13–17 Sept 2004	Changes in Land Use: Consequences on Estuaries and Coastal Zones ECSA38 *with ERF*	Rouen, France
9–10 Sept 2004	Towards an Integrated Knowledge and Management of Estuarine Systems	Lisbon, Portugal
6–8 Apr 2005	Estuaries of South-West England Local Meeting	Plymouth
5–7 Sept 2005	The Ecosystem Approach ECSA 39	HWU Edinburgh
26–29 Nov 2005	Ecosystems in Changing Estuaries	Groningen
27–31 Mar 2006	Seaweed Identification Workshop	HWU Edinburgh
9–12 May 2006	Sustainable Co-development of Enclosed Seas: Our Shared Responsibility ECSA40/EMECS7	Caen, France
11–13 Sept 2006	Marine Governance Workshop	Hull Univ
15–20 Oct 2006	Measuring and Managing Change in Estuaries and Lagoons ECSA41	Venice, Italy
19–23 Mar 2007	Seaweed Identification Workshop	HWU Edinburgh

Date	Topic	Place
16–22 Sept 2007	Estuarine Ecosystems: Structure Function and Management ECSA42	Kaliningrad, Russia
7–9 Feb 2008	Climate Change Impacts on South European Coastal Ecosystems	Lisbon, Portugal
7–11 Apr 2008	Seaweed Identification Workshop	HWU Edinburgh
7–9 Apr 2008	Severn Estuary Local Meeting	Cardiff
29 Sept–3 Oct 2008	Science and Management of Estuaries and Coasts: A Tale of Two Hemispheres ECSA44	Bahia Blanca, Argentina
22–24 Apr 2009	The Thames, Coasts and Estuaries of South-East England Local Meeting	South Bank Univ, London
30 Aug–4 Sept 2009	Estuarine Goods and Services ECSA45	TCD Dublin
13–15 Apr 2010	Coasts and Estuaries of the North-East Irish Sea Local Meeting	Liverpool Univ
26–30 Apr 2010	Numerical Methods in Estuarine and Marine Ecology Workshop	Hull Univ
3–6 May 2010	The Wadden Sea: Changes and Challenges in a World Heritage Site ECSA46	Sylt, Germany
23–27 Aug 2010	European Marine Biology Symposium 45 – ECSA48	HWU Edinburgh
14–24 Sept 2010	Wadden Sea Summer School	Sylt, Germany
22–23 Sept 2010	All at Sea? Synergies between Past and Present Coastal Processes and Ecology *with Quaternary Research Association*	Loughborough Univ
14–19 Sept 2010	Integrative Tools and Methods for Assessing Ecological Quality in Estuarine and Coastal Systems Worldwide ECSA47	Figuera de Foz, Portugal
9–10 Dec 2010	Functioning of Muddy Waters with Focus on the Ems Estuary	Emden, Germany
26–28 Apr 2011	The Humber and the English East Coast Local Meeting	Hull Uhiv
3–7 Apr 2011	Estuarine, Coastal and Oceanic Ecosystems: Breaking Down the Barriers ECSA49	Grahamstown, South Africa
23–27 Sept 2012	Research and Management of Transitional Waters ECSA51	Klaipedia, Lithuania
24–28 Oct 2011	Estuarine and Lagoon System Trajectories *(CEMAGREF)*	Bordeaux, France
16–18 May 2012	Scottish Sea Lochs and Adjacent Waters Local Meeting	SAMS Oban
3–6 June 2012	Today's Science for Tomorrow's Management ECSA50 *with Elsevier*	Venice, Italy
26–28 June 2012	Macronutrients Workshop	Plymouth
21–23 Nov 2012	An Integrated Approach to Emerging Challenges in a World Heritage Site ECSA52–13th International Wadden Sea Symposium	Leeuwarden, Netherlands
8–12 Apr 2013	Problems of Small Estuaries Local Meeting	Swanaea Univ
14–18 Oct 2013	Estuaries and Coastal Areas in Times of Intense Change ECSA53 – *with Elsevier*	Shanghai, China
3 Nov 2013	Estuaries of the World – *ECSA session in EMECS10 conference*	Marmaris, Turkey
6–9 Nov 2013	Changing Hydrodynamics of Estuaries and Tidal River Systems – *ECSA session in 8th International SedNet*	Lisbon, Portugal
8–10 Apr 2014	Estuaries and Coasts of North and Mid-Wales Local Meeting	Bangor Univ
12–18 May 2014	Coastal Systems under Change: Tuning Assessments and Management Tools ECSA54	Sesimbra, Portugal
27–28 May 2015	Restoration of Estuarine Environments: The Example of the Seine Estuary	Le Havre, France

Date	Topic	Place
15–19 Jun 2015	10th Baltic Sea Science Congress *ECSA was co-sponsor*	Riga, Latvia
6–9 Sept 2015	Unbounded Boundaries and Shifting Baselines: Estuarine and Coastal Areas in a Changing World ECSA55 *with Elsevier*	London
5–9 Jul 2016	Estuarine Restoration	Antwerp, Belgium
4–7 Sept 2016	Coastal Systems in Transition ECSA56	Bremen, Germany
15–17 May 2017	Marine Protected Areas: Science: Policy and Management *with Poole Harbour Study Group*	Poole
16–20 Oct 2017	Where Land Meets the Ocean: The Vulnerable Interface *with State Key Laboratory of Estuarine and Coastal Research*	Shanghai, China
3–6 Sept 2018	Changing Estuaries Coasts and Shelf Systems: Diverse Threats and Opportunities ECSA57 *with Elsevier*	Perth, WA
14–15 Mar 2019	Environmental Status of Estuaries and Coasts in India	Kerala, India
30 Apr–1 May 2019	Forth and Tay Estuaries and Adjacent Coastlines Local Meeting	HWU Edinburgh
18–20 Jun 2019	Wetlands of the Future *with Australian Mangrove and Saltmarsh Network*	Melbourne, Australia
16 Jul 2019	Restoring Estuarine and Coastal Habitat in the North-East Atlantic (REACH) – *ECSA co-sponsor*	NHM London
4–8 Nov 2019	Global Challenges and Estuarine and Coastal Systems Functioning: Innovative Approaches and Assessment Tools – *ECSA co-sponsor*	Bordeaux, France
21 Nov 2019	Oyster Habitat Restoration Webinar REACH – *ECSA co-sponsor*	Internet
20–24 Jan 2020	Future Vision and Knowledge Needs for Coastal Transitional Environments (EUROLAG9)	Venice, Italy
19–21 Nov 2020	Conservation of Mangrove Ecosystems: Synergies for Fishery Potential in India	Kerala, India
19–21 Jan 2021	Ocean Recovery (CMS joint with REACH) *ECSA was co-sponsor*	Internet
28 Apr 2021	Estuaries of South-West England Local Meeting	Internet (Plymouth)
6–10 Sept 2021	Estuaries and Coastal Seas in the Anthropocene ECSA58 *with EMECS and Elsevier*	Internet (Hull)

International recognition came early as the Estuarine Research Federation (ERF) in the USA was founded about the same time as EBBA in the UK in November 1971 and we were invited to co-sponsor the ERF meeting in South Carolina in October 1973. In 1978, 1980 and 1984, collaboration was made with the Dutch Oceanografenclub and Hydrobiologische Vereniging.

From these initial collaborations developed a series of major meetings or international symposia, which are recognised as numbered meetings in Table F1. The first to be specifically numbered was EBSA15 in 1985, a feat achieved by silently counting the earlier international meetings. ECSA58 has now been reached. The selected proceedings of most of the numbered meetings have been published in the EBSA/ECSA journal, *Estuarine Coastal and Shelf Science*, or another relevant journal.

In addition a number of workshops have been held, with two themes apparent. Workshops have either been for taxonomic purpose to give detailed training in species identification, especially for seaweeds and invertebrates, or for techniques such as numerical analysis. These workshops often stemmed from the need for the staff of regulatory agencies to acquire the skills needed for pollution assessment, and show how much EBSA/ECSA has been the catalyst for collaboration between universities and pollution control managers.

The Table of meetings shows clearly how EBSA/ECSA has moved from the local (i.e. UK) to the international (Europe and beyond). It also shows how collaboration between universities and water industries has been sustained. Throughout its 50 years, the association has successfully developed collaboration between many aspects of estuarine science, including physics, hydrology, chemistry, geology, biology and management.

Publications

EBSA had intended to establish its own special brackish and estuarine journal, but Academic Press was about to launch *Journal of Coastal and Marine Science* and was willing to change its name to *Journal of Estuarine and Coastal Marine Science* (it was launched as ECMS), with EBSA-nominated members being added to the editorial board. Professor Ernie Naylor of the Port Erin Marine Lab of Liverpool University, later from UCNW, Menai Bridge, became the first EBSA editor of ECMS. The journal was later renamed *Estuarine Coastal and Shelf Science* (often abbreviated *ECSS*), and ECSA has nominated successive editors (Donald McLusky, Michael Elliot and Steve Mitchell).

ECSA is also now linked to the journal *Ocean & Coastal Management*, which was until recently edited by an ECSA councillor, the late Victor de Jonge. Both of these journals are now managed by Elsevier, who have also become closely linked to our major international meetings (from ECSA50 onwards) both in terms of publications as well as organisation.

EBSA also agreed to produce a members' bulletin, and Peter Liss of the University of East Anglia was the first editor. There were initially three bulletins per year, and over the years the print quality and style massively improved; it is now distributed online. Successive editors were Simon Aston, Donald McLusky, Mike Elliott, Jim Wilson, Jean-Paul Ducrotoy and Patrick Meire. In addition, there is now an e-newsletter edited by Anita Franco. The selected proceedings of most of the numbered EBSA/ECSA meetings have been published in the association's own journal, *Estuarine Coastal and Shelf Science*, or another relevant journal or as separate books.

As an occasional series, ECSA published 'Coastal Zone Topics: Process, Ecology & Management', which was designed to improve the flow of information on costal resources, processes and management, based on the papers presented at ECSA's local meetings. Five volumes were published between 1995 and 2003.

EBSA sponsored the publication of three practical handbooks. The first dealt with aspects of estuarine hydrography and sedimentation (edited by K.R. Dyer, 1979), the second with estuarine chemistry (edited by P.C. Head, 1985) and the third with biological surveys (edited by J.M. Baker and W.J. Wolff, 1987). The first has recently been totally rewritten and republished (edited by R.J. Uncles and S.B. Mitchell, 2017).

In collaboration with the Linnean Society, ECSA has been closely involved in the publication of the series 'Synopses of the British Fauna'. These books focus on the identification of particular groups of species, and arose in large part from ECSA workshops, where draft versions of the text were often trialled.

Office bearers

ECSA is entirely run by volunteers who have enthusiastically contributed to the development of estuarine science. It is managed by an elected group of councillors. The ECSA Council is chaired by a president who usually serves a three-year term of office (Table F2). The council is supported by a secretary and treasurer as well as other office bearers (membership secretary, meetings secretary, and publications secretary).

Table F2 Principal Officers of EBSA/ECSA 1971–2021

Presidents		Secretaries	
1972	Ronald Bassindale	1972–86	Prof. Richard S.K. Barnes
1973	Hugh C. Gilson MBE	1986–91	Dr Neville Jones
1974–76	Prof. Jack Allen FRSE	1991–2000	Dr Donald McLusky
1977–79	Prof. Alfred Steers CBE	2000–05	Prof. Mike Elliott
1980–82	Dr John Corlett	2005–11	Dr Jim Wilson
1983–85	Roland S. Glover FRSE	2011–14	Dr Mark Fitzsimmons
1986–88	Prof. Ernest Naylor	2014–21	Dr Gillian Glegg
1989–91	Prof. Keith Dyer		
1992–95	Prof. Alasdair McIntyre	**Treasurers**	
1996–98	Prof. John McManus	1972–78	Dr Colin Little
1999–2001	Dr Neville Jones	1978–2021	Prof. Martin Wilkinson
2002–4	Dr Alan Gray OBE		
2005–7	Prof. Mike Elliott		
2008–10	Dr Reg Uncles		
2011–14	Prof. Geoff Millward		
2015–17	Prof. Kate Spencer		
2018–20	Prof. Axel Miller		
2021	Dr Sally Little		

Research grants

As part of its role to encourage young scientists to undertake estuarine research ECSA has developed a number of grant schemes. The Charles Boyden fund for small grants gives funds for small research projects or for attendance at our major meetings. Charles Boyden was the first meetings secretary of ECSA from 1974 to 1976, before he moved to New Zealand. His premature death in 1979 is marked by this scheme. The Peter Jones Memorial Award commemorates a long-time ECSA councillor, and there is an annual award for the best publication arising from a recent postgraduate student. Awards are also given for the best oral and poster presentation at ECSA symposia. A full list of award winners is on the ECSA website.

Conclusion

Initially, ECSA was focused on the estuarine environment of the British Isles, but it has expanded its remit so that it now encompasses the estuarine (or transitional waters) environment throughout Europe as well as Africa, Asia and Australasia. ECSA has always encouraged an interdisciplinary approach to the study of estuaries and coasts, ranging from the biological through to physical and chemical studies. It has also fostered links between the academic world of research and the world of applications to estuarine management such as pollution control.

As the years have progressed since EBSA was founded, many of the estuaries of Britain, Europe and much of the rest of the world have indeed become cleaner, and we all hope that the work of EBSA/ECSA has contributed to this both by bringing academics and managers together through our meetings and conferences, as well by publishing the many studies that have now been undertaken on our estuaries, brackish waters and coasts. Our knowledge of estuaries at our 50th anniversary is immense compared with our knowledge at the beginning.

References

Baker, J.M. and Wolff, W.J. (1987) *Biological Surveys of Estuaries and Coasts*. Cambridge: Cambridge University Press.

Barnes, R.S.K. and Green, J. (eds) (1972) *The Estuarine Environment*. Barking: Applied Science Publishers. https://doi.org/10.4319/lo.1973.18.2.0348

Dyer, K.R. (ed.) (1979) *Estuarine Hydrography and Sedimentation*. Cambridge: Cambridge University Press.

Green, J. (1968) *The Biology of Estuarine Animals*. London: Sidgwick & Jackson.

Head, P.C. (ed.) (1985) *Estuarine Chemistry*. Cambridge: Cambridge University Press.

Lauff, G.H. (ed.) (1967) *Estuaries*. Washington, DC: American Association for the Advancement of Science.

Little, C. and Barnes, R.S.K. (2019) Genesis and birth of a transitional association. *ECSA Bulletin* 60: 3–4.

Muus, B.J. (1967) The fauna of Danish estuaries and lagoons. *Meddelelser fra Danmarks Fiskeri-Og Havundersogelser* new series 5: 1–316.

Naylor, E. 2002. An historical assessment of the achievements of ECSA over its first thirty years. *Estuarine, Coastal and Shelf Science* 55: 807–13. https://doi.org/10.1006/ecss.2002.1030

Remane, A. and Schlieper, C. (1958) *Die biologie des brackwassers*. Stuttgart: E. Schweizerbart'sche verlagsbuchhandlung.

Uncles, R.J. and Mitchell, S.B. (2017) *Estuarine and Coastal Hydrography and Sediment Transport*. Cambridge: Cambridge University Press. https://doi.org/10.1017/9781139644426

CHAPTER 1

Morphodynamics of Tropical Atlantic River Mouths and their Adjacent Shorelines

HELENE BURNINGHAM, SILVIA PALOTTI POLIZEL and AWA BOUSSO DRAMÉ

Abstract

The tropical, coastal plain hinterlands of the Atlantic comprise large rivers, with plentiful sediment supply, that meet swell-dominated wave climates at the coast, capable of modifying river mouth deposits into a range of deltaic sedimentary frameworks. Land-use, land-cover changes and other anthropogenic activities across the river basins at these latitudes (e.g. deforestation, agriculture, large-scale water management) have impacted shorelines through modifications to the discharge and sediment supply to the coast. The tropical coastal river systems of West Africa and east Brazil encompass a range of morphologies that reflect the relative roles that marine and fluvial processes play. Along both the west and east oceanic margins, these river mouth shorelines are sensitive to climate change influencing fluvial discharge and wave climate variability. Modifications in wave climate and hydrological management (i.e. dam activities) can affect the biogeomorphology of river mouths through changes in sediment supply to the coast, and sediment redistribution along the adjacent deltaic shorelines. In future decades, climate change and sea-level rise have the potential to trigger a range of changes in these processes that would impact the morphodynamics of river mouths and associated delta shorelines. Understanding the hydro- and morphodynamics associated with these coastal landforms is crucial to deal with such future challenges and evaluate adaptation measures.

Keywords: river mouth, hydrological regime, shoreline change analysis, delta, estuary, inlet, catchment water management

Correspondence: h.burningham@ucl.ac.uk

Introduction

River mouths are some of the most productive environments in the world, where the interplay of marine and fluvial processes supply sediments and nutrients from different sources, encouraging the development of a heterogeneous framework of sedimentary environments that support a range of habitats and species. The depositional systems that develop in these contexts can also be very important in providing several ecosystem services

Helene Burningham, Silvia Palotti Polizel and Awa Bousso Dramé, 'Morphodynamics of Tropical Atlantic River Mouths and their Adjacent Shorelines' in: *Challenges in Estuarine and Coastal Science*. Pelagic Publishing (2022).

to local coastal and hinterland communities, particularly in terms of regulating services (e.g. protection) and provision services (e.g. food).

The geomorphology of river mouths can be highly varied and is determined by the relative influence of river, wave and tide processes (Wright 1977), which may lead to tides extending far into river valleys or wave modification of a river course, for example. The potential for complex interactions between these processes has underpinned much of the work classifying the depositional systems that develop in river mouth contexts (e.g. Galloway 1975; Boyd et al. 1992). Deltas and estuaries are the key sedimentary and geomorphological frameworks that develop around river mouths. Although there are several different definitions that address specific geological, hydrological and biochemical features and processes, at a simple geomorphological level, the differentiation between deltas and estuaries reflects the scale of the zone of interaction and relative dominance of tidal and river currents. Estuarine river mouths are those where tides extend landward into the river valley, where currents exert some control on sediment transport and where there is some degree of mixing of the fresh (river) and saline (marine) water bodies. Deltaic river mouths extend seaward of the former or structural river valley through the accumulation of riverine sediment, and tidal interactions are largely limited to only the most seaward inlet of the deltaic system. Wave processes are largely confined to the seaward-most part of the river mouth and are more important in the shaping of mouth morphology and supply of sediments to adjacent shorelines. These systems therefore have the potential to vary significantly in both time and space.

Accommodation space (the potential sedimentary infill zone contained by the antecedent valley or coastal plain topography) and sediment supply are also key factors in influencing the morphology and dynamics of depositional structures formed, both at the river mouth and across the wider coast. For example, in coastal plain environments and catchments delivering large sediment fluxes, river mouth deposits can extend far across the shoreface (cross-shore and alongshore). As such, the role of catchment management, and specifically the construction of dams, has been highlighted by many as a key driver of river mouth and coastal erosion over recent decades (Syvitski et al. 2005; Fagherazzi et al. 2015) and a fundamental control on coastal sediment budgets (Warrick 2020). Wider human impacts include enhanced subsidence linked to groundwater abstraction and petroleum mining, and when combined with sea-level rise and shifts in weather patterns associated with climate change, there is a clear concern over the sustainability of river mouths in the future, bringing significant potential risks to millions of people (Syvitski 2008).

These risks and the sensitivity of coastal communities to future environmental change are more pronounced within the tropics where there is a greater proportion of less economically developed countries. Climate change and climate variability are the significant drivers of societal challenges in the tropics, where the increased frequency of hurricanes, droughts and floods impacts shoreline populations in both direct (storm surge and destructive wind) and indirect (changes in rainfall patterns and delivery of water across catchments) ways. In the tropical Atlantic, the Inter Tropical Convergence Zone (ITCZ) atmospheric system that connects South America to West Africa moves north and south each year and drives the wet and dry seasons on each continent, meaning that the climate of the east and west Atlantic is interlinked. Decadal variability in this, driven by the North–South Atlantic thermohaline gradient, impacts rainfall most significantly in South America and West Africa (Foltz et al. 2019). Several of the tropical river systems entering the Atlantic Ocean contribute significant freshwater that drives the salinity variability, influencing the large-scale Atlantic Meridional Overturning Circulation that controls much of the climate around the North Atlantic Ocean, but significantly also controls the Atlantic Multidecadal Variability and hence rainfall in West Africa (Jahfer, Vinayachandran and Nanjundiah 2020). Tropical river systems are also particularly important geomorphologically, in delivering relatively more sediment to the oceans than their temperate counterparts, and thereby performing a significant denudation function across continental land masses (Milliman and Farnsworth 2013).

The climate interlinkage across the tropical Atlantic has important implications on the future sustainability of river mouth systems in relation to the way that water is managed across catchments, and the associated controls on sediment flux, particularly when combined with the additional pressures of increasing coastal populations and sea-level rise. In this chapter, we review the key geomorphological characteristics of tropical Atlantic river mouths to understand more specifically the possible sensitivities to climate and climate change. We also present an analysis of morphology, hydrology and shoreline dynamics of two case examples from east and west Atlantic – the São Francisco in Brazil, South America, and the Senegal on the Senegal–Mauritania border, Africa. We illustrate the impact that anthropogenic interventions have had on these systems, and reflect on the future challenges for communities in these regions.

Overview of the approach

River mouths of the tropical Atlantic were assessed in terms of their current geomorphology. The Caldwell et al. (2019) global coastal and river delta dataset was subsampled to consider just those river mouths located within the tropics entering the Atlantic. A systematic review of these was undertaken in the context of high-resolution satellite imagery (ESA Sentinel-2 and through Google Earth Pro) to identify the presence of key geomorphological features within the river mouth environment (including depositional features such as beach ridges and spits, and structural control such as reefs or engineering). Within this database, Caldwell et al. (2019) identified deltaic river mouths based on the presence of distributary networks and/or extension of a subaerial deposit from the main shoreline. Here, we extend this to identify estuarine river mouths, but owing to the lack of clear visible criteria for this, we used the literature to inform this classification; that is, if the river is described anywhere in the published literature as an estuary, it is recorded as such in this analysis. This database was also joined to the global delta and river mouth dataset of Nienhuis et al. (2020) to supplement the marine process context (wave height, tidal range, sea-level rise, bathymetric slope) derived by Caldwell et al. (2019) with fluvial process context (basin area and river discharge).

The two specific river mouths of São Francisco (Brazil) and Senegal (West Africa) are reviewed in more detail to evaluate the relative change in fluvial and marine contexts over recent decades on either side of the Atlantic. Coastal dynamics were analysed following a change analysis of shorelines derived from satellite imagery between the 1980s and 2020, and compared with change since 2010. Catchment water level and discharge data, and the European Re-Analysis (ERA5) hindcast wave data were evaluated over the same timeframe to capture river and marine processes.

Tropical Atlantic river mouths

The catchments of rivers entering the tropical Atlantic lie almost entirely within tropical climate regions (Fig. 1.1a); despite arid climates being present across much of the north and south of the tropical zone of Africa, it is perhaps not surprising that these areas do not comprise significant river catchments. The arid, hot, steppe of Eastern Brazil covers a significant area of the São Francisco catchment, overlapping part of an area known as Polígono das Secas, which experiences irregular rainfall and is characterised by socioeconomic underdevelopment (Soares, 2013). Tropical South America and western

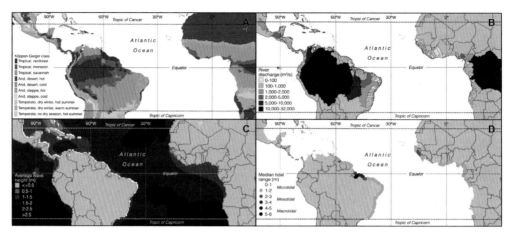

Figure 1.1 Tropical Atlantic contexts for river mouths: (a) Köppen-Geiger climate classification, (b) delineation of drainage basins and their relative river discharge, (c) mean annual significant wave height (1980–2020), and (d) median tidal range. Data sourced from Beck et al. (2018) [a], Nienhuis et al. (2020) [b], ERA5 [c], Caldwell et al. (2019) [d]

Africa are both dominated by near continental-scale basins (the Amazon and the Congo (Fig. 1.1b)) that exert well-established controls on tropical Atlantic circulation and climate variability (Materia et al. 2012; Jahfer, Vinayachandran and Nanjundiah 2020). These two basins therefore indirectly exert significant controls over all tropical Atlantic river systems through their function in controlling climate variability. In terms of marine processes, annual average wave heights (Fig. 1.1c) are an equivalent magnitude to the median tidal range (Fig. 1.1d). Both are minimum in the Caribbean and Gulf of Mexico, but this does not capture the storm-wave climate associated with the prevalence for hurricanes in this region. For much of the west coast of Africa, annual average wave heights are less than 1.5 m, but rise to 2–2.5 m at the north and south extents of the tropics. The east coast of South America experiences a more spatially consistent annual average of 1.5–2 m except in the vicinity of the more shallow and sheltered mouth region of the Amazon. Tidal range is for the most part microtidal throughout the tropics but increases to low mesotidal in West Africa (Guinea and Guinea-Bissau) and meso/macrotidal in the Amazon region.

Given the discrete nature of spatial variation in tidal range, wave climate and river discharge across the tropics (beyond some specific regions (e.g. Amazon, Congo, Guinea) there are large stretches of coastline experiencing very similar conditions), it could be assumed that river mouths would exhibit similar geomorphology. River mouths previously identified as deltas and estuaries (Fig. 1.2a/b) are found throughout the tropics; around 39% of river mouths have been described as deltas, 28% as estuaries and 54% as neither. But interestingly, around 13% of river mouths are concurrently referred to as both deltaic and estuarine (Fig. 1.3a). For instance, the Senegal and the São Francisco river mouths are recognised in the literature as both estuary and delta. This duality often reflects the presence of delta-front estuary systems (Fairbridge 1980) – estuaries that develop within the seaward reaches of large-river delta frameworks (e.g. Bianchi and Allison 2009). But in some cases, this might highlight an Anthropocene shift from regressive to transgressive systems, incorporating (a) human intervention in catchment hydrology and sediment flux, reducing the potential for progradational processes (Syvitski et al. 2009; Nienhuis et al. 2020) and (b) sea-level rise and coastal subsidence, leading to flooding of river valleys and development of transgressive coastal sequences (Dalrymple et al. 1992).

The presence of specific landforms and landform features at river mouths, including barriers and distributaries, can provide some basis for understanding the recognition of a

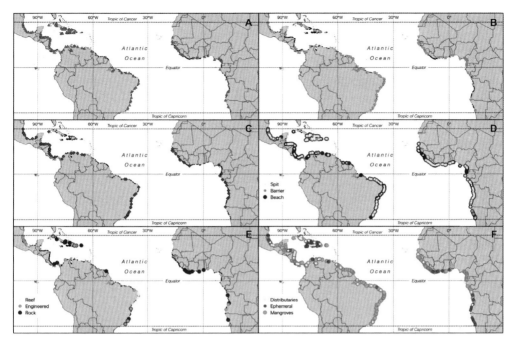

Figure 1.2 Presence of specific geomorphological features at tropical Atlantic river mouths: (a) delta (Caldwell et al. 2019), (b) estuary (derived from a literature review), (c) beach ridges, (d) beaches, barriers and spits, (e) structural control at the mouth (reef, rock and engineering) and (f) those that comprise distributaries, show evidence of being ephemeral and where mangroves are present. The small black dots in each sub-figure are river mouth locations (Caldwell et al. 2019) where the features are not present.

river mouth as deltaic or estuarine. Systematic review of the presence of such features is shown in Fig. 1.2c–f, which again displays widespread occurrence of a range of river mouth characteristics. There is limited direct association between the presence of wave-formed landforms such as spits and beach ridges and higher energy wave climates, and the analysis supports the view that it is the relative, not absolute strength of different processes, in combination with sediment supply, that determines the development of specific landforms. River mouths illustrating evidence of being ephemeral are largely confined to Central America and parts of the African coast. But even this feature is a more complex reflection of river flow variability and persistence in the wave climate. Multivariate analysis of the presence and absence of river mouth landforms relative to key process parameters such as wave height and river discharge (Fig. 1.3b) shows that there are no dominant gradients driving the distribution of this geomorphology and no specific geographic pattern. Despite a lack of evidence in the geographic distribution, wave height is more strongly associated with the presence of wave-formed landforms such as beaches, barriers and spits, but is also associated with the presence of ephemeral river mouths and a lack of developed mangrove system. Tidal range is more strongly associated with the presence of distributaries. River mouths that have neither been defined as estuaries nor deltas are more strongly associated with smaller basins, lower river discharge and the presence of bedrock control. River mouths defined as estuaries and/or deltas tend to be linked to larger river systems; interestingly, no specific landforms are associated with the river mouth definitions, suggesting that this geomorphology is not an underpinning factor in the classification of river mouths within the tropics.

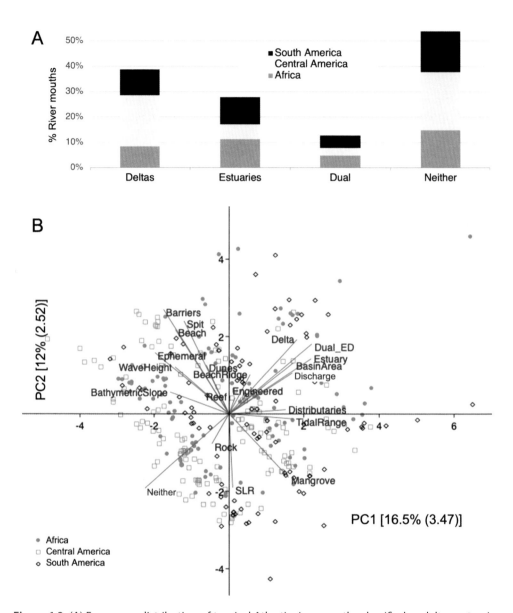

Figure 1.3 (A) Frequency distribution of tropical Atlantic river mouths classified as deltas, estuaries, both (dual) or neither, (B) principal component (PC) analysis of different geomorphological features at river mouths explored in the context of marine and river variables (wave height, bathymetric slope, tidal range, sea-level rise derived from Caldwell et al. (2019); river basin area and discharge from Nienhuis et al. (2020)), quoting percent explained and component scores for the first two components

Tropical coastal river systems of east Brazil and West Africa

As already demonstrated, there is a strong climate link between South America and West Africa, and a broad agreement that climate change impacts will be significant for both regions in terms of changes in rainfall patterns. Coupled with the legacy of successive catchment management interventions, rivers and their mouth environments are already responding to forcing associated with the Anthropocene (Brown et al. 2017).

Case example: São Francisco River

The São Francisco River's source is in the State of Minas Gerais at an altitude of 1,800 m (Guimarães 2010; Rangel and Dominguez 2019), and the river flows for 2,486 km until it reaches the Atlantic Ocean, in north-east Brazil. The São Francisco watershed occupies an area of 639,219 km^2, representing 7.5% of the national territory (Pereira et al. 2007; Medeiros et al. 2014). Owing to its size and diversity of environments, this watershed is divided into four physiographic regions: upper, middle, lower-middle and lower (Bettencourt et al. 2016) (Fig. 1.4a). The São Francisco River is an important and strategic river for the north-east region, as it provides 70% of the superficial waters available for the region, supplying a population of 18.2 million (Castro and Pereira 2019). As noted previously, São Francisco watershed covers different climatic zones, transitioning from humid to arid, meaning that both evapotranspiration ($c.$ 900 mm yr^{-1}) and precipitation ($c.$ 1,000 mm yr^{-1}) play important roles in catchment hydrology; rainfall is mainly controlled by the South American Monsoon System and the ITCZ (Dominguez and Guimarães 2021) and varies from 1,400 mm in the upper reaches to 350 mm in the coastal valley (Castro and Pereira 2019). Human interventions in the São Francisco valley started early in the seventeenth century with the development of routes to connect the coast to the interior. Intensification of intervention in the mid-1940s reflected increased need for water management for irrigated agriculture and energy production (IBGE, 2009).

The São Francisco River delta is wave dominated, and is one of the most prominent delta complexes on the Brazilian coast. Covering over 900 km^2, the delta region is not densely populated (Piaçabuçu is the largest settlement with only 17,203 inhabitants (in 2010; IBGE 2021). The delta was formed during the Quaternary period with the construction of beach ridges and associated development of secondary sedimentary environments including mangroves and dunes (Dominguez 1996; Bittencourt et al. 2007). The deltaic shoreline is mainly formed of sand, with a predominance of medium sands in the beaches near the São Francisco River mouth (Dominguez et al. 2016); beach, spit, dune, mangrove, barrier and beach ridge environments are located around the river mouth, and all demonstrate dynamic behaviour over the recent multi-decadal timescale. The system experiences a semidiurnal microtidal regime (1.74 m spring tidal range) (Bittencourt et al. 2007; Dominguez and Guimarães 2021), and the wave climate is characterised by more frequent (60%) larger waves (1–3 m) from the south-east and less frequent (~35%) smaller (1–2 m) waves from east-south-east (Fig. 1.4b).

The hydrological regime of the São Francisco watershed is illustrated at three hydrometric stations (Fig. 1.4c). Dams were first constructed in the 1940s for the purpose of generating electricity (Castro and Pereira 2019), and over the decades since, a series of dams working in cascade have been installed: Paulo Afonso (1949), Três Marias (1962), Moxotó (1977), Sobradinho (1979), Itaparica (1988) and Xingó (1994) (Dominguez and Guimarães 2021). These dams control 98% of the watershed and in combination have reduced the discharge to the river mouth by 35%, regulating the original seasonal discharge to a steady flow of around 2,000 m^3 s^{-1} (Knoppers et al. 2005). There was a significant reduction in the magnitude of discharge at Piranhas and Traipu following completion of the Xingó dam in 1994; discharge control has been intensified since 2012 with a critical decrease of the flows during droughts (Melo, Araújo Filho and Carvalho 2020), and this is evident downstream

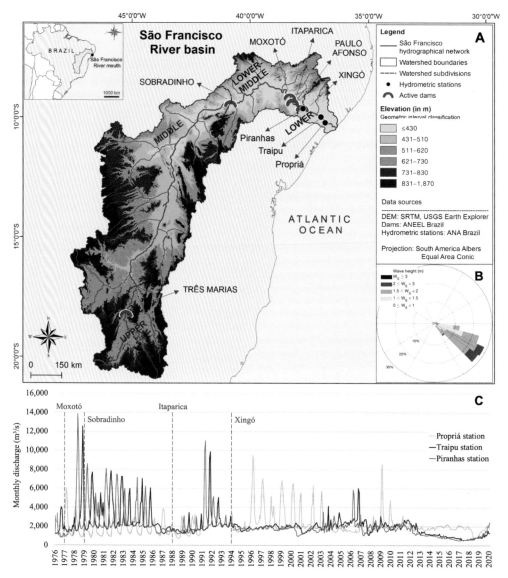

Figure 1.4 (a) Location of the São Francisco River in Brazil and its respective watershed with subdivisions (upper, middle, lower-middle and lower), also showing the location of dams and hydrometric stations; (b) wave rose for the 1984–2020 wave climate for the area adjacent to the São Francisco River mouth; (c) monthly discharges of the São Francisco River for the period 1976–2020 at the Piranhas (bright blue line), Traipu (grey line) and Propriá (cyan line) hydrometric stations, located, respectively, 176.8 km, 87.7 km and 55.7 km upstream from the river mouth. Red dashed lines indicate the beginning of operation of each hydroelectric plant constructed along the São Francisco River in this period. Discharge data from ANA for 1976–2020 (National Water Agency – Hidroweb – https://www.snirh.gov.br/hidroweb/serieshistoricas)

at Propriá. Discharge reduction has also impacted the circulation and salt balance in the estuarine reaches of the São Francisco. With less freshwater reaching the coast, saline intrusion has migrated upstream, meaning seawater influences the river up to 12 km from the mouth on a spring high tide (Paiva and Schettini 2021).

The delta shoreline has experienced significant changes in recent decades, with large stretches of coast experiencing either persistent advance or retreat (Fig. 1.5). Shoreline dynamics directly around the river mouth dominate the overall picture, where significant reshaping of this shoreline has generated shoreline change rates of over 50 m yr^{-1}. Over the multi-decadal timescale, erosional behaviour is exhibited in both margins of the mouth, but over the recent decade, the south margin has experienced notable progradation. Scales of shoreline change decrease as you move away from the river mouth, but increase again around the Parapuca channel (a back-barrier distributary) mouth that has migrated southward by 10 km over the 36-year period analysed. The stretch of shoreline between the São Francisco and Parapuca mouths, exhibiting retreat at the north end and advance at the south end, implies shoreline rotation. It is possible that sediment eroded from the shoreline south of the São Francisco mouth has been supplied to the north of the Parapuca mouth, facilitating this anticlockwise rotation. The orientation of the shoreline on this southern aspect of the delta fan allows waves, predominately from the south-east, to arrive at the shoreline with a slight onshore tendency to the south-west, meaning that they can facilitate alongshore sediment transport southward away from the main river mouth.

Over the last ten years, shoreline change has been more variable, illustrating a predominance of shoreline advance along the beach-dune coasts beyond the river and channel mouth regions. The greater magnitudes of change over the short term suggest that changes occur episodically, and that over the longer term the net effect is smaller, meaning

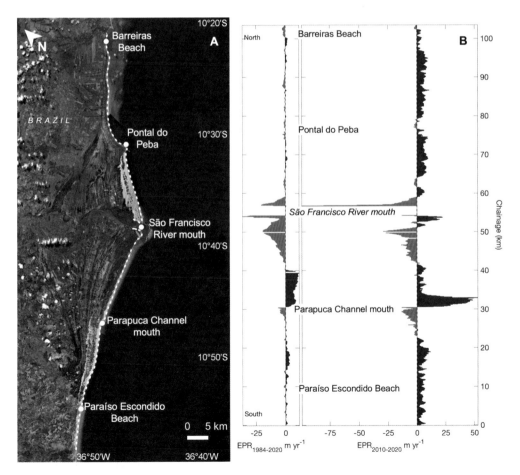

Figure 1.5 Shoreline change analysis for the São Francisco River delta, showing (a) the São Francisco delta shoreline and key locations; (b) End Point Rate (EPR) calculations for a long-term (1984–2020) and a short-term (2010–20) period of analysis

there is some cyclicity in behaviour. The north/east shoreline of the delta shows less alongshore variability over the shorter term, with a prevalence for shoreline advance; to the south/west, the patterns over multi-decadal and multi-annual timescales are similar, beyond the shift from retreat to advance on the south margin of the mouth, but the magnitudes are notably different, particularly in the more recent enhanced progradation of Paraíso Escondido Beach. The shoreline rotation between the São Francisco and Parapuca mouths is enhanced, but over a shorter alongshore distance as the shoreline advanced immediately to the south of the São Francisco mouth, in response to the onshore welding of river mouth bars. Interestingly the pivot of anticlockwise shoreline rotation between the São Francisco and Parapuca mouths does not change between the two timescales; a 1.5 km stretch of coastal barrier here, 250–500 m wide, comprising transverse dunes backed by mangrove and the Parapuca channel, has experienced only minor change in shoreline position over the last 40 years, irrespective of timescale within that period. But the rotation of the São Francisco and Parapuca stretch of coast is more extreme over the recent decade, largely because of increased accretion north of the Parapuca mouth.

This increase in magnitude of change around the river and channel mouths over the short term likely reflects the tendency for localised spit development and sediment bypassing around the mouth through bar development, migration and shoreline attachment. It is possible that these dynamics in the recent decade have increased with reduction in the São Francisco discharge following 2010, which has been consistently suppressed since then. Dominguez and Guimarães (2021) point out that river flow regulation with the maintenance of constant discharge (below 2,000 m^3/s) and absence of peak flows might not be sufficient to maintain the delta shoreline. The national water agency also decreed in 2017 to significantly reduce the minimum discharge volumes from the dams located in the lower parts of the catchment (Sobradinho and Xingó) (ANA 2017), which are now delivering less than a third of the discharge than was expected to maintain the shoreline and contain erosion processes considering the river inputs.

Case example: Senegal River

Among the largest estuaries in Africa, the Senegal River has a cross-border watershed between Senegal, Mauritania and Mali (Fig. 1.6a), sourced initially in the Fouta Djallon mountains in Guinea (Descroix et al. 2020). The river flows through different subtropical climates from Guinea (>1,000 mm of rainfall), through Sudanian (900–1,000 mm) to reach the Atlantic with a Sahelian climate (c. 350 mm), at the confluence between harmattan winds and Atlantic trade winds (Ministère de l'agriculture 1995). Like the São Francisco, the Senegal river mouth is described as both delta and estuary, and comprises an integrated suite of beach, dune, barrier and spit environments. The delta system was primarily constructed during the post-Nouakchottian period (5,500 years ago) (Michel 1973), and reshaping, valley infilling and meander development occurred during the Tafolien marine transgression (4,000–2,500 years ago) (Ameryckx et al. 1967; Thiam 2012). The shift toward estuarine functioning may have started with the Sahelian drought of the 1960s–80s that significantly lowered water levels in the Senegal River, and reduced its competency to transport sediments downstream. Human interventions such as damming policies undertaken to mitigate water availability in the valley, followed by breaching in the early 2000s of the Langue de Barbarie barrier that fronts the estuary and delta system, have possibly reinforced the development of estuarine processes. Tidal regime in the Senegal estuary is semidiurnal and microtidal. The wave climate is predominately from the north-north-east (Fig. 1.6b).

The Senegal River has a hydrological year that starts in May and a simple tropical hydrological regime that peaks in autumn. Fig. 1.6c shows water level variability in the Senegal River at Diama (upstream dam, upper estuary) and Saint-Louis (in the estuary). The Sahelian drought is evidenced in the records from both locations over the 1986–92

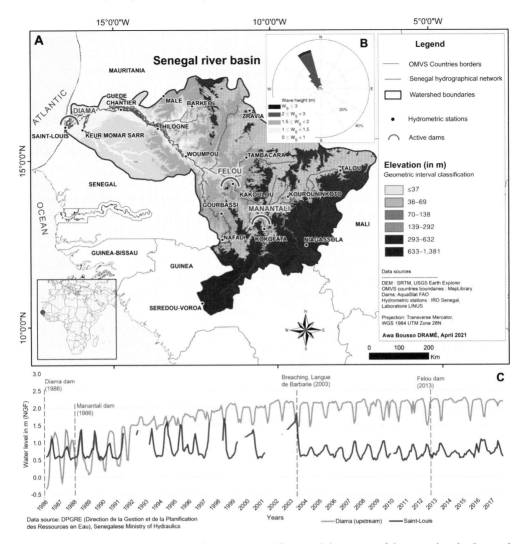

Figure 1.6 (a) Location of the Senegal River in West Africa, and the extent of the cross-border Senegal watershed, also showing the location of dams and hydrometric stations; (b) wave rose for the 1985–2020 wave climate for the area adjacent to the Senegal River mouth; (c) water levels recorded at Saint-Louis (in the Senegal estuary) and upstream of the Diama dam for the period 1986–2020

period. However, the differences in river levels from these two stations outline the spatial variabilities of the Senegal river. Water levels are indeed more constant in Saint-Louis from 1986 to 1990 than in Diama, where an upward trend is evident, with a marked shift in 1992. This reflected implementation of new damming policies from the managing authority of the Senegal River basin that were largely in reaction to hydroclimatic variabilities. The Diama dam was put into service in 1986 (22 km upstream of Saint-Louis) with an initial function to stop saline intrusion in the lower estuary (Dumas et al. 2010; Diedhiou et al. 2020). However, local water use conflicts related to irrigated agriculture, market gardening and urban consumption (Ndiaye, 2003) changed that first goal to supply water to the valley for agricultural purposes year-round (Cogels, Coly and Niang 1997; Taïbi et al. 2007; Mietton et al. 2005). The Manantali dam was completed in 1988 (more than 1,000 km upstream) to produce hydroelectricity (Faye, 2018). Following 1990, the impact of the Diama dam (located in the estuary) is evident in the Saint-Louis water level record, which became artificial and less regular than upstream. The influence on catchment hydrology that these, and more recent interventions (Felou dam, 2013) have had has also led to significant impacts on sediment delivery from the upper basin to the coast, which is now reduced but also more significantly regulated throughout the year (Kane 2005).

In addition to these policy-based water management decisions, there have also been responsive interventions in the Senegal River, most notably at the coast on the Langue de Barbarie sand spit. This southward-extending spit diverts the course of the river to the south, protecting the Senegal estuary from the Atlantic, but in doing so, retaining river waters within the lower reaches of the Senegal; in early 2003, the diversion was over 32 km. Flood alerts in early October 2003 led to a rapid decision to breach the Langue de Barbarie in order to reduce the flood risk at Saint-Louis. The initial 4 m wide and 1.5 m deep breach quickly became the new mouth of the Senegal River (Barry and Kraus 2009; Durand, Anselme and Thomas 2010; Weissenberger et al. 2016), inducing extreme and rapid morphological changes in the estuary, the first being a reduction in the diversion to just 8 km.

The Langue de Barbarie spit system represents the southern half of the Senegal delta shoreline, and much of the seaward shoreline of the Senegal estuary. In 1985, the spit diverted the Senegal mouth 23 km to the south; progressive extension of the spit continued until the breach in 2003 at which point the inlet was 32 km to the south. Mouth dynamics

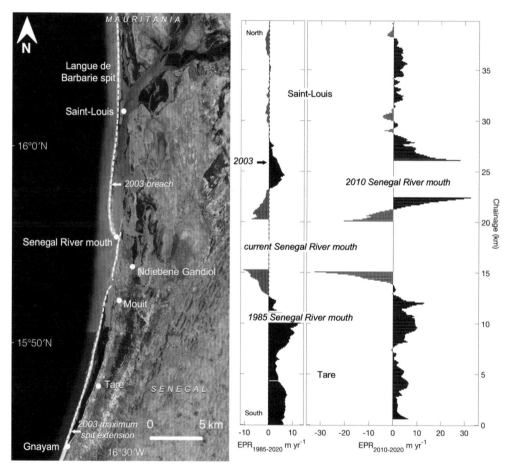

Figure 1.7 Shoreline change analysis for the Senegal River delta, showing (a) the Senegal delta shoreline and key locations; (b) End Point Rate (EPR) calculations for a long-term (1985–2020) and a short-term (2010–20) period of analysis

play a dominant role in shoreline dynamics (Fig. 1.7), particularly in terms of barrier fragmentation and disintegration; following the breach in 2003, the mouth expanded through erosion of the south margin, to 500 m wide by early 2004, to over 2.5 km by 2010, and almost 4.5 km in 2020. Rates of shoreline change are marked by locations and size of the mouth; erosion of the southern margin expands the mouth width and accretion is experienced on the north margin, allowing the spit to extend the barrier system into the space of the former mouth. There have been important changes in the shoreline elsewhere along the barrier. Around Saint-Louis and further north, the rates of change over the longer term are small relative to the rest of the shoreline, but indicate a persistent retreat. In the last decade, this has largely been reversed, with much of this shoreline exhibiting advance. To the south of the region, the signature is persistently accretional, illustrating equivalent rates of advance over both timescales.

The patterns identified here are well reported, and several studies have recognised the progressive return of the mouth location and behaviour to that before breaching (Bergsma et al. 2020). This is being driven by the dominant north-westerly long-range swells from the mid- to high latitudes. The expansion in the river mouth is possibly aided, however, by swell from the south-west, influenced by the South Atlantic Southern Annular Mode (Almar, Kestenare and Boucharel 2019), but is likely more associated with the enhanced tidal exchange that the Senegal River has experienced since the artificial breaching. The recent configuration of the Langue de Barbarie permits a more efficient release of fluvial flood waters, but has allowed an increased saline intrusion and tidal influence; the hydrodynamics of the river are now clearly estuarine rather than deltaic. Before the breach, tides were imperceptible during the flood season and had a maximum range of 0.3 m (2000–1); following the breach, tides became detectable all year long, with a maximum range of 0.93 m (2004–5) (Mietton et al. 2005).

Discussion

Deltas are deposition-dominated systems that develop through the accumulation of sediment that usually results in the seaward advance of the shoreline. As sedimentary frameworks, they comprise a range of depositional landforms, and the nature of both the delta system and the landforms it comprises is a product of fluvial and marine processes, and the sediment supplied and transported by both. In their construction phase, deltas are largely formed in environments experiencing higher discharges and sediment loads that supplies their progradation. Changes in the relative dominance of fluvial and marine processes, and more importantly the availability of sediment, can result in shift to a supply-limited system, and the development of non-deltaic conditions within the delta framework. It is therefore not surprising that delta systems comprise a range of dynamic landforms, and have the potential, in response to climate change and anthropogenic intervention, to operate as estuaries, particularly in their seaward reaches.

The sites analysed here within the tropical Atlantic demonstrate the consequence of human intervention, mainly owing to dam building, on delta function. Dams are a common feature within the tropics where strongly seasonal weather patterns, which are often modulated over multi-annual to decadal timescales in response to ocean- to global-scale teleconnections, drive significant changes in rainfall and availability of freshwater. The construction of dams is a well-established approach to water management in order to regulate water supply to communities and regions throughout the year. But these structures modify natural river processes and interfere in the supply of sediment to the coast, significantly reducing river flows and their seasonal and interannual peaks. In the São Francisco and Senegal deltas, the alteration in river hydrology over the last 40 years appears to have led to a shift to estuarine conditions within the lower delta (changing river-tide dominance) and a change in shoreline dynamics alongshore from the delta mouth (reduced sediment supply to the adjacent coast).

The impacts of human interventions are also intensified by climate change. Considering the influence of extreme climatic events in the São Francisco River basin, de Jong et al. (2018) have demonstrated that rainfall in the region has been lower than normal since 1992, with the streamflow in 2015–17 at least 60% below the long-term average; their modelling work has highlighted that rainfall reductions projected for 2100 can be anticipated by 2050. Changes in rainfall patterns also have consequences to the West African coast. For instance, Debenay, Pages and Guillou (1994) illustrated how these changes across Senegal resulted in a significant change to the relative influence of fluvial and tidal processes in the Casamance River estuary. Here, decreasing rainfall associated with larger-scale cycles in atmospheric circulation led to extended incursion of saline conditions into previously freshwater environments, with important and rapid impacts on river and estuary ecology. A similar situation is taking place in the Senegal through a combination of river flow reduction, artificial barrier breaching and increase in tidal incursion. Discharge regulation in the São Francisco and Senegal rivers has affected their function as a sediment source to the coast. Dominguez and Guimarães (2021) have suggested that contemporary fluvial inputs from the São Francisco River are sufficient to sustain but not advance the shoreline, and that the last major event of delta progradation was 1,000 years ago. For the Senegal River, the dams have decreased the quantity and type (grain size) of sediment transported downstream, forcing a significant reduction in the competency of the sediment transport system, and ultimately the supply of new material to the coast.

Analyses of shoreline dynamics have revealed here that the vicinity of contemporary river mouths is the area at most risk of erosion. It is possible that mouth margins are exhibiting front-line responses to reduced sediment supply. The São Francisco and Senegal mouths have both experienced expansion over the last 40 years. In the 1980s, the São Francisco and Senegal mouths were 1 km and 400 m wide respectively; in 2020 they were 1.75 km and 4.5 km respectively. The mouth margins of these systems are formed in littoral sediments, as beaches, barriers, spits and beach ridges, which are all responsive to erosional processes. Estuary inlet cross-sections have a well-established positive correlation (power law) with tidal prism (O'Brien, 1931), and an expansion in inlet width usually implies an increase in the degree of tidal incursion. In both the São Francisco and Senegal, mouth expansion has taken place over a period characterised by increased regulation of and decreased discharge in river flow; but the longer-term consequence of these changes is through the development of estuarine hydrodynamics, which are already imposing significant challenges such as flooding in the densely populated Saint-Louis on the Senegal estuary.

The adjacent shorelines of the São Francisco and Senegal deltas have also experienced spatial and temporal variability, displaying stretches of retreat and advance over different timeframes. In most cases, the magnitudes of change have been larger for the most recent decade. Some parts of these shorelines have also exhibited a tendency for rotation, a behaviour that is often linked to seasonal reversals in the direction of wave forcing, but here presents a longer-term behaviour that is evident over longer timescales. Again, this is possibly a consequence of decreasing sediment supply, but might also reflect climate change impacts on wave forcing. McCloskey, Bianchette and Liu (2013) have demonstrated that a positive North Atlantic Oscillation might produce more tropical cyclones along the US east coast, resulting in more north-westerly swell waves reaching the Senegalese coasts. Various studies are demonstrating that wave climates in the tropical Atlantic are changing as a consequence of climate change, largely in terms of increased variability and energy, and it is clear that this exerts some control on the dynamics of delta shorelines. For instance, Dada et al. (2016) have demonstrated that increasing wave energy, and subtle shifts in the dominant wave directions in the wave climate of the Gulf of Guinea, western Africa, have increased rates of alongshore sediment transport along the Niger delta shoreline. But similar to the points made here, they recognise that climate variability and human modifications within the Niger catchment and coast make it difficult to directly attribute specific changes in processes to shoreline response and change in delta configuration. Indeed, the morphodynamics of tropical river mouths present clear examples of complex systems where non-linearities in shoreline dynamics most likely reflect a combination of the changes in both fluvial and marine forcing over multi-annual to decadal timescales.

Conclusions

Tropical Atlantic river mouths are subject to similar climatic regimes and wave conditions, but they comprise a variety of different landforms across a range of different contexts that do not explicitly align with their definition as delta or estuary. Their morphodynamics is directly linked to catchment water management (influencing fluvial discharge and sediment loads) and marine processes, which vary geographically, but are also changeable in time. Greater scales of morphological change are found in the vicinity of these river mouths than along their adjacent shorelines, and with an increase in broader environmental changes, the intensity of modification seems to happen in a shorter period. The millennia-scale development of the São Francisco and Senegal river deltas is now contrasting with the decadal-scale expansion of erosional processes around the river mouths that is facilitating the development of estuarine conditions within these deltaic sedimentary frameworks. Reduction in fluvial sediment supply to the coast is being balanced, in terms of shoreline sediment budgets, by erosion of the delta shoreline and growth of coastal-erosion derived sediment supply. The complexity of tropical river systems and the interplay of marine and fluvial processes in their adjacent shorelines open several opportunities for investigation. Climate variability, mainly of rainfall (flood and droughts), and catchment management (rivers modified by dams, not only influenced by natural forces but human intervention as well) may also contribute to accelerate the modification in these areas, making them hotspots for additional studies to deal with ongoing and future challenges/adaptations to which these tropical regions are susceptible/subjected. Moreover, changes in policy that regulate the water discharge and the deforestation patterns in the watershed, land-use and land-cover changes, and climate change impacts (including intensification of droughts, frequency of storms and sea-level rise) will together bring significant challenges to these environments and the communities that depend on them.

References

Almar, R., Kestenare, E. and Boucharel, J. (2019) On the key influence of remote climate variability from tropical cyclones, north and south Atlantic mid-latitude storms on the Senegalese coast (West Africa). *Environmental Research Communications* 1 (7): 071001. https://doi.org/10.1088/2515-7620/ab2ec6

ANA (Agência Nacional de Águas) (2017) *Resolução N° 2.081, de 04 de Dezembro de 2017 – Documento n° 00000.080754/2017–91*. Available at: https://arquivos.ana.gov.br/resolucoes/2017/2081-2017.pdf (Accessed: 26 April 2021).

Barry, K.M. and Kraus, N.C. (2009) Stability of blocked river mouth on west coast of Africa: inlet of Senegal River Estuary. US Army Corps of Engineers,

Coastal and Hydraulics Laboratory (U.S.) Engineer Research and Development Center (U.S.) report ERDC/CHL TR-09-20, 56pp.

Bergsma, E.W., Sadio, M., Sakho, I., Almar, R., Garlan, T., Gosselin, M. and Gauduin, H. (2020) Sand-spit evolution and inlet dynamics derived from space-borne optical imagery: is the Senegal-river inlet closing? *Journal of Coastal Research* 95 (SI): 372–6. https://doi.org/10.2112/SI95-072.1

Bettencourt, P., Fulgêncio, C., Grade, M., Alcobia, S., Monteiro, J.P., Oliveira, R., Leitão, J.C., Leitão, P.C., Fernandes, P.A., de Sousa, S. and Brites, S. (2016) Plano de recursos hídricos da bacia hidrográfica do rio São Francisco. *Recursos Hídricos* 37 (1): 73–80. https://doi.org/10.5894/rh37n1-cti3

Bianchi, T.S. and Allison, M.A. (2009) Large-river delta-front estuaries as natural 'recorders' of global environmental change. *Proceedings of the National Academy of Sciences* 106 (20): 8085–92. https://doi.org/10.1073/pnas.0812878106

Bittencourt, A.C.D.S.P., Dominguez, J.M.L., Fontes, L.C.S., Sousa, D.L., Silva, I.R. and da Silva, F.R. (2007) Wave refraction, river damming, and episodes of severe shoreline erosion: the São Francisco river mouth, Northeastern Brazil. *Journal of Coastal Research* 23 (4, 234): 930–8. https://doi.org/10.2112/05-0600.1

Boyd, R., Dalrymple, R. and Zaitlin, B.A. (1992) Classification of clastic coastal depositional environments. *Sedimentary Geology* 80 (3–4): 139–50. https://doi.org/10.1016/0037-0738(92)90037-R

Brown, A.G., Tooth, S., Bullard, J.E., Thomas, D.S., Chiverrell, R.C., Plater, A.J., Murton, J., Thorndycraft, V.R., Tarolli, P., Rose, J. and Wainwright, J. (2017) The geomorphology of the Anthropocene: emergence, status and implications. *Earth Surface Processes and Landforms* 42 (1): 71–90. https://doi.org/10.1002/esp.3943

Caldwell, R.L., Edmonds, D.A., Baumgardner, S., Paola, C., Roy, S. and Nienhuis, J.H. (2019) A global delta dataset and the environmental variables that predict delta formation on marine coastlines. *Earth Surface Dynamics* 7 (3): 773–87. https://doi.org/10.5194/esurf-7-773-2019

Castro, C.N.D. and Pereira, C.N. (2019) Revitalização da bacia hidrográfica do rio São Francisco: histórico, diagnóstico e desafios. Brazil: IPEA (Instituto de Pesquisa Econômica Aplicada). Available at: http://www.ipea.gov.br/portal/images/stories/PDFs/livros/livros/190724_livro_revitalizacao_hidrografica.pdf (Accessed: 26 April 2021).

Cogels, F.X., Coly, A. and Niang, A. (1997) Impact of dam construction on the hydrological regime and quality of a Sahelian lake in the River Senegal basin. *Regulated Rivers: Research & Management* 13 (1): 27–41. https://doi.org/10.1002/(SICI)1099-1646(199701)13:1<27::AID-RRR421>3.0.CO;2-G

Dada, O.A., Li, G., Qiao, L., Asiwaju-Bello, Y.A. and Anifowose, A.Y.B. (2018) Recent Niger Delta shoreline response to Niger River hydrology: conflict between forces of nature and humans. *Journal of African Earth Sciences*, 139: 222–31. https://doi.org/10.1016/j.jafrearsci.2017.12.023

de Jong, P., Tanajura, C.A.S., Sánchez, A.S., Dargaville, R., Kiperstok, A. and Torres, E.A. (2018) Hydroelectric production from Brazil's São Francisco River could cease due to climate change and inter-annual variability. *Science of the Total Environment* 634: 1540–53. https://doi.org/10.1016/j.scitotenv.2018.03.256

Debenay, J.P., Pages, J. and Guillou, J.J. (1994) Transformation of a subtropical river into a hyperhaline estuary: the Casamance River (Senegal) – paleogeographical implications. *Palaeogeography, Palaeoclimatology, Palaeoecology* 107 (1–2): 103–19. https://doi.org/10.1016/0031-0182(94)90167-8

Descroix, L., Faty, B., Manga, S.P., Diedhiou, A.B., Lambert, L.A., Soumaré, S., Andrieu, J., Ogilvie, A., Fall, A., Mahe, G. and Sombily Diallo, F.B. (2020) Are the Fouta Djallon highlands still the water tower of West Africa? *Water* 12 (11): 2968. https://doi.org/10.3390/w12112968

Diedhiou, R., Sambou, S., Kane, S., Leye, I., Diatta, S., Sane, M.L. and Ndione, D.M. (2020) Calibration of HEC-RAS model for one dimensional steady flow analysis – a case of Senegal River estuary downstream Diama Dam. *Open Journal of Modern Hydrology* 10 (3): 45–64. https://doi.org/10.4236/ojmh.2020.103004

Dominguez, J.M.L. (1996) The Sao Francisco strandplain: a paradigm for wave-dominated deltas? *Geological Society, London, Special Publications* 117 (1): 217–31. https://doi.org/10.1144/GSL.SP.1996.117.01.13

Dominguez, J.M.L., Bittencourt, A.C.D.S.P., Santos, A.N. and do Nascimento, L. (2016) The sandy beaches of the states of Sergipe-Alagoas. In: A. Short and A. Klein (eds) *Brazilian Beach Systems*, Coastal Research Library 17, pp. 281–305. Cham: Springer. https://doi.org/10.1007/978-3-319-30394-9_11

Dominguez, J.M.L. and Guimarães, J.K. (2021) Effects of Holocene climate changes and anthropogenic river regulation in the development of a wave-dominated delta: the São Francisco River (eastern Brazil). *Marine Geology* 435: 106456. https://doi.org/10.1016/j.margeo.2021.106456

Dumas, D., Mietton, M., Hamerlynck, O., Pesneaud, F., Kane, A., Coly, A., Duvail, S. and Baba, M.L.O. (2010) Large dams and uncertainties: the case of the Senegal River (West Africa). *Society and Natural Resources* 23 (11): 1108–22. https://doi.org/10.1080/08941920903278137

Durand, P., Anselme, B. and Thomas, Y.F. (2010) L'impact de l'ouverture de la brèche dans la langue de Barbarie à Saint-Louis du Sénégal en 2003: un changement de nature de l'aléa inondation? *Cybergeo: European Journal of Geography* 496. https://doi.org/10.4000/cybergeo.23017

Fagherazzi, S., Edmonds, D.A., Nardin, W., Leonardi, N., Canestrelli, A., Falcini, F., Jerolmack, D.J., Mariotti, G., Rowland, J.C. and Slingerland, R.L. (2015) Dynamics of river mouth deposits. *Reviews of Geophysics* 53 (3): 642–72. https://doi.org/10.1002/2014RG000451

Fairbridge, R.W. (1980) The estuary: its definition and geodynamic cycle. In E. Olausson and I. Cato (eds) *Chemistry and Biogeochemistry of Estuaries*, pp. 1–35. New York: John Wiley & Sons.

Faye, C. (2018) Weight of transformations and major drifts related to major river water projects in Africa: case of the Manantali dam on the Senegal river basin. *Journal of Research in Forestry, Wildlife and Environment* 10 (3): 13–24.

Foltz, G.R., Brandt, P., Richter, I., Rodríguez-Fonseca, B., Hernandez, F., Dengler, M., Rodrigues, R.R., Schmidt, J.O., Yu, L., Lefèvre, N. and Da Cunha,

L.C. (2019) The tropical Atlantic observing system. *Frontiers in Marine Science* 6: 206. https://doi.org/10.3389/fmars.2019.00206

Galloway, W.E. (1975) Process framework for describing the morphologic and stratigraphic evolution of deltaic depositional systems. In: M.L. Broussard (ed) *Deltas: models for exploration*, pp. 87–98. Houston: Houston Geological Society.

Guimarães, J.K. (2010) Evolução do delta do rio são Francisco-estratigrafia do Quaternário e relações morfodinâmicas. PhD thesis. Federal University of Bahia. Available at: http://repositorio.ufba.br/ri/handle/ri/21481 (Accessed: 14 April 2021).

IBGE (Instituto Brasileiro de Geografia e Estatística) (2009) *Vetores estruturantes da dimensão socioeconômica da bacia hidrográfica do Rio São Francisco 2009*. Rio de Janeiro: série Estudos & Pesquisas – Informação Geográfica 6. Available at: https://biblioteca.ibge.gov.br/visualizacao/livros/liv42291.pdf (Accessed: 25 April 2021).

IBGE (Instituto Brasileiro de Geografia e Estatística) (2021) *Banco de Dados – Cidades@*. Available at: https://cidades.ibge.gov.br/brasil/al/piacabucu/panorama (Accessed: 26 April 2021).

Jahfer, S., Vinayachandran, P.N. and Nanjundiah, R.S. (2020) The role of Amazon river runoff on the multidecadal variability of the Atlantic ITCZ. *Environmental Research Letters* 15 (5): 054013. https://doi.org/10.1088/1748-9326/ab7c8a

Kane, A. (2005) Regulation du Fleuve Sénégal et flux de matières particulaire vers l'estuaire depuis la construction du Barrage de Diama. In: A.H. Horowitz and D.E. Walling (eds) *Sediment Budgets 2*, IAHS Publication 292: 279–90.

Knoppers, B., Medeiros, P.R., de Souza, W.F. and Jennerjahn, T. (2006) The São Francisco estuary, Brazil. In: Wangersky P.J. (eds) *Estuaries*, pp. 51–70. The Handbook of Environmental Chemistry, vol 5H. Berlin, Heidelberg: Springer. https://doi.org/10.1007/698_5_026

Materia, S., Gualdi, S., Navarra, A. and Terray, L. (2012) The effect of Congo River freshwater discharge on Eastern Equatorial Atlantic climate variability. *Climate Dynamics* 39 (9): 2109–25. https://doi.org/10.1007/s00382-012-1514-x

McCloskey, T.A., Bianchette, T.A. and Liu, K.B. (2013) Track patterns of landfalling and coastal tropical cyclones in the Atlantic basin, their relationship with the North Atlantic Oscillation (NAO), and the potential effect of global warming. *American Journal of Climate Change* 2: 12–22. https://doi.org/10.4236/ajcc.2013.23A002

Medeiros, P.P., dos Santos, M.M., Cavalcante, G.H., De Souza, W.F.L. and da Silva, W.F. (2014) Características ambientais do Baixo São Francisco (AL/SE): efeitos de barragens no transporte de materiais na interface continente-oceano. *Geochimica Brasiliensis* 28 (1): 65–78. https://doi.org/10.5327/Z0102-9800201400010007

Melo, S.C., Araújo Filho, J.C. and Carvalho, R.M.C.M.O. (2020) Curvas-chave de descargas de

sedimentos em suspensao no Baixo Sao Francisco. *Revista Brasileira de Geografia Física* 13 (3): 1248–62. https://doi.org/10.26848/rbgf.v13.3.p1248-1262

Michel, P. (1973) *Les bassins des fleuves Sénégal et Gambie: étude géomorphologique*. Memoires Orstom 63. Paris: Orstom.

Mietton, M., Dumas, D., Hamerlynck, O., Kane, A., Coly, A., Duvail, S., Baba, M.L.O. and Daddah, M. (2005) Le delta du fleuve Sénégal Une gestion de l'eau dans l'incertitude chronique. *Incertitude et Environnement*, 321–36. hal-00370662.

Milliman, J.D. and Farnsworth, K.L. (2013) *River Discharge to the Coastal Ocean: a Global Synthesis*. Cambridge: Cambridge University Press.

Ministère de l'agriculture (1995) Senegal – Rapport de pays pour la conférence technique internationale de la FAO sur les ressources phytogénétiques. Food and Agriculture Organization of the United Nations.

Ndiaye, E.H.M. (2003) Le fleuve Sénégal et les barrages de l'OMVS: quels enseignements pour la mise en œuvre du NEPAD?. *VertigO – la revue électronique en sciences de l'environnement* 4 (3). https://doi.org/10.4000/vertigo.3883

Nienhuis, J.H., Ashton, A.D., Edmonds, D.A., Hoitink, A.J.F., Kettner, A.J., Rowland, J.C. and Törnqvist, T.E. (2020) Global-scale human impact on delta morphology has led to net land area gain. *Nature* 577 (7791): 514–18. https://doi.org/10.1038/s41586-019-1905-9

O'Brien, M.P. (1931) Estuary tidal prisms related to entrance areas. *Civil Engineering* 1 (8): 738–9.

Paiva, B.P. and Schettini, C.A. (2021) Circulation and transport processes in a tidally forced salt-wedge estuary: the São Francisco river estuary, Northeast Brazil. *Regional Studies in Marine Science* 41: 101602. https://doi.org/10.1016/j.rsma.2020.101602

Pereira, S.B., Pruski, F.F., Silva, D.D.D. and Ramos, M.M. (2007) Estudo do comportamento hidrológico do Rio São Francisco e seus principais afluentes. *Revista Brasileira de Engenharia Agrícola e Ambiental* 11 (6): 615–22. https://doi.org/10.1590/S1415-43662007000600010

Rangel, A.G.D.A.N. and Dominguez, J.M.L. (2020) Antecedent topography controls preservation of latest Pleistocene-Holocene transgression record and clinoform development: the case of the São Francisco delta (eastern Brazil). *Geo-Marine Letters* 40 (6): 935–47. https://doi.org/10.1007/s00367-019-00609-8

Soares, E. (2013) Seca no Nordeste e a transposição do rio São Francisco. *Geografias* 9 (2): 75–86.

Syvitski, J.P., Vörösmarty, C.J., Kettner, A.J. and Green, P. (2005) Impact of humans on the flux of terrestrial sediment to the global coastal ocean. *Science* 308 (5720): 376–80. https://doi.org/10.1126/science.1109454

Syvitski, J.P. (2008) Deltas at risk. *Sustainability Science* 3 (1): 23–32. https://doi.org/10.1007/s11625-008-0043-3

Syvitski, J.P., Kettner, A.J., Overeem, I., Hutton, E.W., Hannon, M.T., Brakenridge, G.R., Day, J., Vörösmarty, C., Saito, Y., Giosan, L. and Nicholls, R.J. (2009) Sinking deltas due to human activities. *Nature Geoscience* 2 (10): 681–6. https://doi.org/10.1038/ngeo629

Taïbi, A.N., Barry, M.E.H., Jolivel, M., Ballouche, A., Baba, M.L.O. and Moguedet, G. (2007) Enjeux et impacts des barrages de Diama (Mauritanie) et Arzal (France): des contextes socio-économiques et environnementaux différents pour de mêmes conséquences. *Norois. Environnement, aménagement, société* 203: 51–66. https://doi.org/10.4000/norois.1536

Warrick, J.A. (2020) Littoral sediment from rivers: patterns, rates and processes of river mouth morphodynamics. *Frontiers in Earth Science* 8: 355. https://doi.org/10.3389/feart.2020.00355

Weissenberger, S., Noblet, M., Plante, S., Chouinard, O., Guillemot, J., Aubé, M., Meur-Ferec, C., Michel-Guillou, É., Gaye, N., Kane, A. and Kane, C. (2016) Changements climatiques, changements du littoral et évolution de la vulnérabilité côtière au fil du temps: comparaison de territoires français, canadien et sénégalais. *VertigO – la revue électronique en sciences de l'environnement* 16 (3). https://doi.org/10.4000/vertigo.18050

Wright, L.D. (1977) Sediment transport and deposition at river mouths: a synthesis. *Geological Society of America Bulletin* 88 (6): 857–68. https://doi.org/10.1130/0016-7606(1977)88<857:STADAR>2.0.CO;2

CHAPTER 2

Coastal Erosion and Management Challenges in the United Kingdom

KENNETH PYE and SIMON J. BLOTT

Abstract

Coastal erosion is a long-standing issue on some parts of the UK coast, resulting in significant loss of agricultural land, as well as towns and villages in places such as Suffolk and the East Riding of Yorkshire. Erosion of beaches, dunes and shingle ridges on low-lying coasts can also lead to an increased flood risk and significant loss of habitats including intertidal flats and saltmarshes. Current climate change projections suggest that these issues are likely to become worse owing to the combined effects of sea-level rise, higher inshore wave energy associated with greater water depths and possible increases in storminess in some areas. There will be greater pressure on coastal defences and increased risk to coastal towns and infrastructure, including nuclear power stations, oil and gas terminals and transport links. This chapter provides an overview of coastal erosion in the UK, aided by specific examples, and considers the challenges for future management in an era of climate change and increasing pressure for economic development.

Keywords: coastal erosion, agricultural land loss, flood risk, management challenges, climate change impacts

Correspondence: k.pye@kpal.co.uk

Introduction

Coastal erosion involves the loss of sediment or rock from the edge of the land, resulting in landward displacement of the interface between land and sea. This is often associated with landward migration of the intertidal zone and subtidal bathymetric contours in the shallow subtidal zone, from where there may also be loss of sediment volume. Coastal erosion is a global phenomenon that has long attracted scientific attention (Matthews 1913; Komar 1983; Charlier and De Meyer 1998; Pranzini and Williams 2013; Pranzini, Wetzel and Williams 2015; Mentaschi et al. 2018; Williams et al. 2018), but it usually only presents a problem when assets and activities of human interest are impacted. Negative consequences include loss of agricultural land and damage to property, industrial assets, transport infrastructure and ecological habitats, either directly or indirectly through increases in tidal flood risk (Table 2.1). Erosion of legacy landfills and industrial waste tips can also be a source of environmental pollution and have adverse impact on scenic quality (Pope et al. 2011; Cooper et al. 2014; Brand and Spencer 2019). Erosion can also create a public

Kenneth Pye and Simon J. Blott, 'Coastal Erosion and Management Challenges in the United Kingdom' in: *Challenges in Estuarine and Coastal Science*. Pelagic Publishing (2022). © Kenneth Pye and Simon J. Blott. DOI: 10.53061/KLYK5564

Table 2.1 The nature, causes and broadscale consequences of coastal erosion

Area affected	Positive aspects	Negative aspects
Land area (above Highest Astronomical Tide (HAT))	Creation of attractive coastal scenery	Loss of agricultural land
	Release of sediment to feed beaches, dunes and saltmarshes downdrift	Increased sedimentation in ports, harbours, navigation channels and on shellfish layings
	Exposure of rock and sediment sequences of geological interest	Threats to assets including property, transport infrastructure, important habitats
	Exposure of archaeological structures and artefacts	Loss of archaeological and historical sites
	Creation of cliff and cliff-top habitats	Loss of habitats of ecological importance (dunes, saltmarshes, saline lagoons)
		Potential increase in coastal flood risk
Intertidal zone (HAT-LAT)		Loss of recreational beaches
		Loss of saltmarsh, sandflat and mudflat habitat; limitation of aeolian sand supply to dunes; increased risk of cliff erosion
		Loss of shell fishing areas
		Reduced wave energy dissipation and increased threat of damaged to coastal defences
		Increased coastal flood risk
Subtidal areas (below LAT)	Improved navigation depths	Reduced sediment supply to beaches and dunes
	Supply of sediment to areas of lower energy (coastal, estuarine or deeper water)	Reduced wave energy dissipation; increased risk of shoreline erosion and coastal flooding
		Exposure of assets on the seabed (cables, pipelines, footings of windfarm monopiles, etc.)

safety hazard through risk of cliff collapse and landslides (Table 2.1). On the other hand, erosion can be beneficial through the release of sediment to the nearshore system, allowing development of tidal flats, saltmarshes, beaches and dunes. Erosion is responsible for some of the most dramatic coastal scenery in the UK (Steers 1946, 1953; Whittow 2019), exposing key geological sequences and sites of archaeological or historical interest, although these may be short lived (May 2014; Murphy 2014; Westley and McNeary 2014). Erosion also maintains bare slopes and cliffs, which are important habitats for specialist invertebrates, plants and nesting birds (Howe 2015; Rees, Curson and Evans 2015; Table 2.1).

This chapter provides an overview of coastal erosion in the UK, including its extent, causes and the resulting management challenges at a time of increasing development pressure and climate change, drawing on case examples from each of the four nations (Fig. 2.1).

Extent and rates of coastal erosion

Sediments, weakly indurated rocks and landfill deposits have higher susceptibility to coastal erosion than 'hard' rocks, although the latter are not immune (Lee and Clark 2002; Lee 2005; Lim et al. 2010; Table 2.2). Moderately and highly susceptible lithologies occur extensively along the coasts of southern and eastern England, and more locally in Wales, northwest

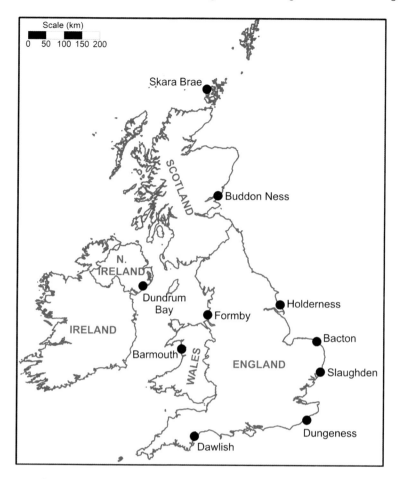

Figure 2.1 Map of the UK showing locations of coastal erosion examples referred to in the text

England, Scotland and Northern Ireland (Table 2.3). Masselink et al. (2020) suggested typical average cliff recession rates of 0.01–0.1 m/yr on hard rock coasts and 0.1–1.0 m/yr on soft rock coasts, although in non-cohesive sediments average rates can exceed 7 m/yr (Brooks and Spencer 2012). Recession is often an episodic process (Cambers 1976; Pethick 1996; Lee 2002; Brooks and Spencer 2010, 2013), and average rates at any location are dependent on the time period considered.

Based largely on an assessment of Ordnance Survey (OS) and British Geological Survey maps, FUTURECOAST (Halcrow 2002) estimated that up to 67% of the coast of England and Wales is at risk of erosion, although a significant proportion of this is protected by defences. The EUROSION (2004) study concluded that approximately 28% of the coast in England and Wales is eroding at rates of >0.1 m/yr. Based on limited further analysis, the National Coastal Erosion Risk Mapping Project (Rogers et al. 2014) suggested that 42% of the coast of England and Wales is at risk from erosion, of which 82% in undefended. McInnes and Moore (2011) reported that around 53% of the coast in England and Wales is characterised by cliffs, of which about 75% are formed in weak rocks vulnerable to weathering, erosion and landslides.

However, considerable uncertainty surrounds these estimates; even the total coastline length is not agreed since estimates are dependent on the resolution of measurement, and hence on the scale of maps, aerial photographs or satellite data used (Mandelbrot

Table 2.2 Susceptibility of rocks and sediments to coastal erosion

Lithology group code	Example lithological groupings	Example lithologies	Susceptibility to wave action	Susceptibility to tide and current action	Susceptibility to wind action	Susceptibility to mass movement	Overall Erosion susceptibility and typical average rates (m/y)
A	Hard igneous and metamorphic rocks	Granite Dolerite Basalt Gneiss	L	L	L	L	L 0.1–0.3
B	Hard sedimentary rocks	Carboniferous Limestone and Coal Measures Jurassic Sandstones Hard Chalk	L	L	L	M	M 0.1–0.5
C	Soft sedimentary rocks	Soft Chalk Tertiary sands, gravels and clays	H	M	L	M	M – H 0.2–3.0
D	Unconsolidated Quaternary sediments	Glacial till, raised beach deposits, sand dunes, saltmarshes	H	M	L – H	M – H	H 0.2–3.0
E	Waste/landfill/made ground	Mining waste, demolition rubble, historical landfill	M – H	L – M	L	L – M	M 0.1–3.0

Table 2.3 Spatial variation of coastal erosion risk in the UK. (The different lithological grouping codes are exemplified in Table 2.2)

Region	Spatial limits	Dominant lithological groupings	Glacial till and periglacial slope deposits	Raised beach deposits	Sand dunes	Sand beaches, ridges and tidal flats	Gravel beaches and ridges	Saltmarshes and mudflats	Landfill and waste
Scotland	Solway to the Tweed	A & B, local D	X	X	X	X	X	x	x
NE England	Tweed to the Humber	B, local A & D	X	x	x	x	x	x	X
E England	Humber to the Thames	C & D	X	x	x	X	X	x	x
SE England	Thames to the Solent	C & D			x	X	X	X	X
SW England	Solent to the Severn Estuary	A, B, C, local D	x	x	X	X	X	X	x
Wales	Severn Estuary to the Dee	A, B, local D	X	x	X	X	X	X	x
NW England	Dee to the Solway	B, D	X	x	X	X	x	X	X
N Ireland	Lough Foyle to Carlingford Lough	A, B, local C	X	X	X	X	x	x	x

X – relatively major; x – relatively minor

1967), what size of islands are included, the landward limit chosen within estuaries and the elevation contour chosen (e.g. mean high water, mean high water springs, edge of the 'land'). The CIA Handbook (CIA 2020) gives the length of the UK coast as 12,429 km, whereas EUROSION (2004) suggested 17,381 km and the OS have reported the coastline length of Great Britain (excluding Northern Ireland, Isle of Man and the Channel Islands) to be 17,820 km (Tabor, 2018). It is also uncertain how much of the UK coast is 'defended', or even 'developed', and therefore at lower risk of actual erosion. A survey commissioned by the National Trust to underpin its Operation Neptune campaign in 1965 noted that large sections of the coast in England, Wales and Northern Ireland had been developed, but identified 5,378 km of "pristine land for permanent preservation" (National Trust, 2015a). A separate survey by the Countryside Commission (1968) reported that only 8.8% of the Yorkshire and Lincolnshire coast was developed, compared with 35% in Kent, Sussex, Hampshire and the Isle of Wight. This figure has since increased to almost 50% (National Trust, 2015a).

The Dynamic Coast project (Hansom, Fitton and Rennie 2017) found that 78% of Scotland's coast is 'hard or mixed', and therefore unlikely to erode at detectable rates (>1 mm/yr), 19% is 'soft/erodible', while 3% (591 km) has man-made defences. A comparison of OS maps surveyed in the 1890s and 1970s indicated a 22% reduction in extent of coastal progradation in Scotland, a 39% increase in extent of erosion and a doubling of average erosion rates from 0.5 to 1.0 m/yr. A comparison of 1970s map surveys with aerial photography flown in around 2012 indicated that 77% of the soft/erodible coast had remained stable, 11% had prograded and around 12% had experienced erosion. The areas worst affected by erosion are areas backed by glacial deposits, Holocene raised beaches and sand dunes. EUROSION (2004) estimated that 19.5% (89 km) of the coastal length in Northern Ireland (approx. 456 km) is vulnerable to coastal erosion, but historically it has been of relatively minor concern in the Province (Amey and HR Wallingford 2018; Cooper, O'Connor and McIvor 2020). The most rapid erosion occurs on sections of composed of soft sediments, notably glacial till, outwash, raised beach and dune deposits (Carter and Bartlett 1990; McKenna, Carter and Bartlett 1992).

Causes of erosion

Coastal erosion resulting from natural processes has occurred for millennia in areas such as east Yorkshire, north Norfolk and Suffolk, and there are many accounts of towns and villages being lost to the sea (Boyle 1889; Sheppard 1912; English 1991; Comfort 1994). In all areas of long-term erosion, tidal currents, waves and wave-driven currents have combined to remove sediment from the intertidal and nearshore zones and to transport it into deeper water or lower energy embayments and estuaries such as the Humber, Wash and Greater Thames. Erosion is likely anywhere there is a negative sediment budget and absence of a wide, high protective beach (Lee 2011). Natural changes in coastal, nearshore and offshore morphology are an important cause of changes in beach sediment budget, and hence drivers of erosion (Burgess, Jay and Nicholls 2007; Blott et al. 2013; Moore and Davis 2014). Human activities have also long been a significant cause of erosion, including dredging for navigation and aggregate extraction, coastal quarrying, the removal of material for ship ballast and roadmaking, and the construction of harbour walls and coastal defences, which modify coastal processes.

In the 1880s, the British Association for the Advancement of Science set up a committee to investigate coastal erosion. The findings (Grantham et al. 1885) confirmed human activities as a major factor. An inquiry into the destruction of the village of Hallsands also identified dredging of shingle for the construction of Devonport docks in the 1890s as the major cause (Harvey 1980). A Royal Commission on Coast Erosion was subsequently set up to investigate the problem (RCCE 1907). The terms of reference were widened in 1908 to include afforestation as a method of erosion control, land 'reclamation', timber production and addressing unemployment (RCCEA 1909). Evidence presented to the inquiry by the Director of the Geological Survey showed that the gain of land area owing to natural accretion, warping and embanking (48,000 acres) was in fact considerably larger than the land area lost to erosion (6,640 acres), although the gain was almost entirely within estuaries while losses were mainly on the open coast. The Commission's final report (RCCEA 1911) concluded that 'while some localities have suffered seriously from the encroachment of the sea, from a national point of view the extent of erosion need not be considered alarming' (p. 159). However, the report also concluded that many works carried out to counter erosion had been of the 'wrong type', and in particular the construction of groynes had often led to 'undue stoppage' of the natural movement of beach material. The report recommended that before works were undertaken there should be detailed investigation to understand each local area. The report also recommended that it should be made illegal to remove any sediment, rock or minerals from the shores of the United Kingdom without prior consent of the Board of Trade, which should also be empowered to control the erection of any works or deposit and to supervise local authorities and organisations concerned with sea defence, with assistance from scientific experts with 'specialist knowledge of sediment movements and engineering design'.

In 1919, responsibilities for coast protection passed to the Ministry of Transport, but central government interest remained limited. The Land Drainage Act of 1930 gave to the Ministry of Agriculture and Fisheries, Drainage Boards, Catchment Boards and Local Authorities the powers formerly held by the Commissioners of Sewers relating to land drainage and 'defence against water', but the interests of these bodies were largely restricted to rivers, estuaries and sections of open coast where there was direct risk of sea flooding. In 1939, a Coast Protection Act was passed that prevented removal of materials from the foreshore, but it was not until 1949 that a new Coast Protection Act finally delivered a framework for a coordinated, centrally regulated approach to coast protection (Roddis 1950). The 1949 Act gave permissive powers to maritime local authorities to undertake works for a range of purposes, including maintenance of the general safety and protection of the borough, protection of the highway, and the creation and protection of public amenities such as public walks and esplanades, open spaces and recreational grounds, ornamental gardens, swimming pools and car parks. The 1949 Act also gave the Coast Protection Authorities powers to seek financial grant-in-aid from the Exchequer to undertake defence works, and to object to works proposed by private landowners that might affect the foreshore.

Over the following half-century, central responsibility for coast protection passed through the Ministry of Transport, the Ministry for Housing and Local Government (1951–70), the Department of Environment (1970–85), the Ministry of Agriculture, Fisheries and Food (1985–2001) and the Department of Environment, Food and Rural Affairs (DEFRA, 2001–present). There have been several adjustments to legislation over this period, culminating in the Flood and Water Management Act 2010 (Environment Agency and Maritime Local Authorities 2010). DEFRA has overall responsibility, with operational delivery delegated to the Maritime Local Authorities and the Environment Agency (EA). Inland and coastal flooding continue to be given higher priority than coastal erosion in view of the perceived higher economic and social cost. The EA recently estimated (2020) that 5.2 million homes and business in England are at risk of flooding but only around 700 properties are vulnerable to coastal erosion over the next 20 years (EA 2020). However, many significant elements of strategically important national infrastructure are located on the coast. In the March 2020 Budget, the government announced an increase in capital funding for flood and coastal defence projects from £2.6bn in the period 2015/16–2020/1 (£434 m/yr) to £5.6bn (£800 m/yr) in the period 2020/1–2026/7. The EA has estimated that this increase in investment will reduce flood risk by 11% but has no plans to monitor progress towards this goal (NAO 2020). Neither the NAO (2020) report or the National Flood and Coastal Erosion Risk Management Strategy for England (EA 2020) make much specific mention of coastal erosion or coast protection.

Case studies

Soft cliff erosion

One of the best examples of soft coastal cliff erosion in the UK is provided by Holderness in east Yorkshire. Most of the coast between Flamborough Head and Kilnsea is formed of glacial till, with areas of fluvio-glacial outwash and sand dunes (Fig. 2.2). Temporal and spatial variations in erosion rates on this coast have been intensively studied (Matthews 1905; Thompson 1923; Valentin 1954, 1971; Newsham et al. 2003; Quinn, Philip and Murphy 2009; Pye and Blott 2015; Hobbs et al. 2019) and are currently monitored by the East Riding of Yorkshire Council. Undefended cliffs 2–17 m high occur along most of the coast, although hard defences have been built at different times to protect the major settlements such as Hornsea, Mappleton and Withernsea, and around the strategically important Easington gas terminal.

Erosion in Holderness is induced by a combination of storm waves, mainly from the north-east, episodic failures of the cliffs and a net southerly drift of sediment away from the area. Maximum average annual rates of cliff recession and total recession distance typically occur immediately downdrift of areas with coastal defences, partly owing to interruptions in the alongshore sediment supply and partly to refraction of breaking waves and increased longshore transport rates around the end of the defences, known as the terminal groyne effect (Brown, Barton and Nicholls 2012; Fig. 2.3). Rates of cliff recession show variation on a quasi-decadal timescale, which is related to the southerly movement of beach ridges and runnels (ords) that are orientated at a slight oblique angle to the cliff toe. This gives rise to 'erosion hot spots' which migrate southwards at a rate of 500–700 m/yr (Pringle 1981, 1985; Pye and Blott 2015). Failure to recognise this resulted in a requirement for emergency cliff protection works to allow completion of the Aldborough Gas Storage Facility in 2010–11. Some failing defences have recently been removed in order to allow natural adjustment of the shoreline, but elsewhere (e.g. at Kilnsea and along Spurn Peninsula) failed defences have simply been abandoned, with consequences for visual amenity and public safety (Lee 2018). In some undefended areas, low-value houses close to retreating cliffs have been demolished and mobile caravan parks relocated inland. This is likely to be necessary on a much wider scale in future.

At Bacton, in north-east Norfolk, cliffs composed of glacial, fluvioglacial and lacustrine sediments have retreated approximately 100 m since 1885, following centuries of coastal retreat, yet a major gas terminal was built in the late 1960s only a short distance from the cliff edge (Fig. 2.4a). The terminal is of national strategic importance, handling a third of all UK gas supplies. It is connected by pipelines to gas fields in the southern and central North Sea and also to Belgium and the Netherlands. Various measures have been undertaken to stabilise the cliffs, including regrading, drainage, construction of timber revetments and groynes, but with limited success. Hard defences have not been built mainly due to concerns about the potential impacts on the neighbouring villages of Bacton and Walcott (HR Wallingford 2003). A storm in December 2013 caused about 10 m of erosion in front of the gas terminal, prompting a search for an alternative solution. A beach mega-nourishment project was undertaken in 2019, taking its cue from the 'sand engine' in the Netherlands (Stive et al. 2013). The Bacton scheme, designed by Royal Haskoning and sponsored by the terminal operators (Shell UK and Perenco), North Norfolk District Council and the EA, cost £19m and involved pumping 1.8m m³ of sand, dredged from licensed offshore

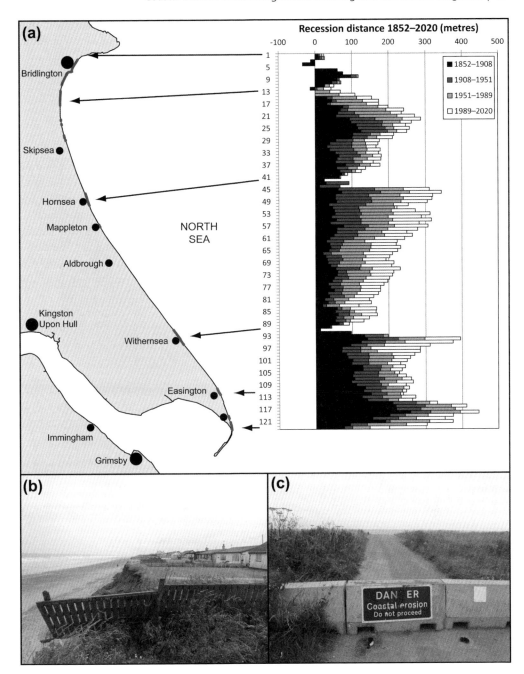

Figure 2.2 (a) Distances of cliff erosion along the Holderness coast since 1852, with defended sections of coastline in 2020 highlighted with a bold red line on the map; (b) properties close to an undefended cliff edge near Skipsea; (c) road to nowhere

extraction areas, to create a wide sand berm in front of the cliffs (Fig. 2.4b). The scheme aims to provide protection for the gas terminal against a 0.01% probability storm event, and to provide additional benefits for neighbouring villages as the sand is dispersed alongshore. However, it remains to be seen if the scheme will deliver its objectives over the design life of 15–20 years.

Figure 2.3 (a) Aerial photograph of southern Withernsea, flown on 27 October 2019, with yellow lines indicating the historical position of the cliff top from OS maps and aerial photographs, and predicted position in 2050 and 2010 based on historical erosion rates of 3.2 and 4.3 m/yr; (b) eroding cliff downdrift of the defences at Withernsea; (c) removal of old defences near Skipsea

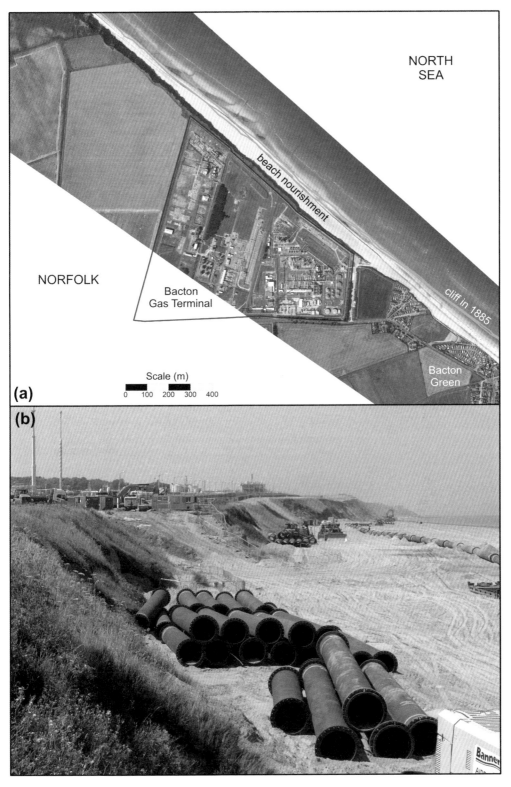

Figure 2.4 (a) Aerial photograph of Bacton on the Norfolk coast, flown 7 August 2020: the position of the top and bottom of the cliffs in 1885 (from OS maps) is indicated by green lines, and the outline of the modern gas terminal is indicated by the red line; (b) the mega-nourishment being undertaken in 2019

Shingle beach and ridge erosion

The coasts of eastern and southern England are notable for shingle beaches and larger shingle formations, including the Cley–Salthouse ridge, Orfordness, Shingle Street, Dungeness and Hurst Castle Spit. A large proportion of the shingle is composed of flint and chert, ultimately derived mostly from the Chalk Group. The shingle deposits provide an important element of the defences against tidal flooding, as well as being internationally important habitats and geomorphological features in their own right. At the present time, the supply of new shingle from the sea floor and eroding cliffs is limited, and many shingle beaches are eroding, requiring periodic nourishment to maintain their flood defence function.

Orfordness in Suffolk (Fig. 2.5a) originated as a spit that extended southwards from Aldeburgh in response to net southerly drift of shingle, forcing the River Alde to divert its course (Steers 1926; Carr 1970). The northern part of the spit around Slaughden and Sudbourne Beach is very narrow, but the shingle widens at Orfordness where a complex of shingle recurves, often cross-cutting, has developed. South of the Ness the shingle narrows again and continues as far as North Weir Point, opposite Shingle Street. The neck of the spit south of Aldeburgh has shown a long-term tendency for narrowing and landward migration, exposing back-barrier sediments on the foreshore. Defences were first constructed at Aldeburgh and Slaughden in the late nineteenth century and were upgraded in the 1930s, 1950s, 1990s and the last few years. In recent decades, recycling of shingle has been undertaken by the EA to maintain the beach and artificial ridge which reduce flood risk around the Alde-Ore estuary. The shingle used for nourishment is taken from a downdrift borrow area opposite Lantern Marshes, where natural processes favour accumulation. Despite regular nourishment, the ridge has become very narrow just south of the defences (Fig. 2.5b) and there is a significant risk of a major washover at the northern end of Lantern Marshes. If not repaired, a breach could develop into a new tidal entrance, with significant implications for the hydrodynamic regime and functioning of the estuary. The area is closely monitored by the EA and dredged shingle is now being placed on the landward side of the ridge with the aim of creating a more resilient ridge morphology. Erosion has also occurred at the Ness itself over the past 50 years, resulting in demolition of the iconic lighthouse in June 2020 before it toppled into the sea (Fig. 2.5c). Erosion in this area presents no immediate threat, although a breakthrough into former gravel workings further north could potentially create a new tidal connection between the sea and estuary in the medium term.

Another large shingle formation occurs at Dungeness on the coast of East Sussex. The present ness feature evolved from a series of spits during the later Holocene (Lewis 1932; Roberts and Plater 2007) and now protects a large area of Romney Marsh from tidal flooding. The southern side of the foreland has experienced long-term recession, with eroded shingle being transported eastwards where it accumulates on the north side of the point (Fig. 2.6a). In the early 1960s, the Central Electricity Generating Board decided to build a Magnox nuclear power station (Dungeness A) behind the eroding shore. It began generating in 1965 and ceased operation on 31 December 2006. A second power station (Dungeness B) was built in the 1970s; its two Advanced Gas-Cooled Reactors began generation in 1983 and 1985, respectively. Both reactors were shut down in 2018 following safety concerns and in June 2021 the operators, EDF Energy, announced plans to shut the plan and enter decommissioning with immediate effect.

Following selection of the site for the A Station, studies by Sir William Halcrow & Partners indicated that the south shore of Dungeness was eroding at 1.1–1.5 m/yr (Maddrell, Osmond and Li 1994). An options appraisal concluded the most cost-efficient method of managing erosion and flood risk was to recycle shingle from the east side of the point, initially at a rate of 15,000 t/yr. This was undertaken between 1965 and 2007, when planning permission expired. The excavated shingle was used to create and maintain an artificial

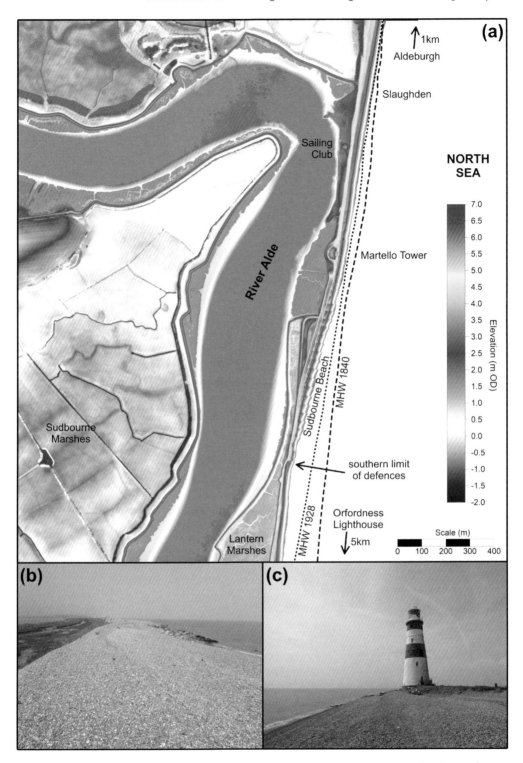

Figure 2.5 (a) LiDAR DTM of the coast at Slaughden, showing the very narrow shingle ridge separating the sea and the Alde estuary behind; (b) Sudbourne Beach in June 2020; (c) Orfordness Lighthouse shortly before its demolition in June 2020

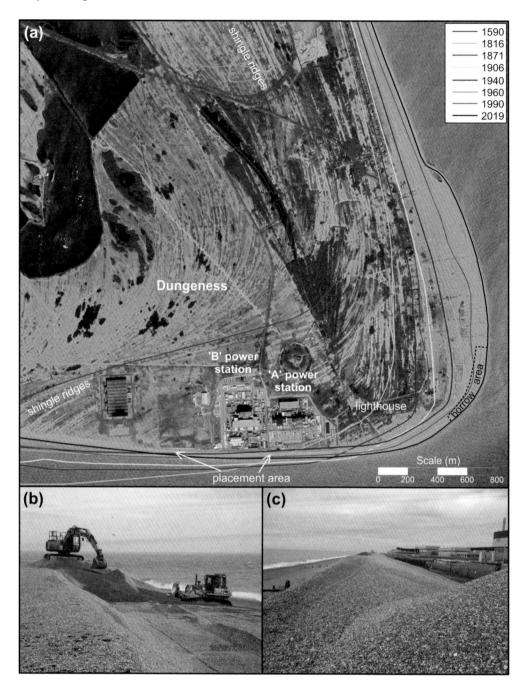

Figure 2.6 (a) Aerial photograph of Dungeness taken in May 2020, with shoreline positions taken from historical maps (1590, 1816, 1871 and 1906), aerial photography (1940, 1960 and 1990) and airborne LiDAR (2019); (b) shingle extraction from the east side of Dungeness; (c) shingle placement in front of the nuclear power stations

berm in front of the power stations that was designed to provide a 1 in 10,000 year (0.01%) standard of protection. Shingle was also placed along the frontage of Lydd Ranges and Denge Marsh, but continued erosion in this area led to a change of policy to managed realignment with shingle reprofiling after 1993 (Maddrell 1996). Following the Fukushima nuclear incident in 2011 and a review of the safety case for the Dungeness stations, EDF and the EA sought planning consent to take up to 30,000 m³ of shingle per year from the borrow area east of the Point until 2023 (Figs 2.6b and c). The berm in front of the B station has since been upgraded to withstand a 0.01% tsunami-type wave. Shingle from the borrow pit is also being used to replenish parts of the beach between Denge Marsh Sewer and the power stations. However, as part of the £40m Lydd Ranges Sea Defence Scheme, which began in July 2021 and is scheduled to continue until July 2025, further managed realignment will be undertaken on this frontage and timber-lined plastic piling inserted behind the reprofiled shingle ridge. Along the western part of Lydd Ranges, 34 new groynes will be constructed, followed by placement of 320,000 t of dredged shingle and construction of a rock armour revetment. Although some of the placed shingle is likely to move eastwards, maintenance of the defensive berm in front of the power stations is likely to require recycling of shingle from the east for the foreseeable future.

Sand dune erosion

Sand dunes form a fringe around many parts of the UK's lowland coasts and in some places comprise systems of considerable areal extent. In both cases the dunes can be important as 'natural' flood defences, as well as being significant as ecological habitats and for a range of economic and recreational activities, including golf and military training (Pye, Saye and Blott 2007). Sand has a low resistance to wave erosion, and dune cliff recession rates of up to 14 m have been recorded during a single storm. One of the largest dune systems occurs on the Sefton coast in north-west England (Fig. 2.7a). Since 1900, the central part of the dune frontage at Formby Point has been eroding at an average rate of up to 3 m/yr, while areas to the north and south have experienced progradation (Pye and Neal 1994; Pye and Blott 2008). The central part of the dune system contains a wide belt of interlinked parabolic dunes, with crests reaching 16 m AOD, which were stabilised with conifers in the late nineteenth and early twentieth centuries and now form an effective barrier against tidal flooding. However, parts of the dunes between Formby Point and Hightown are much lower and narrower, partly the result of sand mining between the 1930s and 1970s. Over-extraction at Cabin Hill near Range Lane on the south side of the Point required the construction of a 2 km long flood bank in the early 1970s to reduce the risk of flooding to the mosslands (low-lying peatlands) behind.

At Formby Point, a combination of coastal erosion and relatively high visitor pressure has given rise to some of the most dynamic coastal dunes presently found in the UK (Fig. 2.7b), bringing significant nature conservation benefits, since areas of mobile sand are now recognised to be of great biodiversity importance. Erosion and dune migration in this area presently do not pose a threat to property or infrastructure, but changes are being carefully monitored by the local Coast Protection Authority (Sefton Council). The most recent Shoreline Management Plan for this area recommends a policy of Managed Realignment up to 2100, recognising that dune fencing, boardwalks and some vegetation planting will be required to manage recreational pressures. Intervention will also be required to remove demolition rubble, which is falling onto the beach near the National Trust's Victoria Road car park (Fig. 2.7c).

Sand dunes also form an important component of the flood defences at Dawlish Warren, located at the mouth of the Exe estuary (Fig. 2.8a). This spit complex has varied in width and position over time (Kidson 1964), and the southern end is protected by hard defences. Groynes and gabions were built along much of the frontage, but were falling into disrepair

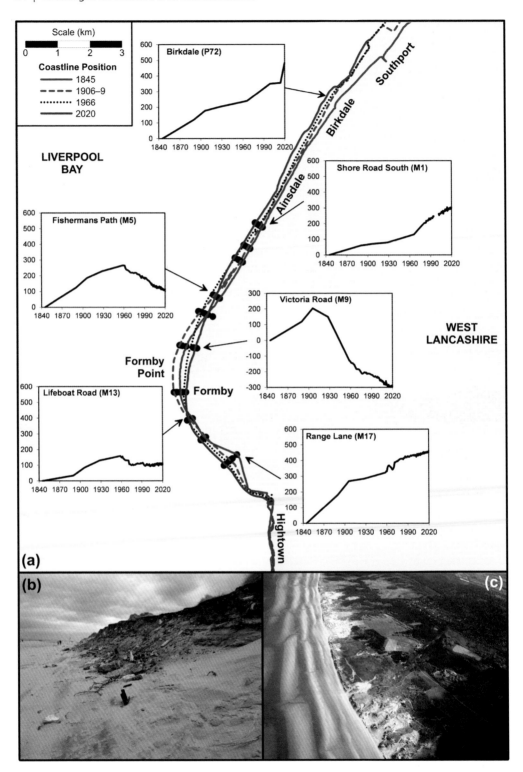

Figure 2.7 (a) Changes in the position of the frontal dune along the Sefton coast near Formby, with graphs showing the erosion or accretion of the dune toe since 1845 (in m); (b) eroding dunes at Victoria Road, showing rubble collapsing from the dune cliff; (c) dynamic sand dunes behind the eroding shore

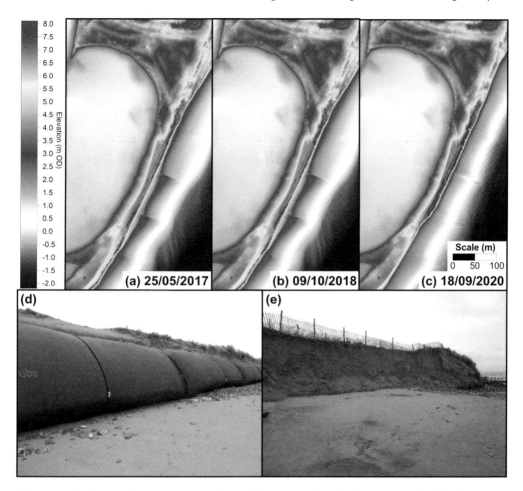

Figure 2.8 (a–c) annual LiDAR surveys of Dawlish Warren, flown before (2017), shortly after (2018) and several years after (2020) beach protection works; (d) photograph of the northern Dawlish Warren frontage showing geotubes exposed by erosion, December 2020; (e) eroding dunes at Dawlish Warren, December 2020

by the early 2000s. A £12m coastal defence scheme, sponsored by the EA and Teignbridge Borough Council, and completed in October 2017, included removal of some groynes and refurbishment of others, the installation of a sand-filled geotube core to strengthen the narrowest part of the frontal dune ridge, and the pumping of 250,000 m³ of dredged sand onto the beach. The width and height of both the frontal dune ridge and the beach initially increased substantially, but as early as the autumn of 2018 the effects of erosion were again becoming evident (Fig. 2.8a). Since then, further erosion has exposed the geotube core and cut back the dune cliffs, exposing the groyne heads (Figs 2.8b and c). The current alignment and position of the reconstructed ridge is evidently out of equilibrium with the prevailing process regime, and further expenditure on major remedial works has been paused pending further studies to better understand the system.

Dundrum Bay in Northern Ireland (Fig. 2.9a) was once entirely fringed by dunes but since the late nineteenth century the western half of the Bay has been progressively 'hardened' by the construction of defences. A concrete wall and promenade was first constructed at Newcastle as the town developed as a visitor destination. In following years, the fronting beach lost much of its sand, exposing a boulder pavement (Fig. 2.9b), partly

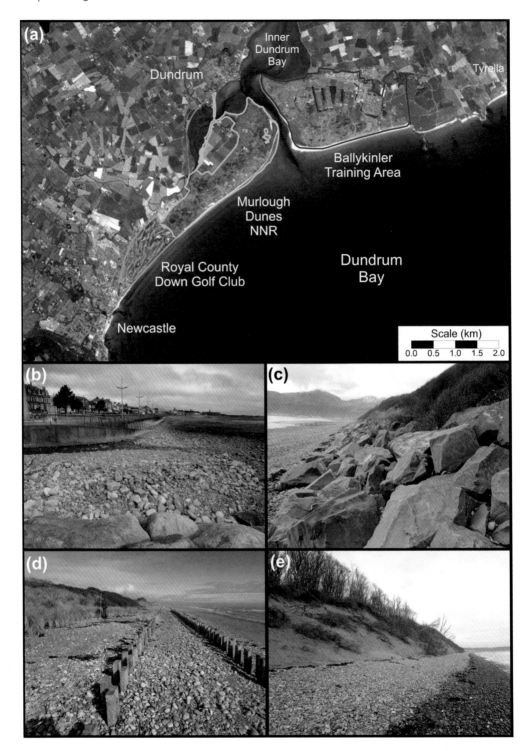

Figure 2.9 (a) Dundrum Bay, Northern Ireland, aerial photograph flown in 2013; (b) Newcastle; (c–d) rock armour and timber protection along the sand dune frontage of the Royal County Down Golf Course, Dundrum Bay; (e) unprotected Murlough dunes (National Trust)

owing to wave reflection off the seawall and partly to a change in bathymetry and wave conditions within the bay that enhanced the eastward littoral drift of sand (Cooper and Navas 2004). Beach lowering and increased frontal dune erosion led the Royal County Down Golf Club to install erosion control measures along its frontage from the 1950s onwards. Early defences included a double row of vertical railway sleepers (Fig. 2.9c), but a rock revetment has since been built in stages along almost the whole frontage (Fig. 2.9d). Dune erosion has now extended as far as the tidal inlet entrance to Inner Dundrum Bay, but the National Trust, which is responsible for the Murlough Dunes National Nature Reserve, has adopted a policy of allowing natural processes to take their course (Fig. 2.9e). The Northern Ireland Environment Agency is also opposed to further hardening of the bay shoreline. To the east of the tidal inlet, the Ballykinler Military Training Area (MTA) frontage is mainly stable or slowly prograding.

In Scotland, the eastern side of Buddon Ness, which is also an MTA, has also experienced long-term erosion associated with landward movement of bathymetric contours within Carnoustie Bay (Fig. 2.10). The Ness represents a foreland composed of raised beach deposits overlain by a variable thickness of dune sand (Hansom 2003). Parabolic dunes have crossed the area from south-west to north-east, and some have been truncated by marine erosion at their eastern ends. In the 1960s, anti-tank blocks and rock armour were placed to defend the northern golf course frontage, and the rock armour was extended southwards to protect the MTA in the 1970s. However, beach levels have continued to drop and the rock armour has slumped in places. An erosional 'bight' has also formed where the rock armour ends. The cable corridor from an offshore wind farm is planned to cross the area and further shoreline stabilisation works are likely to be required in the future.

The Neolithic village of Skara Brae, part of a UNESCO World Heritage Site, is located within sand dunes that fringe the Bay of Skaill on Orkney (Fig. 2.11). The site, consisting of eight clustered houses that were occupied between approximately 3180 and 2500 BCE, was discovered as a result of dune erosion in the mid-nineteenth century. When built, the village almost certainly lay further from the sea. A further storm in 1924 swept away part of one house, prompting action to preserve the site. Excavation began in 1927 and a stone seawall was built to protect against further erosion. The wall has since required repair (Gibson and Bradford 2012) and a risk assessment by Historic Environment Scotland and the Orkney Islands Council in 2019 concluded that the entire Heart of Orkney World Heritage Site, and particularly Skara Brae, is extremely vulnerable to rising sea levels and climate change.

Around the UK there are many locations where rail and road infrastructure is located on erodible dune sand and shingle, or otherwise very close to the sea (Duck, 2015). Defences are sometimes non-existent or consist only of a narrow zone of rock armour. An example is provided by the Cambrian Coast Railway line between Barmouth and Llanaber in west Wales (Fig. 2.12). In places, the single-track line runs behind the top of the beach and during severe storms is prone to blockage by shingle or sand and to washing out of the track ballast. Following the stormy winter of 2013–14, this line was severely damaged in numerous places, and repairs took more than six months.

Conclusions

Tide gauge records provide clear evidence that relative mean sea level has been rising around the UK at an average rate of between 0.5 and 3.5 mm/yr in recent decades, depending on location. Current climate change projections (Lowe et al. 2018; Palmer et al. 2018) suggest this trend will continue for the foreseeable future, associated with warming and a possible higher incidence of weather and sea-level extremes. While there is uncertainty concerning the combined effect of sea-level rise and climate change, deeper water associated with rising sea level and a lack of coastal sediment supply can be expected to lead to greater nearshore

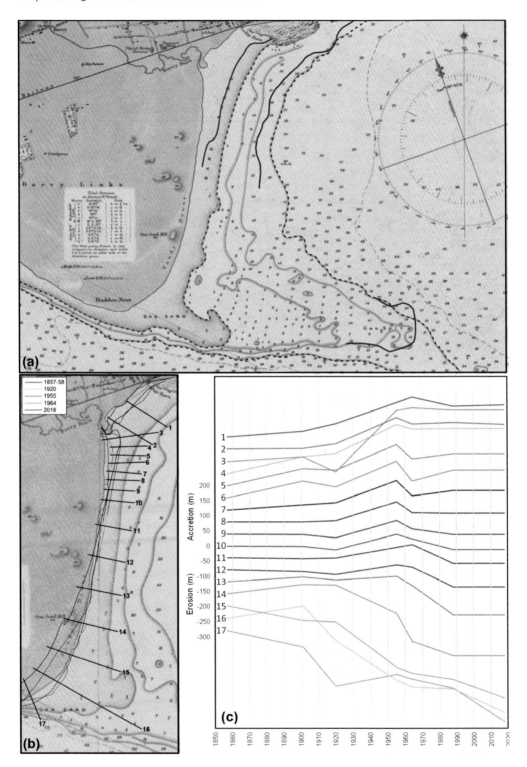

Figure 2.10 (a) Admiralty chart of Buddon Ness surveyed in 1883, with the zero and 18 ft Chart Datum contours picked out by dashed red and blue dashed lines. The equivalent contours in 2003 are shown with solid red and blue lines where data exist; (b) OS six-inch map surveyed in 1901, with MHWS surveyed at different dates shown by coloured lines; (c) temporal pattern of erosion/accretion of the edge of vegetated 'land' at 17 profile locations, interpreted from historical maps and aerial photographs

Figure 2.11 The Neolithic village at Skara Brae: (a) aerial photograph flown 8 September 2019 showing proximity of the village to the eroding shoreline of the Bay of Skaill; (b) ground photograph of one of the huts, which is protected by a seawall first constructed in the 1930s

Figure 2.12 (a) aerial photograph of the coastline north of Barmouth, west Wales, showing the Cambrian Coast railway line, which was blocked by overwash shingle during a series of severe storms during the 2013–14 winter; (b) view north along the track section closed in late winter 2014; (c) view south showing shingle overwash onto the track

wave energy and greater likelihood of beach and coastal erosion (e.g. Brooks and Spencer 2012). The significance of this will depend on the magnitude of warming and sea-level rise that actually occurs, and on the value of economic and social assets at risk in any given area.

Burgess et al. (2007) concluded that, at a national (UK) scale, losses due to coastal erosion on cliffed coasts are not a major concern, although on lowland coasts erosion can have a significant bearing on flood risk. In the future, as now, the most significant erosion and flood related problems will occur at locations where high value assets are at risk. These include nuclear power stations, oil and gas terminals and major centres of population located below potential flood level. The risks to these areas have so far been considered manageable using a combination of soft and hard engineering methods, but delivery of an equivalent or higher standard of protection in the future will inevitably require higher overall expenditure and greater prioritisation of resources. Requirements for large-scale recycling of shingle and sand to maintain adequate protection can only increase as the sea level rises, with or without an increase in storm frequency and magnitude. This will create challenges to identify and exploit adequate sediment source areas without major environmental impacts. Beach nourishment alone is not likely to be sufficient, and there will be a requirement to build and maintain new hard control structures such as groyne fields, offshore breakwaters and revetments, thereby extending the area of 'unnatural' coast.

In most areas where high-value assets are not at risk, significant expenditure on defences will not be justifiable and there will inevitably be greater pressure for abandonment, removal and/or relocation both of defences that have reached the end of their design life and assets that do not achieve the benefit-cost threshold, resulting in short periods of accelerated erosion and 'coastal catch-up' (e.g. Lee 2018). Organisations such as the National Trust have already adopted policies under which new defences will not be constructed unless under exceptional circumstances, and where possible existing defences and assets will be removed to allow natural processes to operate unfettered (National Trust 2015b, c, d). However, the removal of defences, buildings, car parks and waste tips is likely to be an expensive task, and it is unclear how or by whom the costs will be met. Simple abandonment, with resulting negative impacts on environmental quality, including contamination of beaches and coastal waters, will bring its own costs in terms of reputation, environmental standards and public safety, and in most cases should be considered to be an unacceptable option.

References

Amey Consulting and HR Wallingford (2018) Baseline study and gap analysis of coastal erosion risk management NI, report prepared for the Northern Ireland Department of Agriculture, Environment and Rural Affairs and the Department for Infrastructure, p. 102.

Blott, S.J., Duck, R.W., Phillips, M.R., Pontee, N.I., Pye K. and Williams, A. (2013) Great Britain. In E. Pranzini and A.T. Williams (eds) *Coastal Erosion and Protection in Europe*, pp. 173–208. London: Routledge. https://doi.org/10.4324/9780203128558-10

Boyle, J.R. (1889) *Lost Towns of the Humber*. Hull: A. Brown and Son.

Brand, J.H. and Spencer, K.L. (2019) Potential contamination of the coastal zone by eroding historic landfills. *Marine Pollution Bulletin* 146: 282–91. https://doi.org/10.1016/j.marpolbul.2019.06.017

Brooks, S.M. and Spencer, T. (2010) Temporal and spatial variations in recession rates and sediment release from soft rock cliffs, Suffolk coast, UK. *Geomorphology* 124: 26–41. https://doi.org/10.1016/j.geomorph.2010.08.005

Brooks, S.M. and Spencer, T. (2012) Shoreline retreat and sediment release in response to accelerating sea level rise: measuring and modelling cliffline dynamics on the Suffolk Coast, UK. *Global and Planetary Change* 80–1: 165–79. https://doi.org/10.1016/j.gloplacha.2011.10.008

Brooks, S.M. and Spencer, T. (2013) Importance of decadal scale variability in shoreline response: examples from soft rock cliffs, east Anglian coast, UK. *Journal of Coastal Conservation* 18: 581–93. https://doi.org/10.1007/s11852-013-0279-7

Brown, S., Barton, M.E. and Nicholls, R.J. (2012) The effect of coastal defences on cliff top retreat along the Holderness coastline. *Proceedings of the Yorkshire Geological Society* 59: 1–13. https://doi.org/10.1144/pygs.59.1.288

Burgess, K., Jay, H. and Nicholls, R.J. (2007) Drivers of coastal erosion. In C. Thorne, C. Green and E.C. Penning-Rowsell (eds) *Future Flooding and Coastal Erosion Risks*, pp. 267–79. London: Thomas Telford. https://doi.org/10.1680/ffacer.34495.0016

Burgess, K., Jay, H., Nicholls, R.J., Green, C. and Penning-Rowsell, E.C. (2007) Assessment of future coastal erosion risk. In C. Thorne, C. Green and E.C. Penning-Rowsell (eds) *Future Flooding and Coastal Erosion Risks*, pp. 280–93. London: Thomas Telford. https://doi.org/10.1680/ffacer.34495.0017

Cambers, G. (1976) Temporal scales in coastal erosion systems. *Transactions of the Institute of British Geographers* New Series 1: 246–56. https://doi.org/10.2307/621987

Carr, A.P. (1970) The evolution of Orfordness, Suffolk, before 1600 AD: geomorphological evidence. *Zeitschrift für Geomorphologie* New Series 14: 289–300.

Carter, R.W.G. and Bartlett, D. (1990) Coastal erosion in northeast Ireland – Part 1: sand beaches, dune and river mouths. *Irish Geography* 23: 1–16. https://doi.org/10.1080/00750779009478762

Charlier, R.H. and De Meyer, C.P. (1998) *Coastal Erosion*. Berlin: Springer. https://doi.org/10.1007/BFb0011384

CIA (Central Intelligence Agency) (2020) *The CIA World Factbook 2020–2021*. New York: Skyhorse Publishing.

Comfort, N. (1994) *The Lost City of Dunwich*. Lavenham: Terence Dalton.

Cooper, J.A.G. and Navas, F. (2004) Natural bathymetric change as a cause of century-scale shoreline behaviour. *Geology* 32: 513–16. https://doi.org/10.1130/G20377.1

Cooper, J.A.G., O'Connor, M.C. and McIvor, S. (2020) Coastal defences versus coastal ecosystems: a regional appraisal. *Marine Policy* 111, https://doi.org/10.1016/j.marpol.2016.02.021

Cooper, N.J., Thomas, R., Thomson, E. and Wilkinson, J. (2014). Management of landfills and legacy industrial sites on eroding and low-lying coastlines. In W. Allsopp and K. Burgess (eds) *From Sea to Shore – Meeting the Challenges of the Sea Proceedings of the Coasts, Maritime Structures and Breakwaters Conference 2013*, Vol. 1, pp. 755–64. London: ICE Publishing.

Countryside Commission (1968) *The Coasts of England and Wales: Measurements of Use, Protection and Development*. London: HMSO.

Duck, R.W. (2015) *On the Edge: Coastlines of Britain*. Edinburgh: Edinburgh University Press.

English, B. (1991) Ravenser Odd, A Lost East Yorkshire Town. In D.B. Lewis (ed.) *The Yorkshire Coast*, pp. 149–58. Normandy Press, Beverley.

Environment Agency (2020) *National Flood and Coastal Erosion Risk Management Strategy for England*. Bristol: The Environment Agency.

EUROSION (2004) Living with coastal erosion in Europe: sediment and space for sustainability PART II – maps and statistics 29 May 2004. Project commissioned by General Directorate Environment of the European Commission, Contract no. B4–3301/2001/329175/MAR/B3 awarded by the European Commission to the National Institute for Coastal and Marine Management of the Netherlands (RIKZ), and developed in partnership with IGN France International, EADS Systems & Defence Electronics, the French Geological Survey (BRGM), the Autonomous University of Barcelona, the French Institute of Environment and EUCC, The Coastal Union. Available at: https://www.eurosion.org (Accessed 6 January 2021).

Gibson, J. and Bradford, F. (2012) *Rising Tides Revisited: The Loss of Coastal Heritage in Orkney*. Kirkwall: Orkney Archaeology Society.

Grantham, R.B., De Rance, C.E., Redman, J.B., Topley, W., Whitaker, W., Woodall, J.W., Clarke, A., Douglas, J.N., Evans, F.O., Parsons, J., Prestwich, J., Wharton, W.J.L., Easton, E., Valentin, J.S. and Vernon Harcourt, L.F. (1885) Rate of erosion of the sea-coasts of England and Wales. Report of the Committee appointed for the purpose of the inquiry into the Rate of Erosion of the Sea-coasts of England and Wales, and the Influence of the Artificial Abstraction of shingle or other Material in that Action. British Association for the Advancement of Science, 55th Meeting, Aberdeen.

Halcrow (2002) Futurecoast (CD Rom). Report and database produced for DEFRA by Halcrow, British Geological Survey, ABPmer, Queen's University of Belfast and University of Plymouth. Available at: https://www.channelcoast.org/ccooresources/futurecoast (Accessed 7 January 2021).

Hansom, J.D. (2003) Barry Links (Angus (NO 550 320) In V.J. May and J.D. Hansom (eds) *Coastal Geomorphology of Great Britain*, pp. 400–7. Nature Conservation Review 28. Peterborough: Joint Nature Conservation Committee.

Hansom, J.D., Fitton, J.M. and Rennie, A.F. (2017) Dynamic coast – national coastal change assessment: national overview, CREW Report 2014/2. Aberdeen: The James Hutton Institute.

Harvey, J.L. (1980) *The Tragedy of Hallsands Village*. Plymouth: PDS Printers.

Hobbs, P.R.N., Jones, L.D., Kirkham, M.P., Pennington, C.V.L., Morgan, D.J.R. and Dashwood, C. (2019) Coastal landslide monitoring at Aldborough, East Riding of Yorkshire, UK. *Quarterly Journal of Engineering Geology and Hydrogeology* 53: 101–16. https://doi.org/10.1144/qjegh2018-210

Howe, M.A. (2015) Coastal soft cliff invertebrates are reliant upon dynamic coastal processes. *Journal of Coastal Conservation* 19: 809–20. https://doi.org/10.1007/s11852-015-0374-z

HR Wallingford (2003) Overstrand to Walcott strategy study. Cliff Processes. Part II. Technical supporting information, Report EX469. HR Wallingford Ltd, Wallingford, Oxfordshire, February 2003.

Kidson, C. (1964) Dawlish Warren, Devon: late stages in sand spit evolution. *Proceedings of the Geologists' Association* 75: 167–84. https://doi.org/10.1016/S0016-7878(64)80003-1

Komar, P.D. (1983) *Handbook of Coastal Processes and Erosion*. Boca Raton, FL: CRC Press.

Lee, E.M. (2002) Soft cliffs: prediction of recession rates and erosion control techniques. Final report on R& D Project FD2403/1302. DEFRA, London.

Lee, E.M. (2005) Coastal cliff erosion risk: a simple judgement-based model. *Quarterly Journal of Engineering Geology and Hydrogeology* 38: 89–104. https://doi.org/10.1144/1470-9236/04-055

Lee, E.M. (2011) Reflections on decadal-scale response of coastal cliffs to sea level rise. *Quarterly Journal of Engineering Geology and Hydrogeology* 44: 481–89. https://doi.org/10.1144/1470-9236/10-063

Lee, E.M. (2018) Coastal catch-up: how a soft rock cliff evolves when coastal defences fail – Godwin Battery, Kilnsea. Natural England Commissioned Reports 256.

Lee, E.M. and Clark, A.R. (2002) *The Investigation and Management of Soft Rock Cliffs*. London: Thomas Telford. https://doi.org/10.1680/iamosrc.29859

Lewis, W.V. (1932) The formation of Dungeness Foreland. *Geographical Journal* 80: 309–24. https://doi.org/10.2307/1784606

Lim, M., Rosser, N.J., Allison, R.J. and Petley, D.N. (2010) Erosional processes in the hard rock coastal cliffs at Staithes, North Yorkshire. *Geomorphology* 114: 12–21. https://doi.org/10.1016/j.geomorph.2009.02.011

Lowe, J.A., Bernie, D., Bett, P. and 31 others (2018) UKCP18 Science overview report version 2.0. The Met Office, Exeter. Available at: https://www.metoffice.gov.uk/pub/data/weather/uk/ukcp18/science-reports/UKCP18-Overview-report.pdf (Accessed: 7 December 2019).

Maddrell, R.J., Osmond, B. and Li, B. (1994). Review of 30 years beach replenishment experience at Dungeness Nuclear Power Stations. *Coastal Engineering 1994* (Proceedings of the 24th International Conference on Coastal Engineering, 23–8 October, Kobe, Japan): 3548–3563. https://doi.org/10.1061/9780784400890.257

Maddrell, R.J. (1996) Managing coastal retreat, reducing flood risks and protection costs, Dungeness Nuclear Power Station, UK. *Coastal Engineering* 28: 1–15. https://doi.org/10.1016/0378-3839(95)00035-6

Mandelbrot, B.B. (1967) How long is the coast of Britain? Statistical self-similarity and fractional dimension. *Science* 156: 636–8. https://doi.org/10.1126/science.156.3775.636

Masselink, G., Russell, P., Rennie, A., Brooks, S. and Spencer, T. (2020) Impacts of climate change on coastal geomorphology and coastal erosion relevant to the coastal and marine environment around the UK. *Marine Climate Change Impacts Partnership Science Review 2020*: 158–89.

Matthews, E.R. (1905) Erosion on the Holderness coast of Yorkshire. *Minutes of the Proceedings of the Institution of Civil Engineers* 159: 58–78. https://doi.org/10.1680/imotp.1905.16440

Matthews, E.R. (1913) *Coast Erosion and Protection*. London: Charles Griffin & Co.

May, V.J. (2014) Coastal cliff conservation and management: the Dorset and east Devon World Heritage Site. *Journal of Coastal Conservation* 19: 821–9. https://doi.org/10.1007/s11852-014-0338-8

McInnes, R.G. and Moore, R. (2011) *Cliff Instability in Great Britain – A Good Practice Guide*. Birmingham: Halcrow Group, Ltd.

McKenna, J., Carter, R.W.G. and Bartlett, D. (1992) Coast erosion in northeast Ireland Part 2 – cliffs and shore platforms. *Irish Geography* 25: 111–28. https://doi.org/10.1080/00750779209478724

Mentaschi, L., Vousdoukas, M.I., Pekel, J-F., Voukouvalas, E. and Feyen, L. (2018) Global long-term observations of coastal erosion and accretion. *Nature Scientific Reports* 8: 12876. https://doi.org/10.1038/s41598-018-30904-w

Moore, R. and Davis, G. (2014) Cliff instability and erosion management in England and Wales. *Journal of Coastal Conservation* 19: 771–84. https://doi.org/10.1007/s11852-014-0359-3

Murphy, P. (2014) *England's Coastal Heritage: A Review of Progress Since 1997*. Swindon: English Heritage.

NAO (National Audit Office) (2020) Managing Flood Risk. National Audit Office Report HC962, Session 2019–21, 27 November 2020.

National Trust (2015a) *Mapping Our Shores*. Swindon: National Trust.

National Trust (2015b) *Playing Our Part. – National Trust Strategy to 2025*. Swindon: National Trust.

National Trust (2015c) *Shifting Shores – Playing Our Part at the Coast*. Swindon: National Trust.

National Trust (2015d) Shifting Shores – Adapting to Change. Swindon: National Trust.

Newsham, R., Balson, P.S., Tragheim, D.G. and Denniss, A.M. (2003) Determination and prediction of sediment yields from recession of the Holderness coast. In D.R. Green and S.D. King (eds) *Coastal and Marine Geo-Information Systems*, pp. 191–9. Dordrecht: Kluwer. https://doi.org/10.1007/0-306-48002-6_14

Palmer, M., Howard, T., Tinker, J., Lowe, J., Bricheno, L., Calvert, D., Edwards, T., Gregory, J., Harris, G., Krijnen, J., Pickering, M., Roberts, C. and Wolf, J. (2018) UK climate projections science report: marine and coastal projections. UK Met Office, Hadley Centre, Exeter.

Pethick, J.S. (1996) Coastal slope development: temporal and spatial periodicity in the Holderness cliff recession. In M.G. Anderson and S.M. Brooks (eds) *Advances in Hillslope Processes*, Volume 2, pp. 897–917. Chichester: Wiley.

Pope, N.D., O'Hara, S.C.M., Imamura, M., Hutchinson, T.H. and Langston, W.J. (2011) Influence of a collapsed coastal landfill on metal levels in sediments and biota – a portent for the future? *Journal of Environmental Monitoring* 13: 1961–74. https://doi.org/10.1039/c0em00741b

Pranzini, E. and Williams, A.T. (eds) (2013) *Coastal Erosion and Protection in Europe*. London: Routledge. https://doi.org/10.4324/9780203128558

Pranzini, E., Wetzel, L. and Williams, A.T. (2015) Aspects of coastal erosion and protection in Europe. *Journal of Coastal Conservation* 19: 445–59. https://doi.org/10.1007/s11852-015-0399-3

Pringle, A.W. (1981) Beach development and coastal erosion in Holderness, North Humberside. In J. Neale and J. Flenley (eds) *The Quaternary in Britain*, pp. 194–205. Oxford: Pergamon Press.

Pringle, A.W. (1985) Holderness coast erosion and the significance of ords. *Earth Surface Processes and Landforms* 10: 107–24. https://doi.org/10.1002/esp.3290100204

Pye, K. and Blott, S.J. (2008) Decadal-scale variation in dune erosion and accretion rates: an investigation of the significance of changing storm tide frequency and magnitude on the Sefton coast, UK. *Geomorphology* 102: 652–66. https://doi.org/10.1016/j.geomorph.2008.06.011

Pye, K. and Blott, S.J. (2015) Spatial and temporal variations in soft-cliff erosion along the Holderness coast, East Riding of Yorkshire, UK. *Journal of Coastal Conservation* 19: 785–808. https://doi.org/10.1007/s11852-015-0378-8

Pye, K. and Neal, A. (1994) Coastal dune erosion at Formby Point, north Merseyside, England: causes and mechanisms. *Marine Geology* 119: 39–56. https://doi.org/10.1016/0025-3227(94)90139-2

Pye, K., Saye, S.E. and Blott, S.J. (2007) Sand dune processes and management for flood and coastal defence. Parts 1–5. R&D Technical Report FD1302/TR/1/2/3/4/5. Department for Environment, Food and Rural Affairs, London.

Quinn, J.D., Philip, L.K. and Murphy, W. (2009) Understanding the recession of the Holderness Coast, east Yorkshire, UK: a new presentation of temporal and spatial patterns. *Quarterly Journal of Engineering Geology and Hydrogeology* 42: 165–78. https://doi.org/10.1144/1470-9236/08-032

Rees, S., Curson, J. and Evans, D. (2015) Conservation of coastal cliffs in England 2002–2013. *Journal of Coastal Conservation* 19: 761–9. https://doi.org/10.1007/s11852-014-0358-4

Roberts, H.M. and Plater, A.J. (2007) Reconstruction of Holocene foreland progradation using optically stimulated luminescence (OSL) dating: an example from Dungeness, UK. *The Holocene* 17: 495–505. https://doi.org/10.1177/0959683607077034

Roddis, R.J. (1950) *The Law of Coast Protection*. London: Shaw and Sons Ltd.

Rogers, J., Allan, E., Hardiman, N. and Jeans, K. (2014) The national coastal erosion risk management project – from start to finish and beyond. In W. Allsopp and K. Burgess (eds) *From Sea to Shore – Meeting the Challenges of the Sea. Proceedings of the Coasts, Maritime Structures and Breakwaters Conference 2013*, Volume 1, pp. 774–82. London: ICE Publishing.

RCCE (Royal Commission on Coast Erosion) (1907) First report of the Royal Commission appointed to enquire into and to report on certain questions affecting coast erosion and the reclamation of tidal lands in the United Kingdom, Parts I (Report), and II (Minutes of Evidence). London: HMSO.

RCCEA (Royal Commission on Coast Erosion and Afforestation) (1909) Second report (on afforestation) of the Royal Commission appointed to inquire into and to report on certain questions affecting coast erosion, the reclamation of tidal lands, and afforestation in the United Kingdom. London: HMSO.

RCCEA (Royal Commission on Coast Erosion and Afforestation) (1911) Third (and final) report of the Royal Commission appointed to inquire into and to report on certain questions affecting coast erosion, the reclamation of tidal lands, and afforestation in the United Kingdom. London: HMSO.

Sheppard, T. (1912) *Lost Towns of the Yorkshire Coast*. London: A. Brown and Sons.

Steers, J.A. (1926) Orford Ness: a study in coastal physiography. *Proceedings of the Geologists' Association* 37: 306–25. https://doi.org/10.1016/S0016-7878(26)80023-9

Steers, J.A. (1946) *The Coastline of England and Wales*. Cambridge: Cambridge University Press.

Steers, J.A. (1953) *The Sea Coast*. London: Collins. https://doi.org/10.2307/1790640

Stive, M.J.F., de Schipper, M.A., Luijendijk, A.P., Aarninkhof, S.G.J., van Gelder-Maas, C., van de Thiel, J.S.M., de Vries, S., Henriquez, M., Mark, S. and Ranasinghe, R. (2013) A new alternative to saving our beaches from sea-level rise: the sand engine. *Journal of Coastal Research* 29: 1001–8. https://doi.org/10.2112/JCOASTRES-D-13-00070.1

Tabor, M. (2018) UK GEOS Coastal erosion and accretion project. Final report V1.0, April 2018. Ordnance Survey, Southampton.

Thompson, C. (1923) The erosion of the Holderness coast. *Proceedings of the Yorkshire Geological Society* 20: 32–9. https://doi.org/10.1144/pygs.20.1.32

Valentin, H. (1954) Der landverlust in Holderness, Ostengland, von 1852–1952. *Die Erde* 6: 296–315.

Valentin, H. (1971) Land loss at Holderness. In J.A. Steers (ed.) *Applied Coastal Geomorphology*, pp. 116–37. London: Macmillan. https://doi.org/10.1007/978-1-349-15424-1_8

Westley, K. and McNeary, R. (2014) Assessing the impact of coastal erosion on archaeological sites: a case study from Northern Ireland. *Conservation and Management of Archaeological Sites* 16: 185–211. https://doi.org/10.1179/1350503315Z.00000000082

Whittow, J.B. (2019) *The Edge of the Land: Memories of One Person's Enchantment with the Coast*. Salisbury: Riverside Publishing Solutions Ltd.

Williams, A.T., Rangel-Buitrago, N., Pranzini, E. and Anfuso, G. (2018) The management of coastal erosion. *Ocean Coastal Management* 156: 4–20. https://doi.org/10.1016/j.ocecoaman.2017.03.022

CHAPTER 3

Challenges of Sea-level Rise on Estuarine Tidal Dynamics

DANIAL KHOJASTEH, JAMIE RUPRECHT, KATRINA WADDINGTON, HAMED MOFTAKHARI, AMIR AGHAKOUCHAK and WILLIAM GLAMORE

Abstract

Evidence indicates that climate change-induced sea-level rise (SLR) will influence estuarine tidal dynamics. Altered tidal dynamics, in turn, may have a profound impact on urban, rural and natural estuarine environments. A detailed understanding of the challenges posed by the dynamic temporal and spatial responses of estuaries to SLR is required to inform coastal zone planning decisions and management. This chapter presents the mechanisms by which SLR can influence tidal dynamics primarily via changes in friction, geometry, resonance and reflection. These variations can then influence the tidal prism, tidal currents, tidal asymmetry and sediment transport. Feedback loops, compounding and non-linear interactions between these processes may result in significant uncertainties regarding future risk scenarios. For instance, while a rising high tide is likely to increase the incidence of nuisance flooding and contribute to compound flooding risks, the altered low tide will have a direct impact on the resilience of intertidal and benthic environments. Further, variations in the tidal range and currents may affect navigation, tidal energy schemes and sediment dynamics. The rate of both SLR and anthropogenic developments in the coastal/ estuarine zone are expected to accelerate up to and beyond 2100, with land management decisions likely to result in further compounding interactions. In urbanised estuaries, hard engineering structures may replace natural flood mitigation capacity, resulting in the loss of other ecosystem services and changes to the geomorphological characteristics of the estuary. To alleviate negative impacts, a holistic approach to assessing the cumulative impacts of SLR and anthropogenic activities at an estuary-wide scale is recommended.

Keywords: estuary, intertidal habitat, estuary management, compounding effects, climate change

Correspondence: w.glamore@unsw.edu.au

Introduction

Climate change-induced sea-level rise (SLR) is one of the greatest challenges threatening the sustainable management of coastal regions worldwide (Vitousek et al. 2017). Global mean sea level (GMSL) has risen by approximately 0.11–0.16 m in the last century (Kulp

Danial Khojasteh, Jamie Ruprecht, Katrina Waddington, Hamed Moftakhari, Amir AghaKouchak and William Glamore, 'Challenges of Sea-level Rise on Estuarine Tidal Dynamics' in: *Challenges in Estuarine and Coastal Science*. Pelagic Publishing (2022). © Danial Khojasteh, Jamie Ruprecht, Katrina Waddington, Hamed Moftakhari, Amir AghaKouchak and William Glamore. DOI: 10.53061/VYYD7091 and Strauss 2019), mainly owing to increased atmospheric carbon dioxide concentrations and higher global air-surface-water temperatures caused by anthropogenic forcing such as the burning of fossil fuels. This has led to the melting of glaciers and polar ice sheets as well as the thermal expansion of warming oceans. Recent studies reveal that present-day SLR is accelerating (Weeman and Lynch 2018), and the Intergovernmental Panel on Climate Change (IPCC) projections estimate GMSL will increase up to four times, with a likely range of 0.29–1.10 m by 2100 relative to 1986–2005 levels, leading to serious environmental, economic and social implications in coastal regions (Oppenheimer et al. 2019). For instance, without adaptation measures Hinkel et al. (2014) estimated that by 2100, under current SLR projections, up to 5% of the global population could be flooded each year, with expected annual losses of up to 10% of global gross domestic profit. Nuisance flooding (also known as 'sunny-day' flooding), owing to increasingly higher high tides in some coastal estuaries, in the USA is currently three to nine times more frequent than it was 50 years ago (1960–2010) (Moftakhari et al. 2015; Lindsey 2021), and the chronic impact of these events has significant cumulative costs (Moftakhari et al. 2017a). Despite reported uncertainties in the rate of change in future SLR projections, sea levels will continue to rise beyond 2100 (Oppenheimer et al. 2019) as positive feedback loops in the earth-system exceed global equilibrium thresholds (Bamber et al. 2019; Horton et al. 2020).

Real-world impacts of SLR are disproportionately distributed around the globe, varying significantly at regional and local scales. This is exacerbated by compounding anthropogenic drivers, including population growth, urbanisation, industrialisation and global food/energy demand, which are typically concentrated in coastal regions (Haigh et al. 2020; Talke and Jay 2020; Khojasteh et al. 2021). In some coastal regions, such as low-lying islands in the South Pacific Ocean and subsiding coastal cities, such as Jakarta in Indonesia, local rates of SLR are up to four times higher than the global average (Nicholls et al. 2021), increasing regional vulnerability to SLR and posing immediate coastal planning and adaptation management challenges (Moftakhari et al. 2017b).

Coastal zone planning decisions and management considerations that address SLR impacts require an understanding of the physical processes driving mean and extreme water levels in coastal waterways (Khojasteh et al. 2021). Estuaries are located at the interface of the river catchment and the open coast, with both environments contributing to observed water levels. Oceanic water levels are characterised by four components: mean sea level (MSL), tides (astronomic tides plus overtides), non-tidal residuals (including anomalies owing to ocean density changes, offshore wind setup and climatic phenomena such as El Nino) and waves (Serafin and Ruggiero 2014), as illustrated in Fig. 3.1. Of these, the oceanic tides (herein referred to as tides) are the most predictable (refer to Gregory et al. (2019) for details). Tides are shallow-water waves defined by characteristics (amplitudes and phases) that are affected by non-astronomical factors (Haigh et al. 2020) as they respond, largely via non-linear interactions, to changes in MSL through barotropic effects (such as frictional effects and altered tidal resonance (Talke and Jay 2020) and a variety of internal processes (Idier, Dumas and Muller 2012; Devlin et al. 2017).

Over the past two centuries, many regions across the world have experienced significantly altered tidal dynamics beyond the expected range of natural variations (Müller, Arbic and Mitrovica 2011; Arns et al. 2020). Among these regions, changes to tidal dynamics are most apparent in rivers, estuaries and embayments (Talke and Jay 2020). Haigh et al. (2020) identified several mechanisms that influence tidal dynamics, including changes in friction, geometry (i.e. depth, width, length, entrance condition), river inflow, resonance, reflection, tectonic/continental drift and ocean stratification. However, a thorough understanding of the implications that arise from SLR owing to its influence on estuarine tidal dynamics is required to assist decision-makers who are tasked with building resilience in estuarine/deltaic systems.

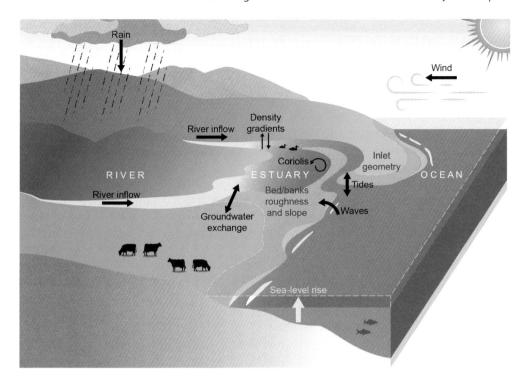

Figure 3.1 A conceptual schematic of an estuarine system including important factors that control hydrodynamics

To provide insights into these implications, this chapter initially describes the estuarine tidal dynamics under SLR. It then highlights the major implications of altered tidal dynamics in urban, rural and natural environments, including more frequent nuisance flooding and failure of drainage infrastructures, as well as altered navigation, tidal renewable energy and sediment dynamics. Finally, the chapter presents a range of challenges for consideration in estuaries under present-day, near future and far future conditions.

Formation of tides

Oceanic tides are generated by gravitational forces between the earth, moon and sun, as well as rotational forces of the earth–moon–sun system. The complex interactions between the earth, moon and sun cause natural fluctuations in the tidal range (high and low tides), with most regions experiencing two high and two low tides each day (semi-diurnal), while some regions have one high and one low tide each day (diurnal). Fortnightly 'spring' tides occur when the orbit of the moon is closer to the earth, increasing the gravitational pull and tidal range to create higher highs and lower lows. Further details regarding tidal dynamics are described in Pugh and Woodworth (2014) and Gregory et al. (2019).

Tidal dynamics and sea-level rise

Several physical processes can influence the tidal range within estuaries, including inertia-related effects, convergence, resonance (reflection) and friction (van Rijn 2011; Haigh et al. 2020). Inertia-related effects are often negligible, except in shallow macrotidal estuaries (Holleman 2013). Converging banks and resonance primarily cause tidal amplification in the landward direction, while friction and diverging banks result in upstream tidal attenuation (Jay 1991; Holleman 2013). However, SLR can influence these processes by changing the water depth (h) and hence altering the tidal range. This is most often due to a reduction in the friction factor (f):

$$f = 8C_D V/3\pi h \tag{1}$$

where C_D is the drag coefficient and V is the tidal current velocity. For instance, in a converging estuary, the tidal range can be amplified if the convergence length, L $\ll$ c/f, and dampened if L $\gg$ c/f, where $c = \omega / \kappa$ is the tidal wave speed, ω denotes the tidal frequency and κ is the wave number (Ralston et al. 2019). Increasing the water depth under SLR (or dredging, for example, as in the Hudson River estuary (Ralston et al. 2019)) can alter both the convergence length and the friction factor. SLR was found to increase the tidal range on the landward side of the Chesapeake estuary owing to an increased water depth and diminished friction (Du et al. 2018). Likewise, SLR was shown to increase the tidal range in the Delaware River estuary if overland flooding is prevented owing to an increase in the convergence length and reduced friction (Lee, Li and Zhang 2017). In contrast, if adjacent lands are allowed to be inundated, SLR would decrease the tidal range of the Delaware River estuary, owing to the addition of new shallow waters that cause more energy dissipation within the system (Lee, Li and Zhang 2017).

Propagating tidal waves can be reflected due to changes in the geometry of an estuary (van Rijn 2011). If a tidal wave intersects a non-porous vertical obstacle, standing waves are generated that can lead to a local amplification in the tidal range. As SLR can change the depth and width of an estuary, it could also alter the tidal reflection patterns. In addition, tidal range amplification and standing waves can occur in estuaries with lengths equal to the resonance length – i.e., one-quarter or three-quarters of a wavelength $\lambda = cT = \sqrt{gh}T$, where T is the tidal period and g the gravitational constant. In these systems, the natural frequency of the estuary is nearly synchronised with the tidal frequency. As SLR changes water depth, an estuary can transition away from a resonance state (tidal range attenuation), or move closer to a resonance state (tidal range amplification) (Haigh et al. 2020). The latter scenario is more common, with the effects of tidal range amplification most noticeable in the upstream areas of an estuary (e.g., Chesapeake estuary (Zhong, Li and Foreman 2008)).

As the shallow water wave speed is proportional to water depth ($c \propto \sqrt{h}$), the trough of the tidal wave travels slower than the crest. This process can cause tidal asymmetry (deformation), where there is an imbalance in the strength of the ebb and flood velocities. As SLR increases water depth, it may alter the tidal asymmetry pattern and sediment dynamics (Khojasteh et al. 2021).

The rise and fall of the tides at the mouth of an estuary influence the water volume (referred to as the tidal prism) penetrating into the estuary, causing the water to move horizontally and creating tidal currents (Carballo, Iglesias and Castro 2009). The altered tidal levels under SLR can influence the exchange of the water volume in and out of the estuary, hence modifying the tidal prism (Khojasteh et al. 2020). For example, by inducing an increase in tidal range and wetted area, SLR has been predicted to increase the tidal prisms of the Choctawhatchee and Perdido Bays by 44% and 52%, respectively (Passeri et al. 2015). In turn, these variations increase the current velocity at the inlets of the Perdido and Choctawhatchee Bays by 4 cm/s (Passeri et al. 2015), affecting tidal asymmetry and sediment transport dynamics (Aubrey and Speer 1985; Dronkers 1986; Guo et al. 2019).

SLR will also influence the tidal dynamics (e.g. tidal range, currents, asymmetry) of estuarine systems, with potential implications for urban, rural and natural environments, as will be discussed later in this chapter.

Compound effects

Altered fluvial flow regimes may also contribute to extreme estuarine water-level patterns. Tidal theory indicates that tides and river inflow can interact through a quadratic bed friction function that diminishes and distorts the tidal wave as discharge increases (Jay 1991; Lanzoni and Seminara 1998). This quadratic bed stress, $\tau_b = \rho C_D |U| U$, can influence wave amplitude and phase as well as energy exchange between frequencies (Flinchem and Jay 2000), where ρ is water density and U is dimensional total velocity (incorporating river and tidal flows) (Parker 1991). Such non-linear interactions between tides and river inflow to estuaries may result in significant uncertainties regarding the prediction of extreme estuarine water levels under various hydrologic scenarios. For instance, a high fluvial flow to the lower estuary may both increase the mean water level and dampen tides (Hoitink and Jay 2016). Non-tidal residuals may also interact with oceanic tides via non-linear mutual phase and amplitude alterations (Proudman 1955; Rossiter 1961). A growing body of literature suggests that ignoring non-linear compounding processes in estuarine systems will lead to the mischaracterisation of flood risk (Moftakhari et al. 2017a; Couasnon et al. 2020).

An example of estuarine water-level variability is provided in Fig. 3.2. This analysis depicts a 6.87 mm/yr increase in MSL at Wyndham, Australia, from the mid-1980s to the mid-2010s. In this case, a statistically significant (at 5%) positive rank correlation exists between MSL and non-tidal residuals. Consequently, a higher average water level at this location would increase the likelihood of storm-surge reaching higher elevations and altering extreme water levels. There also exists a statistically significant positive rank correlation between MSL and S_2 tides (principal semidiurnal solar tidal constituent). This means that in the future, SLR can be expected to result in a larger tidal amplitude under the same gravitational forces. This has potential consequences for coastal populations and assets associated with tidal rivers and estuaries, requiring further attention (Arns et al. 2017; Moftakhari et al. 2017b).

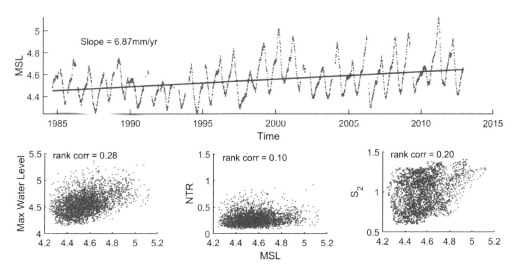

Figure 3.2 Historical trends in mean sea level (MSL) and relationships with estimated maximum water level, non-tidal residual (NTR) and S_2 tide at Wyndham, Australia (Lat: 15.4533, Lon: 128.1017). Units are metres

Implications of tidal dynamics in urban, rural and natural environments

Tidal planes may not increase uniformly in response to SLR owing to the complex interactions between ocean water levels, estuary geometry and sediment dynamics. In estuarine environments, there may be modest changes in low-tide water levels but a large change in tidal amplitude (e.g. the high-tide water level could increase significantly). Conversely, the tidal range could be attenuated through a substantial increase in low-tide water levels, and minimal changes to high-tide water levels. Changes to the low tide would have a direct impact on the efficiency of drainage systems and the resilience of intertidal and benthic environments. The rising high tide may increase the incidence of nuisance flooding and contribute to compound flooding risks. The variations in the tidal range and currents have the potential to affect navigation, tidal energy and/or sediment dynamics. These implications are further discussed below.

Implications of an altered high tide

The spatial and temporal variability of extreme estuarine water levels (i.e. high-tide level) is expected to change under accelerating SLR (Goodwin et al. 2017), with significant effects on infrastructure (Moftakhari et al. 2017a). SLR shifts the high-tide level to higher elevations and increases the chance that historical thresholds will be surpassed, even during non-extreme surge activities (Moftakhari et al. 2015; Vandenberg-Rodes et al. 2016). While changes will be partially due to altered climate forcing, non-linear interactions among different components of extreme water level may also play a major role. Non-linear modulations of high tidal levels owing to a higher MSL directly reduces the difference between the high tidal datum and the elevation of the flood protection crest (the available 'freeboard'), resulting in increased nuisance (a.k.a. sunny-day or tidal) flooding (Moftakhari et al. 2018). It also indirectly influences the risk of compound flooding as the reduced barotropic gradient leads to lower fluvial flow capacity towards the estuarine outlet. Thus, altered high tides caused by SLR can pose a serious threat to the population and assets located in low-lying coastal regions by increasing the frequency and intensity of flooding. They can also adversely affect the functionality of natural environments, such as wetlands, leading to the loss of these ecosystems through altered inundation dynamics and/or patterns (Muñoz et al. 2021; Sadat-Noori et al. 2021).

Implications of an altered low tide

Intertidal and benthic ecosystems, including mangrove and saltmarsh wetlands, mudflats, sandy beaches, rocky shorelines, seaweed and seagrass beds are biologically diverse environments that are well recognised for their extensive ecosystem services (Barbier et al. 2011; Lebreton et al. 2019). On the landward side, intertidal environments are under pressure from anthropogenic developments (Sadat-Noori et al. 2021). The intensification of agriculture and aquaculture, the expansion of urban infrastructure, industrial developments, sea defences and ports have all led to global wetland loss estimated to be in excess of 50% since the beginning of the eighteenth century, with coastal wetland loss accelerating in the twentieth century (Davidson 2014). These pressures are now being compounded on the seaward side by a rising low tide, a threat that has often been overlooked in previous studies (Khojasteh et al. 2020). A rising low tide has the potential to inundate the existing intertidal zone and the resulting increase in water depth will likely reduce sunlight penetration (e.g. light availability) required for photosynthesis in the lower littoral and benthic environments (Mangan et al. 2020). In many areas, this is already resulting in 'coastal squeeze', which is the narrowing of viable habitat areas as the low tide encroaches upslope while there is limited opportunity for landward migration (Sadat-Noori et al. 2021; Kirwan and Megonigal 2013). The ability of these habitats to adapt through migration or accretion is being tested by the SLR rate. Based on such pressures, it is inevitable that some ecosystems will respond to these changes better than others. As a consequence, certain ecosystems are expected to be impacted by the resultant loss of biodiversity (Rayner et al. 2021).

The efficiency of surface drainage systems is also affected by the tailwater conditions at the point of discharge. In a tidal estuary, the low tide represents the physical limit to which these systems can naturally drain. As the low tide rises, the capacity of the drainage system is reduced and, where the low tide becomes too high to permit adequate natural drainage (and/or in areas subject to chronic high-tide (nuisance) flooding), it must be augmented, or even replaced, by pumped drainage systems. In many areas, however, the cost of pumping, including the provision of attenuation storage and the construction, ongoing operation and maintenance of pump stations cannot be justified or sustained on economic or environmental grounds. Raising the land level is an alternative strategy to pumping, but is complicated by existing development, the availability and impacts of obtaining suitable fill material, and so alternative management strategies will need to be considered, with many low-lying areas possibly surrendered to the rising sea.

Effective drainage is fundamental to the amenity of the anthropogenic environment. When functional, floodplain drainage removes excess surface and subsoil water, preventing damage from inundation and erosion, while also mitigating flood risk, managing groundwater levels and avoiding public health risks and other adverse impacts from prolonged ponding. Waterlogging also reduces trafficability, weakens structural foundations and leads to the failure of septic systems and infiltration of sewerage networks (Knott et al. 2018). Further, waterlogging limits oxygen supply to the soil, altering the soil structure and chemistry, hence impacting plant vitality and reducing crop yields (Manik et al. 2019). Overall, the potential implications of elevated low tide levels from SLR are varied, yet limited research has been undertaken in this area.

Implications of altered tidal range and current velocities

Ports play an important role in global supply-chains, with 80% of world trade transported by sea (Becker et al. 2013). Ports, harbours and estuaries associated with this trade are selected and developed to provide a sheltered anchorage, deep water access and appropriate transport linkages, but the movement, mooring and berthing of vessels may be affected by changes to both tidal ranges and currents (Meyers and Luther 2020). Generally, an altered tidal range under SLR is expected to increase the clearance between the hull of a vessel and the bed of the waterway at low tide, but higher water levels will also reduce the distance to overhead obstacles such as bridges and powerlines. Increased water depths under SLR can generally be expected to improve navigability, whereas altered current velocities may present less flexibility for managing ports and harbours (Meyers and Luther 2020). It is worth noting that changes to tidal currents may impact sedimentation and bathymetry, necessitating new or modified navigation routes and dredging regimes.

Changes in tidal range due to SLR may also influence the performance and output power of tidal range power plants located within estuaries. The captured energy (E) is proportional to the impounded wetted surface area (A) and the square of the head difference (d) between the upstream and downstream sides of the impoundment ($E \propto Ad^2$) (Waters and Aggidis 2016). For instance, SLR increases the tidal range of the main stem of the Chesapeake estuary, while minimally reducing the tidal range of the Patapsco River and Choptank River sub-estuaries (Du et al. 2018). Thus, SLR may have either a beneficial or adverse impact on tidal range energy schemes.

SLR can also influence the tidal current energy by altering the current velocities. As the current energy is proportional to current velocity (U) cubed ($E \propto U^3$), any minor changes to the current velocity would substantially impact the output power and the location of energy hotspots (Tang et al. 2014). For example, in the Northern Gulf of Mexico, Grand Bay experiences an increase in tidal current velocities, while the nearby Chandeleur Islands experience a decrease in tidal current velocities under 2 m of SLR (Passeri et al. 2016). The differences in current velocities have implications for altering the tidal current energy resource within these systems.

Sediment dynamics primarily govern the geomorphology of an estuarine system, and are controlled by tidal range, currents and tidal prism (De Swart and Zimmerman 2009). Flood-dominated systems typically experience basin infilling, while ebb-dominated systems undergo sediment flushing, with tidal asymmetry often used to determine the direction of net sediment transport (Guo et al. 2019). As tidal range and currents are altered under SLR, a variation in flood or ebb domination of the system is likely to occur. For example, SLR reduces the flood domination in the Tamar River estuary by up to 40%, with an accompanying mitigation of the sediment import mechanisms (Palmer, Watson and Fischer 2019). Further, alterations in sediment dynamics and geomorphology would have implications for inlet stability, navigational routes, inland fisheries and recreational activities (Duong et al. 2016).

There is growing evidence that increased saltwater intrusion in estuaries will occur because of altered tidal dynamics under SLR (Chen et al. 2016; Chen et al. 2020; Song et al. 2020). SLR may enhance the transport of saltwater into previously non-tidal areas, resulting in the partial or full loss of freshwater ecosystems (Khojasteh et al. 2021). Further, altered tidal dynamics under SLR can modify hydroperiod (i.e. frequency, depth and duration of inundation), which is a key factor affecting saltmarsh and/or mangrove species dominance by controlling the rate of sediment accretion and root creation (Woodroffe et al. 2016; Rayner et al. 2021). This is of particular significance in Ramsar-listed intertidal wetlands, which are considered as internationally important ecosystems for the preservation of local, regional and global biological diversity.

SLR-induced changes in sediment dynamics or vegetation distribution can cause greater variability in estuarine tidal dynamics through feedback loops. Where the entrance conditions of an estuary change under SLR, the tidal dynamics of the system may vary significantly (Khojasteh et al. 2020). Further, changes to vegetation communities may result in different roughness characteristics and, thereby adjust the tidal dynamics of the estuary (Khojasteh et al. 2021). For instance, in the Ems River estuary, channel dredging, which resembles the effects of SLR, has led to altered tidal asymmetry and the formation of a mud layer. Over time, this has led to reduced friction and mixing, affecting estuarine tidal dynamics through feedback loops (de Jonge et al. 2014). Therefore, it is important to consider the hydro-ecologic and geomorphic feedback loops that occur over various temporal and spatial scales (Khojasteh et al. 2021) to ensure the sustainable management of estuaries under accelerating SLR.

Managing tidal challenges now, up to 2100, and beyond

This chapter outlines the range of challenges from SLR that require consideration in estuaries, some of which are illustrated in Fig. 3.3. These SLR-induced challenges include first-order physical processes such as altered tidal prism, tidal range and current velocity. However, additional challenges are also highlighted, which may be deemed second- or third-order processes. For instance, within this chapter, it is noted that changes to the tidal range are likely to influence the drainage of adjacent low-lying landscapes. This can result in increased onsite ponding and decreased flushing which, in turn, may result in increased

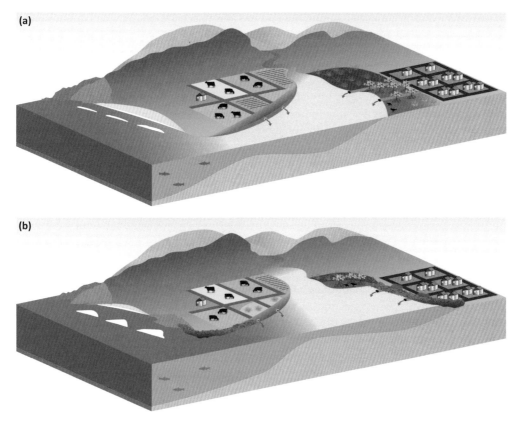

Figure 3.3 A conceptual schematic of an estuarine system under (a) present-day and (b) future scenarios. Under future conditions, an estuary may experience waterlogging and inundation of low-lying lands, reduced drainage and nuisance flooding to urban areas, coastal squeeze, altered fluvial flow regimes, changes in density and salinity gradients, erosion and variations in geomorphology and geometry that may necessitate engineered solutions to protect existing land-uses

water quality problems, mosquito-borne diseases or reduced agricultural productivity. The complex nature of these second- and third-order implications highlights some of the significant challenges faced when managing estuarine systems under accelerating SLR.

The future management of SLR in estuaries is further compounded by the dynamic relationship between variables within an estuary over space and time. As sea levels rise, management decisions will need to consider which areas will be defended and which areas will allow for retreat. It is reasonable to consider that many natural areas will have to retreat as sea levels inundate the landscape and intertidal zones become subtidal. However, this may result in the loss of ecosystem services provided by these habitats. For example, mangroves have been shown to provide substantial ecosystem services in the form of flood mitigation (Barbier 2016; Menéndez et al. 2020), and the potential loss of these habitats owing to inundation may result in the further hardening of upslope structures, to mitigate against flood risk.

In urbanised estuaries, it is anticipated that many sites will choose to defend existing infrastructure from SLR. In these situations, the replacement of intertidal habitat with flood mitigation structures may result in increased tidal velocities and potential scouring of the riverbed. The altered flow paths may, in turn, have broader geomorphological characteristics of the river and result in unforeseen and difficult to manage erosion/ sedimentation concerns. Further research is necessary to understand these challenges and to consider estuary-wide management responses across broad temporal and spatial scales that encompass these dynamic interconnected implications.

It is important to note that this chapter only focuses on the potential implications of altered tidal dynamics under SLR in estuaries. In considering future estuary management plans, additional climate change factors including altered riverine inflows, warming oceanic temperatures, changes in water quality and wind/wave forcing should be taken into account. These changes are often site specific, and in many locations insufficient data (or climate change modelling predictions) is available to undertake a detailed assessment at a local scale (Khojasteh et al. 2021). In these circumstances, broader assessment methods are sometimes applied where the estuarine risk to climate change is based on geomorphic classifications (Dominguez et al. 2019).

In addition to considerations of climate change in estuaries, it should also be noted that anthropogenic developmental pressures will persist. In many locations worldwide, these developmental pressures are a more significant risk to the local environment with substantial changes experienced to (a) inflow volumes/timing (due to changes in land-use, extractions and retentions), (b) water quality (including excessive nutrient and contaminant loads) and (c) habitats. Owing to these compounding pressures, it is apparent that cumulative assessments (e.g. climate change plus anthropogenic activities) should be considered at an estuary-wide scale. The increasingly accepted 'source to sea' concept provides an indication that a holistic approach is feasible by designing an integrated management system for the entire river basin and coastal areas (Granit et al. 2017; Hoagland et al. 2020).

Finally, it is important to note that assessments of SLR in estuaries should consider time periods beyond 2100. While it is widely acknowledged that sea levels are likely to continue to rise after 2100, most planning horizons within estuaries are limited to that year. For instance, under Representative Concentration Pathway 8.6, it is predicted that GMSL will rise by 0.63–1.32 m by 2100 and may rise by 1.67–5.61 m by 2300 (Horton et al. 2020). Planning horizons that acknowledge the shifting baseline over this period and incorporate flexible arrangements based on long-term risk will likely reduce the 'trial and error' approach associated with a 2100 planning horizon. In some way, this is akin to the childhood fable of the Three Little Pigs, where insufficient planning and forecasting resulted in poor on-ground outcomes (e.g. houses made of hay or wood) until eventually all relevant risks were considered and a long-term outcome was determined. With regard to SLR in estuaries, long-term considerations beyond 2100 should be included in decision-making to ensure that the scattered rubble of poor decision-making is not left to slowly become inundated by the rising tides.

Conclusions

For millennia, coastal populations have observed tidal movements as a constant reminder of the periodicity of life. Now, for the first time in living memory for most civilisations, the baseline tidal levels are beginning to shift. As discussed in this chapter, these changes will have various implications on the broader tidal dynamics, including changes to the high and low tide, the tidal range, tidal prism and resonance patterns. In turn, these physical changes will influence the tidal currents, overland drainage, bed stresses, erosional patterns and associated ecosystem distribution and function. In many circumstances these changes are interdependent with feedback loops ever adding to the overall complexity.

The effective management of estuaries under sea-level rise requires an integrated understanding of spatial and temporal tidal dynamics. As an example, highlighted within this chapter, decisions regarding overland inundation and the protection of ecosystem habitat versus the hardening of the flood mitigation defences require careful consideration of social, economic and environmental values. Our ability to understand and adapt to these challenges will determine the legacy of our decision-making for future generations.

References

Arns, A., Dangendorf, S., Jensen, J., Talke, S., Bender, J. and Pattiaratchi, C. (2017) Sea-level rise induced amplification of coastal protection design heights. *Scientific Reports* 7: 1–9. https://doi.org/10.1038/srep40171

Arns, A., Wahl, T., Wolff, C., Vafeidis, A.T., Haigh, I.D., Woodworth, P., Niehüser, S. and Jensen, J. (2020) Non-linear interaction modulates global extreme sea levels, coastal flood exposure, and impacts. *Nature Communications* 11: 1–9. https://doi.org/10.1038/s41467-020-15752-5

Aubrey, D. and Speer, P. (1985) A study of non-linear tidal propagation in shallow inlet/estuarine systems, part I: observations. *Estuarine, Coastal and Shelf Science* 21: 185–205. https://doi.org/10.1016/0272-7714(85)90096-4

Bamber, J.L., Oppenheimer, M., Kopp, R.E., Aspinall, W.P. and Cooke, R.M. (2019) Ice sheet contributions to future sea-level rise from structured expert judgment. *Proceedings of the National Academy of Sciences* 116: 11195–200. https://doi.org/10.1073/pnas.1817205116

Barbier, E.B. (2016) The protective service of mangrove ecosystems: a review of valuation methods. *Marine Pollution Bulletin* 109: 676–81. https://doi.org/10.1016/j.marpolbul.2016.01.033

Barbier, E.B., Hacker, S.D., Kennedy, C., Koch, E.W., Stier, A.C. and Silliman, B.R. (2011) The value of estuarine and coastal ecosystem services. *Ecological Monographs* 81: 169–93. https://doi.org/10.1890/10-1510.1

Becker, A.H., Acciaro, M., Asariotis, R., Cabrera, E., Cretegny, L., Crist, P., Esteban, M., Mather, A., Messner, S. and Naruse, S. (2013) A note on climate change adaptation for seaports: a challenge for global ports, a challenge for global society. *Climatic Change* 120: 683–95. https://doi.org/10.1007/s10584-013-0843-z

Carballo, R., Iglesias, G. and Castro, A. (2009) Numerical model evaluation of tidal stream energy resources in the Ría de Muros (NW Spain). *Renewable Energy* 34: 1517–24. https://doi.org/10.1016/j.renene.2008.10.028

Chen, W., Chen, K., Kuang, C., Zhu, D.Z., He, L., Mao, X., Liang, H. and Song, H. (2016) Influence of sea level rise on saline water intrusion in the Yangtze River Estuary, China. *Applied Ocean Research* 54: 12–25. https://doi.org/10.1016/j.apor.2015.11.002

Chen, W., Mao, C., He, L. and Jiang, M. 2020. Sea-level rise impacts on the saline water intrusion and stratification of the Yangtze Estuary. *Journal of Coastal Research* 95: 1395–400. https://doi.org/10.2112/SI95-269.1

Couasnon, A., Eilander, D., Muis, S., Veldkamp, T.I., Haigh, I.D., Wahl, T., Winsemius, H.C. and Ward, P.J. (2020) Measuring compound flood potential from river discharge and storm surge extremes at the global scale. *Natural Hazards and Earth System Sciences* 20: 489–504. https://doi.org/10.5194/nhess-20-489-2020

Davidson, N.C. (2014) How much wetland has the world lost? Long-term and recent trends in global wetland area. *Marine and Freshwater Research* 65: 934–41. https://doi.org/10.1071/MF14173

De Jonge, V.N., Schuttelaars, H.M., Van Beusekom, J.E., Talke, S.A. and De Swart, H.E. (2014) The influence of channel deepening on estuarine turbidity levels and dynamics, as exemplified by the Ems estuary. *Estuarine, Coastal and Shelf Science* 139: 46–59. https://doi.org/10.1016/j.ecss.2013.12.030

De Swart, H. and Zimmerman, J. (2009). Morphodynamics of tidal inlet systems. *Annual Review of Fluid Mechanics* 41: 203–29. https://doi.org/10.1146/annurev.fluid.010908.165159

Devlin, A.T., Jay, D.A., Zaron, E.D., Talke, S.A., Pan, J. and Lin, H. (2017) Tidal variability related to sea level variability in the Pacific Ocean. *Journal of Geophysical Research: Oceans* 122: 8445–63. https://doi.org/10.1002/2017JC013165

Dominguez, G., Bishop, M., Heimhuber, V., Glamore, W. and Scanes, P. (2019) Ecological responses to climate change. Climate Change in Estuaries – State of the science and framework for assessment, Sydney.

Dronkers, J. (1986) Tidal asymmetry and estuarine morphology. *Netherlands Journal of Sea Research* 20: 117–31. https://doi.org/10.1016/0077-7579(86)90036-0

Du, J., Shen, J., Zhang, Y.J., Ye, F., Liu, Z., Wang, Z., Wang, Y.P., Yu, X., Sisson, M. and Wang, H.V. (2018) Tidal response to sea-level rise in different types of estuaries: the importance of length, bathymetry, and geometry. *Geophysical Research Letters* 45: 227–35. https://doi.org/10.1002/2017GL075963

Duong, T.M., Ranasinghe, R., Walstra, D. and Roelvink, D. (2016) Assessing climate change impacts on the stability of small tidal inlet systems: why and how? *Earth-Science Reviews* 154: 369–80. https://doi.org/10.1016/j.earscirev.2015.12.001

Flinchem, E. and Jay, D. (2000) An introduction to wavelet transform tidal analysis methods. *Estuarine, Coastal and Shelf Science* 51: 177–200. https://doi.org/10.1006/ecss.2000.0586

Goodwin, P., Haigh, I.D., Rohling, E.J. and Slangen, A. (2017) A new approach to projecting 21st century sea-level changes and extremes. *Earth's Future* 5: 240–53. https://doi.org/10.1002/2016EF000508

Granit, J., Liss Lymer, B., Olsen, S., Tengberg, A., Nõmmann, S. and Clausen, T. (2017) A conceptual framework for governing and managing key flows in a source-to-sea continuum. *Water Policy* 19: 673–91. https://doi.org/10.2166/wp.2017.126

Gregory, J.M., Griffies, S.M., Hughes, C.W., Lowe, J.A., Church, J.A., Fukimori, I., Gomez, N., Kopp, R.E., Landerer, F. and Le Cozannet, G. (2019).

Concepts and terminology for sea level: mean, variability and change, both local and global. *Surveys in Geophysics* 40: 1251–89. https://doi.org/10.1007/s10712-019-09525-z

Guo, L., Wang, Z.B., Townend, I. and He, Q. (2019) Quantification of tidal asymmetry and its nonstationary variations. *Journal of Geophysical Research: Oceans* 124: 773–87. https://doi.org/10.1029/2018JC014372

Haigh, I.D., Pickering, M.D., Green, J.M., Arbic, B.K., Arns, A., Dangendorf, S., Hill, D.F., Horsburgh, K., Howard, T. and Idier, D. (2020) The tides they are a-changin': a comprehensive review of past and future nonastronomical changes in tides, their driving mechanisms, and future implications. *Reviews of Geophysics* 58: e2018RG000636. https://doi.org/10.1029/2018RG000636

Hinkel, J., Lincke, D., Vafeidis, A.T., Perrette, M., Nicholls, R.J., Tol, R.S., Marzeion, B., Fettweis, X., Ionescu, C. and Levermann, A. (2014) Coastal flood damage and adaptation costs under 21st century sea-level rise. *Proceedings of the National Academy of Sciences* 111: 3292–7. https://doi.org/10.1073/pnas.1222469111

Hoagland, P., Beet, A., Ralston, D., Parsons, G., Shirazi, Y. and Carr, E. (2020) Salinity intrusion in a modified river-estuary system: an integrated modeling framework for source-to-sea management. *Frontiers in Marine Science* 7 (425). https://doi.org/10.3389/fmars.2020.00425

Hoitink, A. and Jay, D.A. (2016) Tidal river dynamics: implications for deltas. *Reviews of Geophysics* 54: 240–72. https://doi.org/10.1002/2015RG000507

Holleman, C.D. (2013) Dispersion and tidal dynamics of channel-shoal estuaries. D.Phil. thesis. University of California, Berkeley. https://digitalassets.lib.berkeley.edu/etd/ucb/text/Holleman_berkeley_0028E_13799.pdf (Accessed 13 September 2021).

Horton, B.P., Khan, N.S., Cahill, N., Lee, J.S., Shaw, T.A., Garner, A.J., Kemp, A.C., Engelhart, S.E. and Rahmstorf, S. (2020) Estimating global mean sea-level rise and its uncertainties by 2100 and 2300 from an expert survey. *NPJ Climate and Atmospheric Science* 3: 1–8. https://doi.org/10.1038/s41612-020-0126-0

Idier, D., Dumas, F. and Muller, H. (2012) Tide-surge interaction in the English Channel. *Natural Hazards and Earth System Sciences* 12: 3709–18. https://doi.org/10.5194/nhess-12-3709-2012

Jay, D.A. (1991) Green's law revisited: tidal long-wave propagation in channels with strong topography. *Journal of Geophysical Research: Oceans* 96: 20585–98. https://doi.org/10.1029/91JC01633

Khojasteh, D., Glamore, W., Heimhuber, V. and Felder, S. (2021) Sea level rise impacts on estuarine dynamics: a review. *Science of The Total Environment* 780: 146470. https://doi.org/10.1016/j.scitotenv.2021.146470

Khojasteh, D., Hottinger, S., Felder, S., De Cesare, G., Heimhuber, V., Hanslow, D.J. and Glamore, W.

(2020) Estuarine tidal response to sea level rise: the significance of entrance restriction. *Estuarine, Coastal and Shelf Science* 244: 106941. https://doi.org/10.1016/j.ecss.2020.106941

Kirwan, M.L. and Megonigal, J.P. (2013) Tidal wetland stability in the face of human impacts and sea-level rise. *Nature* 504: 53–60. https://doi.org/10.1038/nature12856

Knott, J.F., Daniel, J.S., Jacobs, J.M. and Kirshen, P. (2018) Adaptation planning to mitigate coastal-road pavement damage from groundwater rise caused by sea-level rise. *Transportation Research Record* 2672: 11–22. https://doi.org/10.1177/0361198118757441

Kulp, S.A. and Strauss, B.H. (2019) New elevation data triple estimates of global vulnerability to sea-level rise and coastal flooding. *Nature Communications* 10: 4844. https://doi.org/10.1038/s41467-019-13552-0

Lanzoni, S. and Seminara, G. (1998) On tide propagation in convergent estuaries. *Journal of Geophysical Research: Oceans* 103: 30793–812. https://doi.org/10.1029/1998JC900015

Lebreton, B., Rivaud, A., Picot, L., Prévost, B., Barillé, L., Sauzeau, T., Pollack, J.B. and Lavaud, J. (2019) From ecological relevance of the ecosystem services concept to its socio-political use. The case study of intertidal bare mudflats in the Marennes-Oléron Bay, France. *Ocean & Coastal Management* 172: 41–54. https://doi.org/10.1016/j.ocecoaman.2019.01.024

Lee, S.B., Li, M. and Zhang, F. (2017) Impact of sea level rise on tidal range in Chesapeake and Delaware Bays. *Journal of Geophysical Research: Oceans* 122: 3917–38. https://doi.org/10.1002/2016JC012597

Lindsey, R. (2021) Climate change: global sea level. NOAA. Available at: https://www.climate.gov/news-features/understanding-climate/climate-change-global-sea-level (Accessed 13 September 2021).

Mangan, S., Bryan, K.R., Thrush, S.F., Gladstone-Gallagher, R.V., Lohrer, A.M. and Pilditch, C.A. (2020) Shady business: the darkening of estuaries constrains benthic ecosystem function. *Marine Ecology Progress Series* 647: 33–48. https://doi.org/10.3354/meps13410

Manik, S., Pengilley, G., Dean, G., Field, B., Shabala, S. and Zhou, M. (2019) Soil and crop management practices to minimize the impact of waterlogging on crop productivity. *Frontiers in Plant Science* 10: 140. https://doi.org/10.3389/fpls.2019.00140

Menéndez, P., Losada, I.J., Torres-Ortega, S., Narayan, S. and Beck, M.W. (2020) The global flood protection benefits of mangroves. *Scientific Reports* 10: 4404. https://doi.org/10.1038/s41598-020-61136-6

Meyers, S.D. and Luther, M.E. (2020) The impact of sea level rise on maritime navigation within a large, channelized estuary. *Maritime Policy & Management* 47: 920–36. https://doi.org/10.1080/03088839.2020.1723810

Moftakhari, H., Aghakouchak, A., Sanders, B.F., Matthew, R.A. and Mazdiyasni, O. (2017a) Translating uncertain sea level projections into infrastructure impacts using a Bayesian framework. *Geophysical Research Letters* 44: 11914–21. https://doi.org/10.1002/2017GL076116

Moftakhari, H.R., Aghakouchak, A., Sanders, B.F., Allaire, M. and Matthew, R.A. (2018) What is nuisance flooding? Defining and monitoring an emerging challenge. *Water Resources Research* 54: 4218–27. https://doi.org/10.1029/2018WR022828

Moftakhari, H.R., Aghakouchak, A., Sanders, B.F., Feldman, D.L., Sweet, W., Matthew, R.A. and Luke, A. (2015) Increased nuisance flooding along the coasts of the United States due to sea level rise: past and future. *Geophysical Research Letters* 42: 9846–52. https://doi.org/10.1002/2015GL066072

Moftakhari, H.R., Aghakouchak, A., Sanders, B.F. and Matthew, R.A. (2017b) Cumulative hazard: the case of nuisance flooding. *Earth's Future* 5: 214–23. https://doi.org/10.1002/2016EF000494

Müller, M., Arbic, B.K. and Mitrovica, J. (2011) Secular trends in ocean tides: observations and model results. *Journal of Geophysical Research: Oceans* 116: C05013. https://doi.org/10.1029/2010JC006387

Muñoz, D.F., Muñoz, P., Alipour, A., Moftakhari, H., Moradkhani, H. and Mortazavi, B. (2021) Fusing multisource data to estimate the effects of urbanization, sea level rise, and hurricane impacts on long-term wetland change dynamics. *IEEE Journal of Selected Topics in Applied Earth Observations and Remote Sensing* 14: 1768–82. https://doi.org/10.1109/JSTARS.2020.3048724

Nicholls, R.J., Lincke, D., Hinkel, J., Brown, S., Vafeidis, A.T., Meyssignac, B., Hanson, S.E., Merkens, J.-L. and Fang, J. (2021) A global analysis of subsidence, relative sea-level change and coastal flood exposure. *Nature Climate Change* 11: 338–42. https://doi.org/10.1038/s41558-021-00993-z

Oppenheimer, M., Glavovic, B., Hinkel, J., Van De Wal, R., Magnan, A.K., Abdelgawad, A., Cai, R., Cifuentes-Jara, M., Deconto, R.M. and Ghosh, T. (2019) Sea level rise and implications for low lying islands, coasts and communities: IPCC special report on the ocean and cryosphere in a changing climate.

Palmer, K., Watson, C. and Fischer, A. (2019) Nonlinear interactions between sea-level rise, tides, and geomorphic change in the Tamar Estuary, Australia. *Estuarine, Coastal and Shelf Science* 225: 106247. https://doi.org/10.1016/j.ecss.2019.106247

Parker, B.B. (1991) *Tidal Hydrodynamics*. John Wiley & Sons, New York.

Passeri, D.L., Hagen, S.C., Bilskie, M.V. and Medeiros, S.C. (2015) On the significance of incorporating shoreline changes for evaluating coastal hydrodynamics under sea level rise scenarios. *Natural Hazards* 75: 1599–1617. https://doi.org/10.1007/s11069-014-1386-y

Passeri, D.L., Hagen, S.C., Plant, N.G., Bilskie, M.V., Medeiros, S.C. and Alizad, K. (2016) Tidal hydrodynamics under future sea level rise and coastal morphology in the Northern Gulf of Mexico. *Earth's Future* 4: 159–76. https://doi.org/10.1002/2015EF000332

Proudman, J. (1955) The effect of friction on a progressive wave of tide and surge in an estuary. *Proceedings of the Royal Society of London. Series A. Mathematical and Physical Sciences* 233: 407–18. https://doi.org/10.1098/rspa.1955.0276

Pugh, D. and Woodworth, P. 2014. *Sea-level Science: Understanding Tides, Surges, Tsunamis and Mean Sea-level Changes*. Cambridge: Cambridge University Press. https://doi.org/10.1017/CBO9781139235778

Ralston, D.K., Talke, S., Geyer, W.R., Al-Zubaidi, H.A. and Sommerfield, C.K. (2019) Bigger tides, less flooding: Effects of dredging on barotropic dynamics in a highly modified estuary. *Journal of Geophysical Research: Oceans* 124: 196–211. https://doi.org/10.1029/2018JC014313

Rayner, D., Glamore, W., Grandquist, L., Ruprecht, J., Waddington, K. and Khojasteh, D. (2021) Intertidal wetland vegetation dynamics under rising sea levels. *Science of The Total Environment* 766: 144237. https://doi.org/10.1016/j.scitotenv.2020.144237

Rossiter, J.R. (1961) Interaction between tide and surge in the Thames. *Geophysical Journal International* 6: 29–53. https://doi.org/10.1111/j.1365-246X.1961.tb02960.x

Sadat-Noori, M., Rankin, C., Rayner, D., Heimhuber, V., Gaston, T., Drummond, C., Chalmers, A., Khojasteh, D. and Glamore, W. (2021) Coastal wetlands can be saved from sea level rise by recreating past tidal regimes. *Scientific Reports* 11: 1–10. https://doi.org/10.1038/s41598-021-80977-3

Serafin, K.A. and Ruggiero, P. (2014) Simulating extreme total water levels using a time-dependent, extreme value approach. *Journal of Geophysical Research: Oceans* 119: 6305–29. https://doi.org/10.1002/2014JC010093

Song, Z., Shi, W., Zhang, J., Yuan, D., Wu, Q. and Wang, R. (2020) Impact of sea level rise on estuarine salt water intrusion – a numerical model study for Changjiang Estuarine. *IOP Conference Series: Earth and Environmental Science* 525: 012130. https://doi.org/10.1088/1755-1315/525/1/012130

Talke, S.A. and Jay, D.A. (2020) Changing tides: the role of natural and anthropogenic factors. *Annual Review of Marine Science* 12: 121–51. https://doi.org/10.1146/annurev-marine-010419-010727

Tang, H., Kraatz, S., Qu, K., Chen, G., Aboobaker, N. and Jiang, C. (2014) High-resolution survey of tidal energy towards power generation and influence of sea-level-rise: a case study at coast of New Jersey, USA. *Renewable and Sustainable Energy Reviews* 32: 960–82. https://doi.org/10.1016/j.rser.2013.12.041

Van Rijn, L.C. (2011) Analytical and numerical analysis of tides and salinities in estuaries; part I: tidal wave propagation in convergent estuaries. *Ocean Dynamics* 61: 1719–41. https://doi.org/10.1007/s10236-011-0453-0

Vandenberg-Rodes, A., Moftakhari, H.R., Agha-kouchak, A., Shahbaba, B., Sanders, B.F. and Matthew, R.A. (2016) Projecting nuisance flooding in a warming climate using generalized linear models and Gaussian processes. *Journal of Geophysical Research: Oceans* 121: 8008–20. https://doi.org/10.1002/2016JC012084

Vitousek, S., Barnard, P.L., Fletcher, C.H., Frazer, N., Erikson, L. and Storlazzi, C.D. (2017) Doubling of coastal flooding frequency within decades due to sea-level rise. *Scientific Reports* 7: 1399. https://doi.org/10.1038/s41598-017-01362-7

Waters, S. and Aggidis, G. (2016) Tidal range technologies and state of the art in review. *Renewable and Sustainable Energy Reviews* 59: 514–29. https://doi.org/10.1016/j.rser.2015.12.347

Weeman, K. and Lynch, P. (2018) New study finds sea level rise accelerating. NASA Global Climate Change. Available at: https://climate.nasa.gov/news/2680/new-study-finds-sea-level-rise-accelerating/ (Accessed 13 September 2021).

Woodroffe, C.D., Rogers, K., Mckee, K.L., Lovelock, C.E., Mendelssohn, I. and Saintilan, N. 2016. Mangrove sedimentation and response to relative sea-level rise. *Annual Review of Marine Science* 8: 243–66. https://doi.org/10.1146/annurev-marine-122414-034025

Zhong, L., Li, M. and Foreman, M. 2008. Resonance and sea level variability in Chesapeake Bay. *Continental Shelf Research* 28: 2565–73. https://doi.org/10.1016/j.csr.2008.07.007

CHAPTER 4

Estimating the Residence Time in Estuaries: Methods and Application

PAULA BIROCCHI, JULIANA CORREA NEIVA FERREIRA and MARCELO DOTTORI

Abstract

Hydrodynamic models are widely used to investigate estuarine and coastal processes. We present those models that are used in estuarine regions to study the dispersion/retention of pollutants, larvae and other particulates. The residence time (Rt) is a hydrodynamic parameter that can play an important role in the control of processes for the elimination of substances from an estuary, especially pollutants. The investigation and application of the Rt in estuaries is a challenge because how to determine it remains an open question. The definition of Rt is extensively discussed in the literature, with one possible definition considering Rt to be the average time that particles take to exit the estuary. It can be calculated for any type of material and varies depending on the starting location. Rt represents the timescales of physical transport mechanisms and is often compared with biogeochemical processes. Different methods can be applied to calculate the Rt in estuaries. We show the importance of determining it and summarise the methods established in the literature. We include an estimation of Rt using the ECOM model as applied to the Cananéia-Iguape estuarine lagoon system in south-eastern Brazil. The application of an Rt estimate is a useful approach when facing the challenge of monitoring estuarine and coastal regions.

Keywords: residence time, estuary, hydrodynamic model, Cananéia-Iguape estuarine lagoon system, particle tracking

Correspondence: paula.birocchi@usp.br

Introduction

An estuary can be defined as 'a semi-enclosed coastal body of water, which has a free connection with the open sea, and within which sea water is measurably diluted with freshwater derived from land drainage' (Pritchard 1967). Estuaries are characterised by saline mixing zones between fresh and open-ocean waters, with the dominance of fine sedimentary material that is carried by sea and rivers (Church 1986; McLusky and Elliott 2004). The hydrodynamics and flushing of estuaries are forced by tides, riverine flow, meteorological processes and interactions with bathymetry and morphology (Defne and Ganju 2015). The effects of estuarine circulation – such as the transport and diffusion processes – on residence time (Rt) have been reported by Shen and Haas (2004), Wang et al. (2004) and Liu, Chen and Hsu (2011).

Paula Birocchi, Juliana Correa Neiva Ferreira and Marcelo Dottori, 'Estimating the Residence Time in Estuaries: Methods and Application' in: *Challenges in Estuarine and Coastal Science*. Pelagic Publishing (2022). © Paula Birocchi, Juliana Correa Neiva Ferreira and Marcelo Dottori. DOI: 10.53061/GAYD3051

The Rt is the average time that a dissolved or suspended material resides in an estuary before it is carried out to the open ocean (Liu, Chen and Hsu 2011). The Rt between estuary and open sea can play an important role in the control of pollutant concentrations and the distribution of plankton (Basu and Pick 1996), climatic influences on phytoplankton blooms (Phlips et al. 2012), spatial and temporal variations of dissolved nutrients (Andrews and Muller 1983) and diversity of phytoplankton responses to nutrient loading (Camacho et al. 2015). As an example, an investigation of the annual total nitrogen and total phosphorus budgets for various estuaries around the North Atlantic showed that the net fractional transport of these nutrients through estuaries to the continental shelf is inversely correlated with the log mean Rt of water in the system (Nixon et al. 1996). In this scenario, denitrification is the dominant mechanism responsible for eliminating nitrogen in most estuaries, and the portion of total N input that is denitrified seems to be directly proportional to the log mean water Rt. In general, Rt can be considered a hydrodynamic parameter that represents the timescales of physical transport processes, and is often compared with the timescales of biogeochemical processes, such as the carbon, nitrogen and phosphorus cycles (Rueda, Moreno-Ostos and Armengol 2006; Liu et al. 2010; Wan et al. 2013).

Short Rt suggests a good flushing efficiency and active water exchange with the neighbouring sea, with the pollutants residing in the area for a short time and thus causing minimum adverse effects to the water quality, as evaluated by Zainol, Akhir and Zainol (2021). Numerical modelling can help to estimate eutrophication in estuaries, which is one of the consequences of shrimp farming (Bull, Cunha and Scudelari 2021). As for larval transport and dispersal, quantifying this process is vital to assess local population maintenance, replenishment and resilience for species subjected to exploitation (Shen, Boon and Kuo 1999; Fogarty and Botsford 2007; McManus et al. 2020). A better understanding of larval transport processes is also important for conservation, management and fisheries regulation (Herbert et al. 2011). Larval transport in estuaries can be predicted with a combination of fine-scale hydrodynamic models linked with water quality and individual models of larval behaviour (Sale et al. 2005). Hydrodynamic modelling is also performed for water quality studies, coupled with biogeochemical models (Larson et al. 2005; Ulses et al. 2005; Levasseur et al. 2007). The use of hydrodynamic modelling for aquaculture can lead to the identification of suitable areas for this activity (Zainol, Akhir and Zainol 2021).

The dispersion of contaminants and their pathways within and outside an estuary can be estimated through hydrodynamic patterns and the types of contaminants found in the water and sediment (Azevedo, Bordalo and Duarte 2010; Iglesias et al. 2020). As an example, Birocchi et al. (2021) used a 3D hydrodynamic model to estimate bacteria concentration and their dispersion in a channel in the south-east region of Brazil. Pollutants can generate public health problems (Griffin et al. 2001), which demonstrate the importance of predicting the Rt of these particles with hydrodynamic patterns. In this scenario, a high Rt can increase contamination effects on the organisms of the natural reserve, and the trajectories of floater contaminants driven by tides might be distributed through an entire estuary (Iglesias et al. 2020). Particle trajectories and residual currents can also be used to compute pollutant dispersion, identifying zones in an estuary that are sensitive to environmental pollution and, therefore, protecting these vulnerable regions (Novikov and Bagtzoglou 2006; Balachandran et al. 2008; Ouartassi et al. 2021). Estuarine Rt studies allow the assessment of nutrient exports and imports, quantify the possible effects of changes in nutrient loads on concentration levels, and estimate chlorophyll concentrations and primary production (Monbet 1992; Richardson and Jorgensen 1996; Kelly 1997; Dettman 2001). Generally, Rt is linked to nutrient exchange, dilution, mix and transport, but it also relates to other ecosystem features, indicating interactions between biological processes (Cloern 1982; Officer, Smayda and Mann 1982; Hily 1991; Rueda, Moreno-Ostos and Armengol 2006; de Abreu et al. 2020), including effects in aquaculture (Brooks, Baca and Lo 1999).

In this chapter, we summarise the hydrodynamics models previously used in the literature to estimate Rt in estuarine ecosystems. These methodological approaches provide better information for understanding exchange and transport processes of particles in estuaries, besides evaluating ecological phenomena, such as phytoplankton blooms. Furthermore, the Estuarine Coastal Ocean Model (ECOM), developed by Blumberg and Mellor (1987), which includes a particle tracking module, was used to estimate the Rt in the Cananéia-Iguape estuary as an example. This region is impacted mainly by upstream sources of pollution through the Valo Grande opening, a source of freshwater used by local populations, and the Rt is a good indicator to quantify how long these pollutants remain in the estuary.

Methods

Miller and McPherson (1991) defined the estuarine Rt as the time required to flush out a certain percentage (e.g. 67%, 95%, or 100%) of conservative particles (that do not decay with time) which were uniformly distributed at time zero. If the percentage is set to 100%, the definition of estuarine Rt is similar to the hydraulic retention time required for all the particles in the estuary to be carried out (Liu, Chen and Hsu 2011). Makarynskyy, Zigic and Langtry (2007), Zigic et al. (2009) and Liu, Chen and Hsu (2011) used an estuarine Rt of 67%, which is characterised as the time demanded for 67% of the initial equally distributed particles to be flushed out of the estuary. The Rt can include two other terms that depend on the passive tracer concentration: the mean Rt and the averaged Rt. The mean Rt, according to Zhao, Wei and Zhao (2002), is defined as the time needed for the concentration of the passive tracer in a given region to decay to e^{-1} (37%) of the initial value, while the average Rt (θ) was described by Takeoka (1984) with the following equation:

$$\theta = \int_0^\infty r(t)\, dt \tag{1}$$

where r(t) is C(t)/C(t0) and C is the concentration of the passive tracer. Liu et al. (2004) developed a dispersion model and used it to simulate the exchange processes of passive dissolved conservative matter in Jiaozhou Bay, based on the average Rt approach (Takeoka 1984). In this review, we have focused on the Miller and McPherson (1991) approach, which does not depend on the tracer concentration; we will continue to call it Rt.

The concept of Rt was first used by Craig (1957) to evaluate the Rt of radiocarbon at atmosphere, oceanic mixed layer and deep sea. Craig also stated that Dingle (1954) and Plass (1956) had estimated the Rt of a molecule of carbon dioxide in the atmosphere, although they did not use this term in their studies. The calculation of Rt for particles, first called *average transit time*, was first described by Bolin and Rodhe (1973) in natural reservoirs. These authors also computed other parameters, revealing the Rt as the most relevant.

Liu, Chen and Hsu (2011) applied a lagrangian particle-tracking technique to a three-dimensional, semi-implicit Eulerian–Lagrangian finite-element model (SELFE, Zhang and Baptista 2008) to study the pathways of particulate pollutant dispersion adjacent to the Danshuei River estuarine system. Lagrangian models are a reasonable and convenient set of tools for inspecting and modelling the mixing, transport pathways and estuary–ocean transfer of dissolved pollutants (Burwell et al. 1999; Bilgili et al. 2005). In the Liu, Chen and Hsu (2011) scheme the position of a particle is expressed by:

$$\vec{x}(t + \Delta t) = \vec{x}(t) + \vec{u}\Delta t + z_n\sqrt{2\vec{K}\Delta t} \tag{2}$$

where x is the particle's position, Δt is the time step, u is the velocity vector, K is the eddy diffusion tensor, which can be defined through the hydrodynamic model, and z_n is a random vector that satisfies a normal distribution, based on Press et al. (1994).

According to Defne and Ganju (2015), Rt is an integrative parameter that determines the renewal time for a given water parcel or water body. The authors defined the Rt for each particle as the time elapsed between initial release and exit from the estuary. Defne and Ganju (2015) compared different methods to determine a system-wide mean Rt for a lagoon back-barrier estuary in New Jersey. They used the Lagrangian TRANSport model (LTRANS) (North et al. 2011) for particle tracking to estimate the Rt for the studied estuary. LTRANS runs offline with hydrodynamic model outputs for density, velocity and vertical diffusivity to calculate particle tracks. According to the authors, particle tracking is one numerical technique for quantifying Rt in estuaries; multiple particles are released and tracked until they transit out of the estuary. The change in the total quantity of remaining particles in an estuary is often used as a measure of estimated renewal time for the estuarine water as well (Brooks, Baca and Lo 1999; Abdelrhman 2002; Monsen et al. 2002; Liu et al. 2004). However, the Rt in an estuary usually varies temporally and spatially (Zhang, Wilkin and Schofield 2010), so determining a mean Rt for the entire estuary requires supplementation with an analysis of differential transport of particles within the domain (Defne and Ganju 2015). Defne and Gangu also computed the spatial variation of Rt in each scenario by mapping Rt based on the initial starting location of the particle.

A simple Lagrangian model to simulate Rt is described by:

$$P^{n+1} = P^n + \Delta t \frac{dP^n}{dt} + \frac{\Delta t^2}{2!} \frac{d^2P^n}{dt^2} + O^3 \tag{3}$$

This model was used by Abreu et al. (2020) and, according to the authors, based on the traditional concept, Rt is calculated through the ratio between the compartment volume (m³) and the residual water flow (m³/s) through the compartment. In that study, the transport of components in the Lagrangian model was implemented on the SisBaHiA model (Base System for Environmental Hydrodynamics), which is represented by a number of immaterial particles driven by currents counted through the hydrodynamic model. The location of a particle in the following time (P^{n+1}) is ruled by a second-order expansion in Taylor's series (Taylor, 1715) (which is important for its use in mathematical analysis) based on the known previous position (P^n). De Abreu et al. (2020) also evidenced the importance of including precipitation (rain) in the model that estimates Rt, since this parameter can influence water flow in the area.

Huggett, Purdie and Haigh (2020) have estimated the flushing time for a microtidal estuary, and suggest that the difference between flushing and Rt depends on the approach. Generally, the flushing time considers the time it takes for particles to leave the estuary, including fluvial discharge, and the Rt can be interpreted as the flushing time variability across the estuary (Huggett, Purdie and Haigh 2020). Liu et al. (2004) assumed a different approach, and considered the Rt to be half the flushing time. Huggett, Purdie and Haigh (2020) defined flushing time as the amount of time between release and the point at which only 37% (the *e-folding* flushing time) of the particles remain within the estuary, or the average concentration to be reduced to $e^{-1} = 1/2.7$. These authors used a particle tracking module available in the MIKE 21 Flexible Mesh to estimate the flushing time (DHI 2017). They released particles on spring and neap tide at four different stages of the tidal cycle – flood tide, high tide, ebb tide and low tide – and also used a prism method to compare model results.

The best method for estimating Rt is, therefore, still under investigation. MacCready et al. (2021) estimated Rt using three different methods for the Salish Sea, these being replicable in other estuaries. Estimating the Rt with a box model generally gave higher figures than Rts based on simpler calculations of flushing time. These box model Rts are prolonged partly due to reflux (water mixed down that then heads landward), which means that it takes much longer for the system to flush out conservative tracers, and partly due to lack of complete homogenisation within the volume. According to the authors, the combined use of efflux (water that is mixed up and then heads seaward)/reflux (Cokelet and Stewart, 1985) with Total Exchange Flow (MacCready 2011) is a valuable approach likely to prove fruitful in other estuaries. In addition, the relevance of taking into account efflux/reflux and incomplete tracer homogenisation is expected to be critical in interpreting Rts in many estuaries (MacCready et al. 2021).

Case study

We were particularly interested in using the ECOM, a hydrodynamic numerical model that includes a particle tracking module, to compute the Rt (and flushing time) in the Cananéia-Iguape estuarine lagoon system (CIELS). The ECOM model ran for 55 days from 7 November 2019 to 31 December 2019, with a spin-up of three days (from 4 November to 6 November 2019). The ECOM was designed to simulate sea level, currents, temperature and salinity in aquatic systems that are time dependent in the most realistic way. This model takes into account the hydrostatic approximation, which is a simplification of the equation governing the vertical component of velocity in the ocean and simply says that the pressure at any point in the ocean is due to the weight of the water above it. With an Arakawa C grid configuration, the ECOM solves the momentum and continuity equations of the ocean in a sigma (σ) coordinate system (Blumberg 2010), a very common choice in coastal models with σ varying from –1 to zero, and the bottom is defined by the surface $\sigma = -1$ (Miller, 2007). The ECOM model applied for the CIELS was calibrated and validated in advance (not shown). More details about this model can be found at Blumberg and Mellow (1987) and its applications can be found, for instance, in Birocchi et al. (2021) and Costa et al. (2020).

Brooks, Baca and Lo (1999) offers an example of the use of the ECOM to estimate particle tracking and the residence (flushing) time of neutrally buoyant particles in Cooking Bay, Maine. To develop a map of flushing time, the authors seeded the model with particles released in every grid cell at many different stages of the tidal cycle. They then calculated the tidal-mean Rt. The estimates of the Rt in the CIELS followed the method of Brooks, Baca and Lo (1999), according to the following steps:

- To generate the particle tracking we:

  a) released three particles at the surface in adjacent model grid cells at the Valo Grande opening in the first lunar-hour after the spin-up (three days) of the ECOM model;
  b) plotted the particle trajectories.

- To generate the Rt map we:

  c) released one particle at the surface in every grid cell (removing the boundary points) after the spin-up of the ECOM model (running for 55 days);
  d) calculated the tidal-mean Rt from the mean over 12 individual model runs, with releases at a uniform interval of one lunar hour in the M-2 tidal cycle after the spin-up of the ECOM model;
  e) created a tidal-mean Rt map, as we knew the initial position of the particles and how long they took to leave the estuary. The Rt was the time the particle took to leave the estuary from its initial position.

We used the mean temperature of 25.02 °C measured in 23 stations along the estuary in October 2019 and salinity of 0 to set up the initial condition. At the boundary, we used the values of 25.03 °C and 31.63, for temperature and salinity, respectively, based on the values measured in the nearest station and the radiative boundary condition given by Reid and Bodine (1968). We used horizontally constant local wind data acquired from National Centers for Environmental Prediction (NCEP)/National Center for Atmospheric Research (NCAR) (Kalnay et al. 1996) with a one-hour time interval. We included remote wind influence (updated every 15 minutes) by introducing the subtidal elevation as boundary condition. The tide in the two openings of the system was predicted, using the amplitudes and phases of nine main constituents: Q1, O1, P1, K1, N2, M2, S2, K2 and M3. The tidal components, at the south entrance of the CIELS, were defined from the harmonic analysis performed by Mesquita and Harari (1983) using elevation data measured at the Oceanographic Institute (University of São Paulo) base from 1969 to 1974. At the northern entrance, harmonic analysis was performed using data from a tide gauge installed near the Icapara lighthouse – approximately 3 km from the north mouth – from 24 May 1965 to 12 June 1965. These data are available in the catalogue of the Brazilian tide stations Fundação de Estudos do Mar (FEMAR 1999). We also incorporated constant river discharges for the Mar de Cananéia given by Bérgamo (2000) and, for the Valo Grande opening, daily mean values estimated with Department of Water and Electric Energy (DAAE, in portuguese) data using the GEOBRÁS (1966) method, also applied by Pisetta (2006). Following de Abreu et al. (2020) to include the precipitation in the model to estimate the Rt, we used the hourly precipitation time series from MERGE results from the Centro de Previsão de Tempo e Estudos Climáticos (CPTEC) (Rozante et al. 2010) in our model. This product consists of observed precipitation estimated by satellite.

Applying steps a) and b), the trajectories were computed, and are shown in Fig. 4.1 for three particles (in blue, black and red) released in the first lunar hour on 7 November 2019 at Valo Grande mouth (green dot). All particles left the estuary by the north entrance (Barra

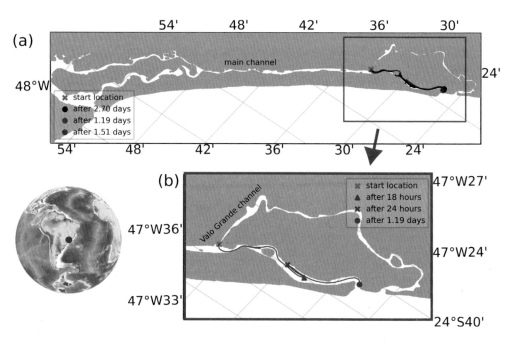

Figure 4.1 The globe map shows the location (red dot) of the Cananéia-Iguape estuary, Brazil. (a) Entire view of the Cananéia-Iguape estuary emphasising its main channel. In this map, it is possible to observe the tracking for three particles (in black, blue and red lines and markers) released on 7 November 2019 in the Valo Grande mouth (green dot) at the first lunar hour. The circle markers show the position and time for the particles when they leave the estuary by the northern entrance (Barra de Icapara). (b) Zoomed map of the Valo Grande mouth and northern entrance of the estuary. The green point corresponds to the location of the particle at the moment of release. The red points and line correspond to the locations and tracking for one of the particles 18 hours after release (triangle marker), after 24 hours (x marker) and after 1.19 days (circle marker), where the last corresponds to the moment when the particle leaves the estuary by the northern entrance.

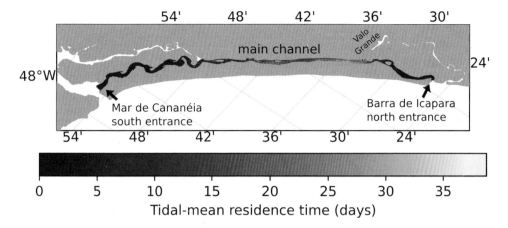

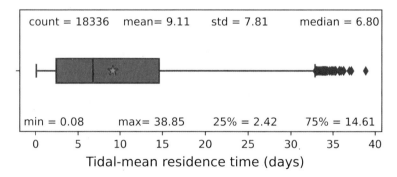

Figure 4.2 Tidal-mean Rt (in days) for the main channel in the Cananéia-Iguape estuary for neutral surface particles released in each grid cell. The average is developed as the mean over 12 individual model runs with releases at a uniform interval of one lunar hour in the M-2 tidal cycle on 7 November 2019. Relatively low values of Rt (<1 day) are evident near the northern (Barra de Icapara) and the southern (Mar de Cananéia) entrances, but particles remain in the inner main channel for ten days or longer.

de Icapara) 1.19, 1.51 and 2.70 days, respectively, after their release (Fig. 4.1a). It is possible to visualise the semi-diurnal tidal influence in the movement of the particles, mainly when we focus on a unique particle (Fig. 4.1b) moving backward and forward in the estuary over a period of 6 hours, between 18 and 24 hours after the release (see red markers).

Applying steps c), d) and e), the tidal-mean Rt is computed and presented in Fig. 4.2. We can observe in the vicinity of the north and south entrances a relatively short tidal-mean Rt of less than a day and, in the inner part of the main channel, values varying from 10 days to 38.85 days. The inner part of the channel, near the south region of Valo Grande channel, is where we found the maximum values for the tidal-mean Rt, reaching 38.85 days (Fig. 4.3).

Figure 4.3 Boxplot for the tidal-mean residence time estimated for the CIELS (in days). The x-axis represents the tidal-mean residence time found in the estuary. 'Count' means the total number of particles released in the estuary, 'mean' is the average represented by the red star filled in with green and 'std' is the standard deviation. The 'median', 'min' and 'max' are the middle, minimum and maximum values found in the estuary. '25%' is the first quartile (Q1) with 25th percentile, or the middle number between the smallest number and the median of the values, and '75%' is the third quartile (Q3) with 75% percentile, or the middle value between the median and the maximum value of the results. Diamond-shaped points are data values that are far apart from the other results, called outliers in a boxplot.

We released 18,336 particles into the estuary and we found values of 9.11 and 7.81 days for the mean and the standard deviation of the tidal-mean Rt, respectively, showing that Rt can vary considerably throughout the system. We also estimated the minimum Rt value equal to 0.08 days, showing that in the vicinities of the northern and southern entrances, the particles quickly leave the estuary. The middle (or median) value detected for the tidal-mean Rt in the CIELS was 6.80 days, with 25% (first quartile) of the total results higher than 2.42 days and 75% (third quartile) of them lower than 14.61 days (Fig. 4.3).

Brichta and Gaeta (2000) have estimated the freshwater Rt in the CIELS for June and December. They found Rt values of 14 days for June and 7 days for December, which were highly influenced by the fluvial discharge. The Rt doubled in June in comparison with December owing to its lower river discharge at both Cananéia and Ribeira de Iguape watersheds, with values of 19 m/s^3 and 333.32 m/s^3, respectively, for June, and 47.79 m/s^3 and 833.37 m/s^3, respectively, for December.

Conclusion

The definition of Rt can vary depending on the approach implemented. ECOM was chosen to compute the time required to flush particles from the Cananéia-Iguape estuarine lagoon system (Brazil) and to calculate the tidal-mean Rt taking into consideration precipitation. ECOM was configured to take into account the particles released after its spin-up, to calculate the tidal-mean Rt. The mean and maximum values for the tidal-mean Rt in the Cananéia-Iguape estuarine system were 9.11 days and 38.85 days, respectively, showing that this property can vary considerably in an estuary. It is important to mention that in this chapter the tidal-mean Rt was first estimated for the main channel of the Cananéia-Iguape estuary, and this approach is one of many ways in which Rt could be calculated here.

Estimation of Rt is a useful approach for water quality monitoring in coastal and estuarine regions. Maintaining water quality in accordance with the legislation in order to guarantee a healthy ocean is a challenge in many Brazilian estuaries. An Rt estimate could be included in Brazilian environmental policies as a way of detecting the self-purification time of coastal and estuarine regions. This is a big challenge in Brazil, where environmental laws are poorly applied, despite being some of the most comprehensive in the world. In the Cananéia-Iguape estuary, for example, the Rt approach could detect areas with high Rt, which consequently retain more pollutants, and could negatively affect the region's oyster aquaculture – the largest production of oysters in the state of São Paulo (Pereira, Henriques and Machado 2003; Ristori et al. 2007).

The Rt for individual Brazilian estuaries is not well known, although de Abreu et al. (2020) and da Silva et al. (2019) have recently tried to estimate the Rt in the Amazonian and Patos lagoon estuaries, respectively. Rt is closely connected to other ecosystem features, including biogeochemical processes (carbon, sulphur, nitrogen, phosphorus, oxygen and others), dilution and self-purification, which control variations in water quality (de Abreu et al. 2020). The application of Rt permits hydrodynamic patterns to be related to ecological and environmental mechanisms; and Rt studies can therefore bring about vital improvements in the conservation of marine biodiversity. Moreover, these studies can be used in other correlated subjects, such as sanitation and ecological and environmental engineering (de Abreu et al. 2020). Besides being considered a good hydrodynamic parameter, the study of Rt in an estuary allows ecological aspects, including the dispersion of pollutants and biogeochemical cycles, to be better understood.

References

Abdelrhman, M. (2002) Modeling how a hurricane barrier in New Bedford Harbor, Massachusetts, affects the hydrodynamics and residence times. *Estuaries* 25 (2): 177–96. https://doi.org/10.1007/BF02691306

Andrews, John C. and Hans Müller (1983) Space-time variability of nutrients in a lagoonal patch reef. *Limnology and Oceanography* 28 (2): 215–27. https://doi.org/10.4319/lo.1983.28.2.0215

Azevedo, I.C., Bordalo, A.A. and Duarte, P.M. (2010) Influence of river discharge patterns on the hydrodynamics and potential contaminant dispersion in the Douro estuary (Portugal). *Water Research* 44 (10): 3133–46. https://doi.org/10.1016/j.watres.2010.03.011

Balachandran, K.K., Reddy, G.S., Revichandran, C., Srinivas, K., Vijayan, P.R. and Thottam, T.J. (2008) Modelling of tidal hydrodynamics for a tropical ecosystem with implications for pollutant dispersion (Cohin Estuary, Southwest India). *Ocean Dynamics* 58 (3): 259–73. https://doi.org/10.1007/s10236-008-0153-6

Basu, B.K. and Pick, F.R. (1997) Phytoplankton and zooplankton development in a lowland, temperate river. *Journal of Plankton Research* 19 (2): 237–53. https://doi.org/10.1093/plankt/19.2.237

Bérgamo, A.L. (2000) Características da hidrografia, circulação e transporte de sal: Barra de Cananéia, sul do Mar de Cananéia e baía do Trapandé. PhD thesis. Oceanographic Institute, University of São Paulo.

Bilgili, A., Priehl, J., Lynch, D.R., Smith, K.W. and Swift, M.R. (2005) Estuary/ocean exchange and tidal mixing in a Gulf of Maine Estuary: a Lagrangian modelling study. *Estuarine Coastal Shelf Science* 65 (4): 607–24. https://doi.org/10.1016/j.ecss.2005.06.027

Birocchi, P., Dottori, M., Costa, C.D.G.R. and Leite, J.R.B. (2021) Study of three domestic sewage submarine outfall plumes through the use of numerical modeling in the São Sebastião channel, Sao Paulo state, Brazil. *Regional Studies in Marine Science*: 101647. https://doi.org/10.1016/j.rsma.2021.101647

Blumberg, A.F. and Mellor, G.L. (1987) A description of a three-dimensional coastal ocean circulation model. *Coastal and Estuarine Sciences* 4. https://doi.org/10.1029/CO004p0001

Blumberg, A. (2010) Stevens ECOM users manual October 2010 – Appendix II.

Bolin, B. and Rodhe, H. (1973) A note on the concepts of age distribution and transit time in natural reservoirs. *Tellus* 25 (1): 58–62. https://doi.org/10.1111/j.2153-3490.1973.tb01594.x

Brichta, M. and Gaeta, S.A. (2000) Biomassa e produção autotrófica planctônica no complexo estuarino-lagunar Iguape-Cananéia, São Paulo. Thesis. Instituto Oceanográfico – USP, São Paulo.

Brooks, D.A., Baca, M.W. and Lo, Y.T. (1999) Tidal circulation and residence time in a macrotidal estuary: Cobscook Bay, Maine. *Estuarine, Coastal and Shelf Science* 49: 647–65. https://doi.org/10.1006/ecss.1999.0544

Bull, E.G., Cunha, C.D.L. and Scudelari, A.C. (2021) Water quality impact from shrimp farming effluents in a tropical estuary. *Water Science and Technology* 83 (1): 123–36. https://doi.org/10.2166/wst.2020.559

Burwell, B., Vincent, M., Luther, M. and Galperin, B. (1999) Modelling residence times: Eulerian vs. Lagrangian. In L. Spaulding and H.L. Butler (eds) *Proceedings of the Sixth International Conference of Estuarine and Coastal Modelling*, pp. 995–1009. New Orleans, LA.

Camacho, René A. et al. (2015) Modeling the factors controlling phytoplankton in the St. Louis Bay Estuary, Mississippi and evaluating estuarine responses to nutrient load modifications. *Journal of Environmental Engineering* 141 (3): 04014067. https://doi.org/10.1061/(ASCE)EE.1943-7870.0000892

Church, Thomas M. (1986) Biogeochemical factors influencing the residence time of microconstituents in a large tidal estuary, Delaware Bay. *Marine Chemistry* 18 (2–4): 393–406. https://doi.org/10.1016/0304-4203(86)90020-4

Cloern, James E. (1982) Does the benthos control phytoplankton biomass in south San Francisco Bay. Marine Ecology Progress Series. *Oldendorf* 9 (2): 191–202. https://doi.org/10.3354/meps009191

Cokelet, E.D. and Stewart, R.J. (1985) The exchange of water in fjords: The efflux/reflux theory of advective reaches separated by mixing zones. *Journal of Geophysical Research* 90 (C4): 7287–7306. https://doi.org/10.1029/JC090iC04p07287

Costa, C.G., Leite, J.R.B., Castro, B.M., Blumberg, A.F., Georgas, N., Dottori, M. and Jordi, A. (2020) An operational forecasting system for physical processes in the Santos-Sao Vicente-Bertioga Estuarine System, Southeast Brazil. *Ocean Dynamics* 70 (2): 257 71. https://doi.org/10.1007/s10236-019-01314-x

Craig, H. (1957) The natural distribution of radiocarbon and the exchange time of carbon dioxide between atmosphere and sea. *Tellus* 9 (1): 1–17. https://doi.org/10.3402/tellusa.v9i1.9078

da Silva, D.V., Oleinik, P.H., Costi, J., de Paula Kirinus, E. and Marques, W.C. (2019). Residence time patterns of Mirim Lagoon (Brazil) derived from two-dimensional hydrodynamic simulations. *Environmental Earth Sciences* 78 (5): 1–11. https://doi.org/10.1007/s12665-019-8162-y

de Abreu, C.H.M., Barros, M. de L.C., Brito, D.C., Teixeira, M.R. and da Cunha, A.C. (2020) Hydrodynamic modeling and simulation of water residence time in the estuary of the lower Amazon river. *Water (Switzerland)* 12 (3). https://doi.org/10.3390/w12030660

Defne, Zafer and Ganju, Neil K. (2015) Quantifying the residence time and flushing characteristics of a shallow, back-barrier estuary: application of hydrodynamic and particle tracking models. *Estuaries and Coasts* 38 (5): 1719–34. https://doi.org/10.1007/s12237-014-9885-3

Dettmann, Edward H. (2001) Effect of water residence time on annual export and denitrification of nitrogen in estuaries: a model analysis. *Estuaries* 24 (4): 481–90. https://doi.org/10.2307/1353250

DHI, MIKE (2017) MIKE 21 & MIKE 3 Flow model FM, hydrodynamic and transport module, scientific documentation. DHI Water & Environment.

Dingle, A.N. (1954) The carbon dioxide exchange between the North Atlantic Ocean and the atmosphere. *Tellus* 6: 342–50. https://doi.org/10.3402/tellusa.v6i4.8759

FEMAR, F.d.E.d.M. (1999) Catálogo de estações maregráficas brasileiras. Technical report.

Fogarty, M.J. and Botsford, L.W. (2007) Population connectivity and spatial management of marine fisheries. *Oceanography* 20 (3): 112–23. https://doi.org/10.5670/oceanog.2007.34

GEOBRÁS (1966). Complexo Valo Grande Mar Pequeno Rio Ribeira de Iguape. Serviço do Vale do Ribeira. Technical report, Department of Water and Electric Energy (DAEE). Secretaria dos Serviços e Obras Públicas. Government of the state of São Paulo.

Griffin, D.W., Lipp, E.K., McLaughlin, M.R. and Rose, J.B. (2001) Marine recreation and public health microbiology: quest for the ideal indicator. *BioScience* 51 (10): 817–25. https://doi.org/10.1641/0006-3568(2001)051[0817:MRAPHM]2.0.CO;2

Herbert, R.J., Willis, J., Jones, E., Ross, K., Hubner, R., Humphreys, J., Jensen, A. and Baugh, J. (2012) Invasion in tidal zones on complex coastlines: modelling larvae of the non-native Manila clam, *Ruditapes philippinarum*, in the UK. *Journal of Biogeography* 39 (3): 585–99. https://doi.org/10.1111/j.1365-2699.2011.02626.x

Hily, Christian (1991) Is the activity of benthic suspension feeders a factor controlling water quality in the Bay of Brest?. Marine Ecology Progress Series. *Oldendorf* 69 (1): 179–88. https://doi.org/10.3354/meps069179

Huggett, R.D., Purdie, D.A. and Haigh, I.D. (2020) Modelling the influence of riverine inputs on the circulation and flushing times of small shallow estuaries. *Estuaries and Coasts*. https://doi.org/10.1007/s12237-020-00776-3

Iglesias, I., Almeida, C.M.R., Teixeira, C., Mucha, A.P., Magalhães, A., Bio, A. and Bastos, L. (2020) Linking contaminant distribution to hydrodynamic patterns in an urban estuary: the Douro estuary test case. *Science of the Total Environment* 707: 135792. https://doi.org/10.1016/j.scitotenv.2019.135792

Kalnay, E., Kanamitsu, M., Kistler, R., Collins, W., Deaven, D., Gandin, L., Iredell, M., Saha, S., White, G., Woollen, J. and others (1996) The NCEP/NCAR 40-year reanalysis project. *Bulletin of the American Meteorological Society* 77 (3): 437–72. https://doi.org/10.1175/1520-0477(1996)077<0437:TNYRP>2.0.CO;2

Kelly, John R. (1997) Nitrogen flow and the interaction of Boston Harbor with Massachusetts Bay. *Estuaries* 20 (2): 365–80. https://doi.org/10.2307/1352350

Larson, M., Bellanca, R., Jönsson, L., Chen, C. and Shi, P. (2005) A model of the 3D circulation, salinity distribution, and transport pattern in the Pearl River Estuary, China. *Journal of Coastal Research* 215: 896–908. https://doi.org/10.2112/03-105A.1

Levasseur, A., Shi, L., Wells, N.C., Purdie, D.A. and Kelly-Gerreyn, B.A. (2007) A three-dimensional hydrodynamic model of estuarine circulation with an application to Southampton Water, UK. *Estuarine, Coastal and Shelf Science* 73 (3–4): 753–67. https://doi.org/10.1016/j.ecss.2007.03.018

Liu, S., Butler, D., Memon, F.A., Makropoulos, C., Avery, L. and Jefferson, B. (2010) Impacts of residence time during storage on potential of water saving for grey water recycling system. *Water Research* 44 (1): 267–77. https://doi.org/10.1016/j.watres.2009.09.023

Liu, Wen-Cheng, Chen, Wei-Bo and Hsu, Ming-Hsi (2011) Using a three-dimensional particle-tracking model to estimate the residence time and age of water in a tidal estuary. *Computers & Geosciences* 37 (8): 1148–61. https://doi.org/10.1016/j.cageo.2010.07.007

Liu, Z., Wei, H., Liu, G. and Zhang, J. (2004) Simulation of water exchange in Jiaozhou Bay by average residence time approach. *Estuarine, Coastal and Shelf Science* 61 (1): 25–35. https://doi.org/10.1016/j.ecss.2004.04.009

MacCready, P. (2011) Calculating estuarine exchange flow using isohaline coordinates. *Journal of Physical Oceanography* 41 (6): 1116–24. https://doi.org/10.1175/2011JPO4517.1

MacCready, P., McCabe, R.M., Siedlecki, S.A., Lorenz, M., Giddings, S.N., Bos, J., Albertson, S., Banas, N.S. and Garnier, S. (2021) Estuarine circulation, mixing, and residence times in the Salish Sea. *Journal of Geophysical Research: Oceans* 126 (2): e2020JC016738. https://doi.org/10.1029/2020JC016738

Makarynskyy, O., Zigic, S. and Langtry, S. (2007) A numerical modelling study of the proposed increase in fish production. In *Proceedings of the II International Conference of Natural Environment: Critical Problems of Ecology and Hydro-meteorology, Merging of Education and Science*, Odessa, Ukraine.

McLusky, Donald S. and Michael Elliott (2004) *The Estuarine Ecosystem: Ecology, Threats and Management*. Oxford: Oxford University Press. https://doi.org/10.1093/acprof:oso/9780198525080.001.0001

McManus, M.C., Ullman, D.S., Rutherford, S.D. and Kincaid, C. (2020) Northern quahog (Mercenaria mercenaria) larval transport and settlement modeled for a temperate estuary. *Limnology and*

Oceanography 65 (2): 289–303. https://doi.org/10.1002/lno.11297

Mesquita, A.R. and Harari J. (1983) Tides and tide gauges of Cananéia and Ubatuba – Brazil (lat. 24 o S). *Relatório Interno do Instituto Oceanográfico da Universidade de São Paulo* 11: 1–14.

Miller, R.L. and McPherson, B.F. (1991) Estimating estuarine flushing and residence times in Charlotte Harbor, Florida, via salt balance and a box model. *Limnology and Oceanography* 36 (3): 602–12. https://doi.org/10.4319/lo.1991.36.3.0602

Miller, R.N. (2007) *Numerical Modeling of Ocean Circulation*. Cambridge University Press.

Monbet, Yves (1992) Control of phytoplankton biomass in estuaries: a comparative analysis of microtidal and macrotidal estuaries. *Estuaries* 15 (4): 563–71. https://doi.org/10.2307/1352398

Monsen, N.E., Cloern, J.E., Lucas, L.V. and Monismith, S.G. (2002) A comment on the use of flushing time, residence time, and age as transport time scales. *Limnology and Oceanography* 47 (5): 1545–53. https://doi.org/10.4319/lo.2002.47.5.1545

Nixon, S.W., Ammerman, J.W., Atkinson, L.P., Berounsky, V.N., Billen, G., Boicourt, W.C., Boynton, W.R., Church, T.M., Ditoro, D.M., Elgren, R., Garber, J.H., Giblin, A.E., Jahnke, R.A., Owens, N., Pilson, M. and Seitzinger, S. (1996) The fate of nitrogen and phosphorus at the land-sea margin of the North Atlantic Ocean. *Biogeochemistry* 35: 141–80. https://doi.org/10.1007/BF02179826

North, E.W., Adams, E.E., Schlag, S., Sherwood, C.R., Socolofsky, S. and He, R. (2011) Simulating oil droplet dispersal from the Deepwater Horizon spill with a Lagrangian approach. AGU series: *Monitoring and Modeling the Deepwater Horizon Oil Spill: A Record Breaking Enterprise* 195, pp. 217–26. https://doi.org/10.1029/2011GM001102

Novikov, A. and Bagtzoglou, A.C. (2006) Hydrodynamic model of the lower Hudson River estuarine system and its application for water quality management. *Water Resources Management* 20 (2): 257–76. https://doi.org/10.1007/s11269-006-0320-9

Officer, C.B., Smayda, Theodore J. and Mann, R. (1982) *Benthic Filter Feeding: A Natural Eutrophication Control*. https://doi.org/10.3354/meps009203

Ouartassi, B., Doyon, B. and Heniche, M. (2021) Numerical prediction of oil mineral aggregates dispersion in the estuary of St-Lawrence River. *Journal of Physics: Conference Series* 1743 (1): 012033. https://doi.org/10.1088/1742-6596/1743/1/012033

Pereira, O., Henriques, M. and Machado, I. (2003). Estimativa da curva de crescimento da ostra Crassostrea brasiliana em bosques de mangue e proposta para sua extração ordenada no estuário de Cananéia, SP, Brasil. *Boletim do Instituto de Pesca* 29 (1): 19–28.

Phlips, Edward J. and others (2012) Climatic influences on autochthonous and allochthonous phytoplankton blooms in a subtropical estuary, St. Lucie Estuary, Florida, USA. *Estuaries and Coasts* 35 (1): 335–52. https://doi.org/10.1007/s12237-011-9442-2

Pisetta, M. (2006) Transporte de Sedimentos por Suspensão no Sistema Estuarino-Lagunar de Cananéia-Iguape. Master's thesis. Oceanographic Institute, University of São Paulo.

Plass, G.N. (1956) Carbon dioxide theory of climatic change. *Tellus* 8: 140–54. https://doi.org/10.3402/tellusa.v8i2.8969

Press, W.H., Teukolsky, S.A. and Flanney, B.P. (1994) *Numerical Recipes in Fortran*, 2nd edn. Cambridge: Cambridge University Press.

Pritchard, D.W. (1967) What is an estuary: a physical viewpoint. *American Association for the Advancement of Science* 83: 3–5.

Reid, R.O. and Bodine, B.R. (1968) Numerical model for storm surges in Galveston Bay. *Journal of the Waterways and Harbors Division* 94 (1): 33–58. https://doi.org/10.1061/JWHEAU.0000553

Richardson, Katherine and Barker Jørgensen, Bo (1996) Eutrophication: definition, history and effects. *Eutrophication in Coastal Marine Ecosystems* 52: 1–19. https://doi.org/10.1029/CE052p0001

Ristori, C.A., Iaria, S.T., Gelli, D.S. and Rivera, I.N.G. (2007). Pathogenic bacteria associated with oysters (Crassostrea brasiliana) and estuarine water along the south coast of Brazil. *International Journal of Environmental Health Research* 17 (4): 259–69. https://doi.org/10.1080/09603120701372169

Rozante, J.R., Moreira, D.S., Gonçalves, L.G.G. and Vila, Daniel A. (2010) Combining TRMM and surface observations of precipitation: technique and validation over South America. *Weather and Forecasting* 25: 885–94. https://doi.org/10.1175/2010WAF2222325.1

Rueda, F., Moreno-Ostos, E. and Armengol, J. (2006) The residence time of river water in reservoirs. *Ecological Modelling* 191 (2): 260–74. https://doi.org/10.1016/j.ecolmodel.2005.04.030

Sale, P.F., Cowen, R.K., Danilowicz, B.S., Jones, G.P., Kritzer, J.P., Lindeman, K.C., Planes, S., Polunin, N.V.C., Garry, R.R., Sadovy, Y.J. and Steneck, R.S. (2005) Critical science gaps impede use of no-take fishery reserves. *Trends in Ecology & Evolution* 20 (2): 74–80. https://doi.org/10.1016/j.tree.2004.11.007

Shen, J., Boon, J.D. and Kuo, A.Y. (1999) A modeling study of a tidal intrusion front and its impact on larval dispersion in the James River estuary, Virginia. *Estuaries* 22 (3): 681–92. https://doi.org/10.2307/1353055

Shen, Jian and Haas, Larry (2004) Calculating age and residence time in the tidal York River using three-dimensional model experiments. *Estuarine, Coastal and Shelf Science* 61 (3): 449–61. https://doi.org/10.1016/j.ecss.2004.06.010

Takeoka, H. (1984) Fundamental concepts of exchange and transport time scales in a coastal sea. *Continental Shelf Research* 3: 311–26. https://doi.org/10.1016/0278-4343(84)90014-1

Taylor, B. (1715) Methodus Incrementorum Directa et Inversa [Direct and Reverse Methods of Incrementation] (in Latin). London. p. 21–23 (Prop. VII, Thm. 3, Cor. 2). Translated into English in

Struik, D.J. (1969). *A Source Book in Mathematics 1200–1800*, pp. 329–32. Cambridge, Massachusetts: Harvard University Press. https://doi.org/10.1515/9781400858002

Ulses, C., Grenz, C., Marsaleix, P., Schaaff, E., Estournel, C., Meulé, S. and Pinazo, C. (2005) Circulation in a semi-enclosed bay under influence of strong freshwater input. *Journal of Marine Systems* 56 (1–2): 113–32. https://doi.org/10.1016/j.jmarsys.2005.02.001

Wan, Y., Qiu, C., Doering, P., Ashton, M., Sun, D. and Coley, T. (2013) Modeling residence time with a three-dimensional hydrodynamic model: linkage with chlorophyll a in a subtropical estuary. *Ecological Modelling* 268: 93–102. https://doi.org/10.1016/j.ecolmodel.2013.08.008

Wang, Chi-Fang, Hsu, Ming-Hsi and Kuo, Albert Y. (2004) Residence time of the Danshuei River estuary, Taiwan. *Estuarine, Coastal and Shelf Science* 60 (3): 381–93. https://doi.org/10.1016/j.ecss.2004.01.013

Zainol, Z., Akhir, M.F. and Zainol, Z. (2021) Pollutant transport and residence time of a shallow and narrow coastal lagoon estimated using a numerical model. *Marine Pollution Bulletin* 164: 112011. https://doi.org/10.1016/j.marpolbul.2021.112011

Zhang, W.G., Wilkin, J.L. and Schofield, O.M.E. (2010) Simulation of water age and residence time in New York Bight. *Journal of Physical Oceanography* 40: 965–82. https://doi.org/10.1175/2009JPO4249.1

Zhang, Y.L. and Baptista, A.M. (2008) SELFE: a semi-implicit Eulerian–Lagrangian finite-element model for cross-scale ocean circulation. *Ocean Modelling* 21 (3–4): 71–96. https://doi.org/10.1016/j.ocemod.2007.11.005

Zhao, L., Wei, H. and Zhao, J.Z. (2002) Numerical study on water exchange in Jiaozhou Bay. *Chinese Journal of Oceanology and Limnology* 33: 23–9 (in Chinese, with English abstract).

Zigic, S., Makarynskyy, O., Langtry, S. and Westbrook, G. (2009) A numerical modelling study for the proposed increase in Barramundi production, Cone Bay, Western Australia. In *Proceedings of the 11th Estuarine and Coastal Modelling Conference*, Seattle, WA. https://doi.org/10.1061/41121(388)26

CHAPTER 5

Coastal and Estuarine Physical Processes: Looking Back, Looking Forwards

STEVEN B. MITCHELL and REGINALD J. UNCLES

Abstract

Despite the continuing importance of good-quality field data in estuarine and coastal settings, recent years have seen comparatively fewer field studies. These have instead been replaced by an explosion in the use of numerical models to simulate the physical processes associated with transport by waves and currents, involving many things including the movement of solutes, solvents and floating particles or swimming organisms. Many of these models are now in such common use that their routine application is no longer described in the scientific research literature. Some of the newer modelling approaches, such as machine learning and artificial intelligence, are also starting to see wider use. The question remains as to the future direction of modelling and the ambition of the community to develop tools to inform the agenda for scientists and managers of coastal and estuarine environments. It seems likely that most of these efforts will be devoted to climate change, rising sea levels and of the transport of microplastics. Here we discuss some of the current achievements in modelling the physical processes of estuaries and coasts and consider some of the barriers to future progress, as there are still ways in which continued uncertainties plague the best efforts of modellers to capture the behaviours of particles and solutes, which may be governed by forces other than those strictly related to the flow of water. We also reflect on some pressing questions currently being asked about the coastal environment and how we manage it, especially in relation to sea-level rise, the fate of microplastics and the increasing urbanisation of coastal zones. We finish by asking to what extent the current approaches used by modellers are appropriate for answering key questions in the coming decades.

Keywords: estuaries, modelling, machine learning, artificial intelligence, climate change, microplastics

Correspondence: steve.mitchell@port.ac.uk

Introduction

This short chapter is not a scientific review, nor is it a research paper; rather, it is a general essay describing a personal view of the subject and a summary of some of the progress made since the early 1970s. One of the most important directions of travel during this time has been the use of field data to calibrate and validate numerical models of hydrodynamics, solute and sediment transport in a range of environments. The 1970s saw the publication

Steven B. Mitchell and Reginald J. Uncles, 'Coastal and Estuarine Physical Processes: Looking Back, Looking Forwards' in: *Challenges in Estuarine and Coastal Science*. Pelagic Publishing (2022). © Steven B. Mitchell and Reginald J. Uncles. DOI: 10.53061/EFRH6987 of some key studies in terms of the physics of estuaries and coastal waters, many of which still underpin the development of modelling approaches (e.g. Anderson and Devol 1973; Huthnance 1973, Harris et al. 1979). Early studies to validate attempts at tidal and wave modelling, and the modelling of solute and sediment transport – the latter understood to depend on the time and provenance of the sediment involved (Van der Graff and Van Overveen 1979) – were among the first to be published in the journal *Estuarine and Coastal Marine Science*. Innovations of this type led to new understanding of meteorological effects (Elliott and Wang 1978) and of morphology and morphological models affected by the erosion and deposition of sediment (Greenwood and Mittler 1979), mobilisation and stabilisation by biota (Holland, Zingmark and Dean 1974; Paterson 1989) and the importance of plants such as mangroves in stabilising deposits (Furukawa, Wolanski and Mueller 1997; Mazda et al. 1997). A review of early progress with the measurement and modelling of important estuarine physical variables and processes is given by Uncles (2002), and the Elsevier Treatise on Estuarine and Coastal Science (Wolanski and McLusky 2011) provides numerous reviews on many aspects of the science at that time. A review of more recent progress, focusing on the achievements of fieldwork and innovations in field instrumentation, is given by Uncles and Mitchell (2017), and a summary of the latest (2017) methodologies to measure estuarine and coastal hydrographic variables and sediment transport, including bedload and suspended floc sizes, is given in Mitchell, Uncles and Stephens (2017).

In terms of the development of the literature over the past 50 years, it is interesting to reflect on the burgeoning availability of peer-reviewed literature in the area of coastal and estuarine sediment transport and hydrodynamics. The year 1973 saw the publication of the first issue of the Journal *Estuarine and Coastal Marine Science*, which in 1981 became today's well-known *Estuarine, Coastal and Shelf Science*. Other journals and book series include *Continental Shelf Research, Estuaries and Coasts, Coastal Engineering* and *Marine Geology*, all dedicated to broadening and deepening interest in and knowledge of the physical processes that affect estuaries and coastal areas. In the sections that follow, we will consider the history of the literature in this area since the 1970s, the key drivers in the research questions asked and the likely direction of travel of future researchers based on what we know now about the key challenges for coastal and estuarine scientists.

Looking back – developments over time

Literature development

In terms of coastal engineering, it is instructive to refer to the work of Silvester (1978), who made it clear that despite the complexities of processes affecting hydrodynamics and sediments, some basic principles always apply and that studies should focus on accurate and appropriate data collection and modelling, a point reinforced later by Bruun (1989). In the early years of the Estuarine and Coastal Sciences Association (ECSA), the focus on data collection and understanding its potential and its limitations was of paramount importance (Bruun 1970). Many of the advances in understanding came at a time when instrumentation allowed data collection at a finer resolution, leading to more sophisticated monitoring and modelling of complex systems (Jarvis and Riley 1987; Kostaschuk, Luternauer and Church 1989; Aldridge 1997).

Different types of estuary, then as now, require different approaches to quantify the processes involved (e.g. Stigebrandt 1981; Uncles, Elliott and Weston 1985, 1986), and the complexities of interactions at different time and distance scales also points to the need for continuous development of understanding in this area (Green, Black and Amos 1997). Use of satellite data (Muralikrishna 1985; Marmorino and Smith 2008; Lavender 2017) has also become an essential element in a broad-scale understanding of the processes occurring in surface layers, but while remote sensing products are available to download from the internet, it is difficult to understand the assumptions and uncertainties involved in the data without first understanding the physics behind the measurements (Lavender 2017). The action of waves and tides has also received sharper focus thanks to data retrieved from autonomous recording buoys (Reynolds 1983; Dhoop and Mason 2018) and sediment traps (Broman, Kugelberg and Näf 1990). Increasingly detailed datasets have allowed vastly improved understandings of sediment budgets around coastlines (Van Rijn 1997) and within estuaries (Wolanski, King and Galloway 1995; Li and Zhang 1998). In general terms, though, it has been the progress in modelling the processes that has driven both the understanding of the systems and the direction of travel in terms of monitoring those systems.

Fieldwork development

Early estuarine field studies, particularly during the 1980s and earlier, often utilised water-column profiling of currents, salinity, temperature and, occasionally, suspended solids and/or turbidity from anchored boats (e.g. Bowden and Sharaf el Din, 1966; Uncles and Jordan 1979; Uncles, Elliott and Weston 1985), or time-series data from moored instrumentation, using self-recording propellor-type (e.g. Plessey impellor) or Savonius rotor-type (e.g. Anderra 4) current meters (e.g. Ramster and Howarth 1975). While studies of this kind are still undertaken, they have always tended to be costly and time consuming, with inherent risks of collecting poor data, or no data, as well as presenting in certain circumstances a safety risk for those involved. In poor weather conditions, fieldwork is, and was, often not possible at all – and often these fast wind, large wave or high run-off conditions are of great interest. There was therefore a strong impetus to develop models to provide data for practical estuarine and coastal engineering projects. Alongside model development there has been a significant improvement in terms of the ability to collect and store large amounts of data. Assuming the quality is as good as it always has been, we now have access to much larger datasets than before, making it possible to test a wider range of conditions and giving greater confidence to our understanding of the processes involved, which has aided the developments seen in modelling over the past half-century.

Model development

Fifty years ago, coastal oceanographic hydrodynamic models were under development by individual scientists primarily concerned with specific issues such as storm surge prediction, residual current formation or tidal phenomena and comparisons with measurements (e.g. Heaps 1969; Pingree 1978; Uncles 1981). These were far from being the powerful or sophisticated tools that we use today. By and large, the early models represented the flow of water by means of a modified version of the Navier–Stokes or St Venant equations, taking as input boundary conditions the water surface level for a set spatial grid size using a discrete time step. The models were adapted to make use of an advection–dispersion equation taking a simple turbulence (mixing) formulation into account, together with friction at the bed, and possibly incorporated sources and sinks of material (e.g. sediment) and carried out an appropriate material mass balance for each node of the spatial finite-difference grid.

Model considerations

The processes that govern flow and transport in estuaries and coastal systems are noted for their complexity, and it is well known that there is no single type of system that behaves the same as another. In other words, each system must be modelled based on an understanding of the system itself and the questions being asked, as well on the availability of the data used to drive any model (Chu 1989). Where the processes of interest are complex, so the model must change to meet the challenges involved, for example with wave–current interactions (Zhang et al. 2004) or the specific question of the effects of coastal development (Zigic, King and Lemckert 2005; Kim, Choi and Lee 2006). A model used to understand the sediment budget in a large estuary is not normally well suited to investigate the release of a pollutant from a single-point source, for example, and the level of detail needed for one would not be appropriate for the other and vice versa. One key question relates to the dimensionality of the modelling used. For simpler applications, a one-dimensional model may be appropriate (Nielsen, Rasmussen and Gertz 2005) where there is not much variation over the depth or width of a channel, as in a tidal river subject to dredging and canalisation. Two dimensions may be suitable for wider channels, where the 'second' dimension may be horizontal or vertical (Young and Murray 1988), while three dimensions may be used for detailed studies of variations in three orthogonal directions (Cugier and Le Hir 2002; Torres and Uncles 2011).

In earlier work, attempts to model flows tended to favour a finite difference approach (Falconer and Owens 1990; Lin and Falconer 1995), where the difference in a water level between two points on a grid could be related to the velocity of water between those points, which could in turn be converted into a flow rate. This type of repetitive calculation could be undertaken in one, two or three dimensions, and over time a general pattern of flows could emerge. Success depended on the appropriate selection of time step and grid size, guided usually by a compromise between accuracy and length of time taken to complete a model run. Flows that were unsteady (i.e. time-varying) owing to changing tidal water levels lent themselves to an approach of this sort, with the results from one time step being fed into the following time step (Warren and Bach 1992).

Establishment of a common approach in the field of finite difference and later finite volume models then led to the distinction between Eulerian versus Lagrangian approaches. Here, the argument was whether it was more appropriate to gain information from the model about what was changing at a single point over a given period of time (Eulerian) or what was changing in terms of the experience of a particular particle (or organism) in a given flow field. A consideration of drifter studies may be important in this case (Suara et al. 2017; Uncles et al. 2020). The choice of which approach to use is generally down to what understanding, or outcome, is required from the decision-making process; that is, whether it is more interesting and useful to understand what is happening to a given particle or at a given site. Two-phase approaches (water/sediment) have also been used, with some success (Nguyen et al. 2009).

When modelling particles or organisms subject to uncertainty for a given condition, it is often desirable to introduce a probabilistic approach. A random walk model, where each movement is subject to an element of randomness influenced by a general direction of flow, is appealing (e.g. such a model for sediment particles is described in Kelsey et al. 1994) or where there is some understanding of the general patterns of behaviour of, say, a group of fish, which are at the same time behaving as individuals. A suite of Individual Behaviour Models is available for addressing this kind of question (Berenshtein et al. 2018).

Recent advances in machine learning and artificial intelligence mean that the advantages of the established models of hydrodynamics, solutes and sediments can be coupled with schemes that can learn or develop given a set of information on actual outcomes.

There have been some notable successes in recent years, including predictions for habitat distribution, dissolved oxygen concentrations and turbidity, all of which can be modelled using traditional approaches but where the outcomes can be improved by incorporating known behaviours for given systems over given time periods (Koudenoukpo et al. 2021; Wang et al. 2021).

Key drivers of model development

As in all fields of human endeavour, the questions posed by researchers are informed by the pressures imposed by society to solve certain problems. Perhaps the most pressing of these – climate change – can be summed up in terms of well-known changes to the earth's systems such as increasing ocean temperatures, sea-level rise, increased or decreased freshwater input to estuaries owing to higher or lower rainfall, depending on location, and changes to the water chemistry of coastal areas owing to ocean acidification, for example. As climate change accelerates or is possibly held in check in response to mitigation measures, the questions become increasingly related to those regarding the impact of changes to the hydrodynamics and related transport in coastal areas and estuaries. In addressing the problems that arise from climate change, some of the possible solutions can also be seen to be addressed by a better understanding of the hydrodynamics of coastal areas. Perhaps the most obvious of these is related to the vast potential of tides and waves to generate so-called 'clean' electricity, but here again the uncertainties of the coastal systems involved have precluded more rapid development in this regard.

In tandem with climate change, the problem of increased plastics in oceans and rivers, especially microplastics, has begun to characterise much of the recent literature in terms of monitoring, modelling and mitigating its effects (Tsang et al. 2020). An increased number and various types of small plastic particles have been noted together with a burgeoning range of provenance. Different types of plastic tend to behave differently under different conditions of tides and waves, and have different effects in different environments. The effects of litter and debris are not new problems but have attracted increasing attention owing to economic growth and expansion of human activities.

Eco-hydrological modelling, and the consideration of the interactions between physical processes and the health of ecosystems, is clearly of economic and environmental importance and considering increased awareness of sustainability drivers (van Niekerk et al. 2019). Sustainability of fisheries is of universal importance but especially to coastal communities around the globe, and the challenge is often to try to discover the impact of some type of change and to mitigate its effects.

Continued economic growth also influences the need for new infrastructure in the marine environment, increased port operations, offshore wind farms, deeper and wider dredged channels for shipping and the disposal of spoil, tidal and non-tidal power plants in coastal areas and so on. Often, development is associated with mobilisation or remobilisation of sediment and the release of pollutants that may have been stored in them (Vane et al. 2020). Especially for the case of fine cohesive sediments, the release of certain chemical species must be understood, both in the short term during construction and afterwards.

Perhaps most importantly as far as researchers are concerned is the vastly increased availability of good-quality datasets, coupled with concerted efforts to make these data widely and freely available. The problem for scientists and modellers has now rather become one of 'what to do with the data', rather than of trying to coordinate resources and equipment to gather a few good samples from one or two days. Therefore, the increased availability of information from in situ and satellite monitoring, from model outputs and from autonomous vehicles of various sorts has meant there is a considerable challenge involved in reducing large amounts of available data to the more useful elements.

Looking forward – future direction of modelling effort and key challenges

Despite the ongoing progress in making models more accurate and more usable, there will always be a need for good-quality field data to calibrate models and validate the approach used in them. Thus, while the publications that gained prominence in the 1970s are still important, with their emphasis on monitoring and logging of data to explain individual phenomena and to understand the key drivers, it is more likely that approaches based on modelling will become more widespread (Fringer et al. 2019), with a relatively small element of field monitoring, all of which is now more sophisticated than ever before. We now identify five main areas of focus for future research and discuss the likely direction of travel for researchers in each case.

Response to climate change

It seems highly likely that sea-level rise and changing climate will continue to influence all decision-making in coastal areas for the foreseeable future, regardless of any progress made in terms of reductions to carbon emissions. This being the case, there is a particular need for coastal engineers to understand the likely system response to the changes seen, especially in the case of extreme events (Gong and Shen 2003). The demands of local and national governments and people alike will probably result in higher seawalls and more material for beaches and intertidal areas, as well as the increased use of 'managed retreat' in some areas. However, questions often arise regarding the impact of engineered solutions on the surrounding natural (or less engineered) areas. These factors, in combination with increased subsidence of coastal zones, means that whenever surge events or high tides occur, local solutions will in many cases have to be found quickly.

The concern also extends to the transport of sediments and salinity in estuaries. The problem of increased saline intrusion is well known (Chen et al. 2009; Prandle and Lane 2015; Little, Wood and Elliott 2017), and this affects drinking water supplies further upstream (Gong and Shen 2011), requiring more treatment or increase in shut down periods at municipal water treatment works. The same arguments, albeit with slightly different mechanistic interpretations, occur for the transport of fine sediments farther into the estuary owing to both tidal mechanisms (tidal pumping, for example, Uncles and Jordan 1979) and the up-estuary migration of the freshwater–saltwater interface (Uncles and Stephens 1993). The constantly developing system response to sea-level rise is likely to be of interest to engineering hydrologists too, in the understanding and prevention of flooding or the management of the attendant risk, as well as ongoing geomorphological concerns (Sandbach et al. 2018).

For researchers, this means access to more complete datasets and scenario-testing using models designed to relate the underpinning mechanisms to the phenomena of interest. There is also a need to share best practice in this area through modellers' forums and the sharing of datasets.

Response to port development and the demand for more efficient sea transport

Increases in global trade together with concerns about emissions from air transport have inevitably led to an increased importance of shipping and port operations. This in turn requires capital and maintenance dredging for ports and access channels, and a proper understanding of the implications for water quality and local geomorphology (de Jonge et al. 2014; Xiao et al. 2020). This is a particular concern where contaminated sediments may be present, although it is at least fair to say that our understanding of the transport mechanisms is sufficient to allow the likely fates to be identified at least at a basic level, a fact that has been recognised for some time (Bohlen, Cundy and Tramontano 1979). However, there is still a key knowledge gap in terms of the long-term transport of dredged (i.e. recently excavated) material within and around potentially sensitive coastal zones (Uncles et al. 2020). Long-term monitoring and more sophisticated modelling is likely to be useful to address this challenge.

Microplastics and debris

The emphasis on understanding the movement of microplastics and other floating debris has captured the attention of researchers, policymakers and the public emphatically over the past decade (Pagter et al. 2020; Tsang et al. 2020), and is perhaps the most urgent of the challenges for flow and transport modellers and researchers. It therefore seems likely that this will be a key focus over the next 50 years, or more. The challenges are significant, in terms of the response of different types of microplastics to currents and waves under a variety of forcing factors. Models that have been in place for the last few decades for predicting sediment transport will need to be repurposed to deal with floating particles or debris, considering the different densities of material involved, and the breakdown of these materials into different constituent parts over time, changing their buoyancy properties. All this makes modelling of plastic material particularly challenging, and probably means an increased emphasis on more intelligent modelling strategies such as those discussed below.

Artificial intelligence and machine learning for modelling complex systems

Over the past half-century since the early publications mentioned above, the tools available to modellers have become much more sophisticated and at the same time much more available to the non-specialist. This has meant that to carry out a modelling study for a given coastal project, all that is needed is some basic understanding of the system and the circumstances involved, coupled with a basic ability to load a model on a PC and an ability to follow a standard user guide. The widespread use of models therefore means that relatively few case studies of numerical models nowadays appear in the scientific literature, simply because there is not much new to be learned from their application.

However, this does not mean that there is no longer any interest in developing new models or applying them to new situations; instead, the emphasis is now on ever more complex approaches (Boyd and Weaver, 2021) including also artificial intelligence or machine learning to apply to more complex types of problems. Long-term geomorphological evolution, for example, so long beyond the reach of even the most sophisticated standard flow and transport models, is now more possible owing to the ability of modellers to write code that allows the model to 'learn' the responses of particles, channels and larger systems (Wang et al. 2021). While the mechanistic link involved may be less clear, this does not matter provided that the answer matches the reality. Similarly, the vast computing power now available to research organisations, together with the increased interest in so-called citizen science, also means that mass repetition of the same models is possible via a Monte Carlo simulation or similar approach, meaning that it is possible to see what the solution to a particular problem could be given thousands of different model results.

Implications of physical processes on ecosystems

Of great importance to many scientists is the implications of certain physical processes on the ecosystems that depend on them (Wolanski et al. 2004). In recent advances in eco-hydraulic modelling, some attempts have been made to predict the distribution of sea grasses, algae and associated species of organisms for different levels of inundation in estuaries. Such models are complicated because of the interrelationships between flow, water quality, sediment transport and ecosystems, through effects due to temperature, light levels, levels of solutes and the presence of other material such as debris (Stark et al. 2017). Designs for new infrastructure must always consider the effect on the local flow conditions as these may influence the passage of migratory fish and other floating organisms (Berenshtein et al. 2018). The levels of uncertainty associated with models of this type are significant, and it is important to be able to quantify these by understanding as far as possible the limitations of the models used, with reference to ground truthing and ongoing monitoring. Attempts to understand the flow field and its effects on individual particles or tracers can be of particular importance in such studies, given the importance of understanding the experience of different organisms in their response to changes caused by climate change or other interventions.

Tidal and wave power

The potential for generating wave and tidal power was attracting attention even in the early days of ECSA (Voss, 1979; Shaw 1980), though only a few schemes have yet come to fruition. In terms of research, there are surely innumerable opportunities for researchers of physical processes to make their mark on the literature in this area. Recent studies for example have shown the importance of understanding the impacts of power generation from tidal and wave installations on the surrounding water body (Rtimi, Sottolichio and Tassi 2021). However, in thinking about this area there seems little question that clean and reliable energy from the sea will continue to increase in importance in terms of its contribution to the energy mix.

Conclusions

Fifty years ago, the group of like-minded scientists who set up ECSA may well have wondered at the potential of technology to solve problems of predicting the response of the coastal and estuarine environment to changes caused by development and climate change, each directly or indirectly related to the activities of humankind. Despite some giant leaps forward, thanks to technological advances in monitoring and modelling, there is still a clear need for a concerted and focused approach to be brought to bear over the next 50 years, to ensure the sustainability of the earth's coastal systems.

References

Aldridge, J.N. (1997) Hydrodynamic model predictions of tidal asymmetry and observed sediment transport paths in Morecambe Bay. *Estuarine, Coastal and Shelf Science* 44: 39–56. https://doi.org/10.1006/ecss.1996.0113

Anderson, J.J. and Devol, A.H. (1973) Deep water renewal in Saanich Inlet, an intermittently anoxic basin. *Estuarine and Coastal Marine Science* 1: 1–10. https://doi.org/10.1016/0302-3524(73)90052-2

Berenshtein, I., Paris, C.B., Gildor, H., Fredj, E., Amitai, Y., Lapidot, O. and Kiflawi, M. (2018) Auto-correlated directional swimming can enhance settlement success and connectivity in fish larvae. *Journal of Theoretical Biology* 439: 76–85. https://doi.org/10.1016/j.jtbi.2017.11.009

Bohlen, W.F., Cundy, D.F. and Tramontano, J.M. (1979) Suspended material distributions in the wake of estuarine channel dredging operations. *Estuarine and Coastal Marine Science* 9: 699–711. https://doi.org/10.1016/S0302-3524(79)80004-3

Boyd, S.C. and Weaver, R.J. (2021) Replacing a third-generation wave model with a fetch based parametric solver in coastal estuaries. *Estuarine, Coastal and Shelf Science* 251: 107192. https://doi.org/10.1016/j.ecss.2021.107192

Bowden, K.F. and Sharaf El Din, S.H. (1966) Circulation, salinity and river discharge in the Mersey Estuary. *Geophysical Journal of the Royal Astronomical Society* 10: 383–99. https://doi.org/10.1111/j.1365-246X.1966.tb03066.x

Broman, D., Kugelberg, J. and Näf, C. (1990) Two hydrodynamically stable self-suspended buoyant sediment traps. *Estuarine, Coastal and Shelf*

Science 30: 429–36. https://doi.org/10.1016/0272-7714(90)90007-E

Bruun, P. (1970) Use of tracers in harbor, coastal and ocean engineering. *Engineering Geology* 4: 73–88. https://doi.org/10.1016/0013-7952(70)90004-9

Bruun, P. (1989) Coastal engineering and use of the littoral zone. *Ocean and Shoreline Management* 12: 495–516. https://doi.org/10.1016/0951-8312(89)90027-1

Chen, S., Fang, L., Zhang, L. and Huang, W. (2009) Remote sensing of turbidity in seawater intrusion reaches of Pearl River Estuary – a case study in Modaomen water way, China. *Estuarine, Coastal and Shelf Science* 82: 119–27. https://doi.org/10.1016/j.ecss.2009.01.003

Chu, W. (1989) Remaining problems in the practical application of numerical models to coastal waters. In: (eds. V.C. Lakhan and A.S. Trenhaile) *Applications in Coastal Modeling, Elsevier Oceanography Series* 49, pp. 355–69. Elsevier Inc., Oxford. https://doi.org/10.1016/S0422-9894(08)70132-6

Cugier, P. and Le Hir, P. (2002) Development of a 3D hydrodynamic model for coastal ecosystem modelling. Application to the plume of the Seine River (France). *Estuarine, Coastal and Shelf Science* 55: 673–95. https://doi.org/10.1006/ecss.2001.0875

De Jonge, V.N., Schuttelaars, H.M., van Beusekom, J.E.E., Talke, S.A. and de Swart, H.E. (2014) The influence of channel deepening on estuarine turbidity levels and dynamics, as exemplified by the Ems estuary. *Estuarine, Coastal and Shelf Science* 139: 46–59. https://doi.org/10.1016/j.ecss.2013.12.030

Dhoop, T. and Mason, T. (2018) Spatial characteristics and duration of extreme wave events around the English coastline. *Journal of Marine Science and Engineering* 6 (14): 16. https://doi.org/10.3390/jmse6010014

Elliott, A.J. and Wang, D. (1978) The effect of meteorological forcing on the Chesapeake Bay: the coupling between an estuarine system and its adjacent coastal waters. In: (ed. Jacques C.J. Nihoul) *Hydrodynamics of Estuaries and Fjords*, Elsevier Oceanography Series 23, pp 127–45. Elsevier Inc., Oxford. https://doi.org/10.1016/S0422-9894(08)71275-3

Falconer, R.A. and Owens, P.H. (1990). Numerical modelling of suspended sediment fluxes in estuarine waters. *Estuarine, Coastal and Shelf Science* 31: 745–62. https://doi.org/10.1016/0272-7714(90)90080-B

Fringer, O.B., Dawson, C.N., He, R., Ralston, D.K. and Zhang, Y.J. (2019) The future of coastal and estuarine modeling: findings from a workshop. *Ocean Modelling* 143: 101458. https://doi.org/10.1016/j.ocemod.2019.101458

Furukawa, K., Wolanski, E. and Mueller, H. (1997) Currents and sediment transport in mangrove forests. *Estuarine, Coastal and Shelf Science* 44: 301–10. https://doi.org/10.1006/ecss.1996.0120

Gong, W. and Shen, J. (2003) Response of sediment dynamics in the York River Estuary, USA to tropical cyclone Isabel of 2003. *Estuarine, Coastal and Shelf Science* 84: 61–74. https://doi.org/10.1016/j.ecss.2009.06.004

Gong, W. and Shen, J. (2011) The response of salt intrusion to changes in river discharge and tidal mixing during the dry season in the Modaomen Estuary, China. *Continental Shelf Research* 31: 769–88. https://doi.org/10.1016/j.csr.2011.01.011

Green, M.O., Black, K.P. and Amos, C.L. (1997) Control of estuarine sediment dynamics by interactions between currents and waves at several scales. *Marine Geology* 144: 97–116. https://doi.org/10.1016/S0025-3227(97)00065-0

Greenwood, B. and Mittler, P.R. (1979) Structural indices of sediment transport in a straight, wave-formed, nearshore bar. *Marine Geology* 32: 191–203. https://doi.org/10.1016/0025-3227(79)90064-1

Harris, J.E., Hinwood, J.B., Marsden, M.A.H. and Sternberg, R.W. (1979) Water movements, sediment transport and deposition, Western Port, Victoria. *Marine Geology* 30: 131–61. https://doi.org/10.1016/0025-3227(79)90010-0

Heaps, N.S. (1969) A two-dimensional numerical sea model. *Philosophical Transactions of the Royal Society of London* Series A 265: 93–137. https://doi.org/10.1098/rsta.1969.0041

Holland, A.F., Zingmark, R.G. and Dean, J.M. (1974) Quantitative evidence concerning the stabilisation of sediments by marine benthic diatoms. *Marine Biology* 27: 191–6. https://doi.org/10.1007/BF00391943

Huthnance, J.M. (1973) Tidal current asymmetries over the Norfolk sandbanks. *Estuarine and Coastal Marine Science* 1: 89–99. https://doi.org/10.1016/0302-3524(73)90061-3

Jarvis, J. and Riley, C. (1987) Sediment transport in the mouth of the Eden estuary. *Estuarine, Coastal and Shelf Science* 24: 463–81. https://doi.org/10.1016/0272-7714(87)90128-4

Kelsey, A., Allen, C.M., Beven, K.J. and Carling, P.A. (1994) Particle tracking model of sediment transport. In: K.J. Beven, P.C. Chatwin and J.H. Millbank (eds) *Mixing and Transport in the Environment*, pp. 419–42. John Wiley and Sons Ltd, Chichester.

Kim, T.I., Choi, B.H. and Lee, S.W. (2006) Hydrodynamics and sedimentation induced by large-scale coastal developments in the Keum River Estuary, Korea. *Estuarine, Coastal and Shelf Science* 68: 515–28. https://doi.org/10.1016/j.ecss.2006.03.003

Kostaschuk, R.A., Luternauer, J.L. and Church, M.A. (1989) Suspended sediment hysteresis in a salt-wedge estuary: Fraser River, Canada. *Marine Geology* 87: 273–85. https://doi.org/10.1016/0025-3227(89)90065-0

Koudenoukpo, Z.C., Hamed, Odountan, Olaniran, Ablawa Agboho, Prudenciène, Dalu, Tatenda, Van Bocxlaer, Bert, Janssens de Bistoven, Luc, Chikou, Antoine and Backeljau, Thierry (2021) Using self-organizing maps and machine learning models to assess mollusc community structure in relation to physicochemical variables in a West Africa river-estuary system. *Ecological Indicators* 126: 107706. Elsevier Inc., Oxford. https://doi.org/10.1016/j.ecolind.2021.107706

Lavender, S.J. (2017) Satellite and aircraft remote sensing. In R.J. Uncles and S.B. Mitchell (eds) *Estuarine Hydrography and Sediment Transport*, pp. 316–44. Cambridge: Cambridge University Press. https://doi.org/10.1017/9781139644426.012

Li, J. and Zhang, C. (1998) Sediment resuspension and implications for turbidity maximum in the Changjiang Estuary. *Marine Geology* 148: 117–24. https://doi.org/10.1016/S0025-3227(98)00003-6

Lin, B. and Falconer, R.A. (1995) Modelling sediment fluxes in estuarine waters using a curvilinear coordinate grid system. *Estuarine, Coastal and Shelf Science* 41: 413–28. https://doi.org/10.1016/0272-7714(95)90002-0

Little, S., Wood, P.J. and Elliott, M. (2017) Quantifying salinity-induced changes on estuarine benthic fauna: the potential implications of climate change. *Estuarine, Coastal and Shelf Science* 198: 610–25. https://doi.org/10.1016/j.ecss.2016.07.020

Marmorino, G.O. and Smith, G.B. (2008) Thermal remote sensing of estuarine spatial dynamics: effects of bottom-generated vertical mixing. *Estuarine, Coastal and Shelf Science* 78: 587–91. https://doi.org/10.1016/j.ecss.2008.01.015

Mazda, Y., Wolanksi, E., King, B., Sase, A., Ohtsuka, D. and Magi, M. (1997) Drag forces due to vegetation in mangrove swamps. *Mangroves and Salt Marshes* 1: 193–9. https://doi.org/10.1023/A:1009949411068

Mitchell, S.B., Uncles, R.J. and Stephens, J.A. (2017) Suspended particulate matter: sampling and analysis. In R.J. Uncles and S.B. Mitchell (eds) *Estuarine and Coastal Hydrography and Sediment Transport*, pp. 179–210. Cambridge: Cambridge University Press. https://doi.org/10.1017/9781139644426.008

Muralikrishna, I.V. (1985) Utility of proposed sensors for coastal engineering studies. *Advances in Space Research* 5: 111–14. https://doi.org/10.1016/0273-1177(85)90262-5

Nguyen, K.D., Guillou, S., Chauchat, J. and Barbry, N. (2009) A two-phase numerical model for suspended-sediment transport in estuaries. *Advances in Water Resources* 32: 1187–96. https://doi.org/10.1016/j.advwatres.2009.04.001

Nielsen, M.H., Rasmussen, B. and Gertz, F. (2005) A simple model for water level and stratification in Ringkøbing Fjord, a shallow, artificial estuary. *Estuarine, Coastal and Shelf Science* 63: 235–48. https://doi.org/10.1016/j.ecss.2004.08.024

Pagter, E., Frias, J., Kavanagh, F. and Nash, R. (2020) Varying levels of microplastics in benthic sediments within a shallow coastal embayment. *Estuarine, Coastal and Shelf Science* 243: 106915. https://doi.org/10.1016/j.ecss.2020.106915

Paterson, D.M. (1989) Short-term changes in the erodibility of intertidal cohesive sediments related to the migratory behaviour of epipelic diatoms.

Limnology and Oceanography 34: 223–34. https://doi.org/10.4319/lo.1989.34.1.0223

Prandle, D. and Lane, A. (2015) Sensitivity of estuaries to sea level rise: vulnerability indices. *Estuarine, Coastal and Shelf Science* 160: 60–8. https://doi.org/10.1016/j.ecss.2015.04.001

Ramster, J.W. and Howarth, M.J. (1975) A detailed comparison of the data recorded by Aanderaa Model 4 and Plessey M021 recording current meters moored in two shelf-sea locations, each with strong tidal currents. *Sonderdruck aus der Deutschen Hydrographischen Zeitschrift* 28: 1–25. https://doi.org/10.1007/BF02226654

Reynolds, M. (1983) A simple meteorological buoy with satellite telemetry. *Ocean Engineering* 10: 65–76. https://doi.org/10.1016/0029-8018(83)90040-9

Rtimi, R., Sottolichio, A. and Tassi, P. (2021) Hydrodynamics of a hyper-tidal estuary influenced by the world's second largest tidal power station (Rance estuary, France). *Estuarine, Coastal and Shelf Science* 250: 107143. https://doi.org/10.1016/j.ecss.2020.107143

Sandbach, S.D., Nicholas, A.P., Ashworth, P.J., Best, J.L., Keevil, C.E., Parsons, D.R., Prokocki, E.W. and Simpson, C.J. (2018) Hydrodynamic modelling of tidal-fluvial flows in a large river estuary. *Estuarine, Coastal and Shelf Science* 212: 176–88. https://doi.org/10.1016/j.ecss.2018.06.023

Shaw, T.L. (1980) *An Environmental Appraisal of Tidal Power Stations: with Particular Reference to the Severn Barrage*. London: Pitman Publishing Ltd.

Silvester, R. (1978) What direction coastal engineering? *Coastal Engineering* 2: 327–49. https://doi.org/10.1016/0378-3839(78)90030-3

Stark, J., Smolders, S., Meire, P. and Temmerman, S. (2017) Impact of intertidal area characteristics on estuarine tidal hydrodynamics: a modelling study for the Scheldt Estuary. *Estuarine, Coastal and Shelf Science* 198A: 138–55. https://doi.org/10.1016/j.ecss.2017.09.004

Stigebrandt, A. (1981) A mechanism governing the estuarine circulation in deep, strongly stratified fjords. *Estuarine, Coastal and Shelf Science* 13: 197–211. https://doi.org/10.1016/S0302-3524(81)80076-X

Suara, K., Chanson, H., Borgas, M. and Brown, R.J. (2017) Relative dispersion of clustered drifters in a small micro-tidal estuary. *Estuarine, Coastal and Shelf Science* 194: 1–15. https://doi.org/10.1016/j.ecss.2017.05.001

Torres, R. and Uncles, R.J. (2011) Modeling of Estuarine and Coastal Waters. *Elsevier Reference Module in Earth Systems and Environmental Sciences: Treatise on Estuarine and Coastal Science*, Volume 2, pp. 395–427. Elsevier Inc., Oxford. https://doi.org/10.1016/B978-0-12-374711-2.00216-3

Tsang, Y.Y., Mak, C.W., Liebich, C., Lam, S.W., Sze, E.T.P. and Chan, K.M. (2020) Spatial and temporal variations of coastal microplastic pollution in Hong Kong. *Marine Pollution Bulletin* 161: 111765. https://doi.org/10.1016/j.marpolbul.2020.111765

Uncles, R.J. (1981) A numerical simulation of the vertical and horizontal M2 tide in the Bristol Channel and comparisons with observed data. *Limnology and Oceanography* 26: 571–7. https://doi.org/10.4319/lo.1981.26.3.0571

Uncles, R.J. and Jordan, M.B. (1979) Residual fluxes of water and salt at two stations in the Severn Estuary. *Estuarine and Coastal Marine Science* 9: 287–302. https://doi.org/10.1016/0302-3524(79)90042-2

Uncles, R.J. and Mitchell, S.B. (2017) Estuarine and Coastal Hydrography and Sediment Transport. In R.J. Uncles and S.B. Mitchell (eds) *Estuarine and Coastal Hydrography and Sediment Transport*, pp. 1–34. Cambridge: Cambridge University Press. https://doi.org/10.1017/9781139644426.002

Uncles, R.J. and Stephens, J.A. (1993) The freshwater–saltwater interface and its relationship to the turbidity maximum in the Tamar Estuary, UK, *Estuaries* 16: 126–41. https://doi.org/10.2307/1352770

Uncles, R.J., Elliott, R.C.A. and Weston, S.A. (1985) Observed fluxes of water, salt and suspended sediment in a partly mixed estuary. *Estuarine, Coastal and Shelf Science* 20: 147–67. https://doi.org/10.1016/0272-7714(85)90035-6

Uncles, R.J., Elliott, R.C.A. and Weston, S.A. (1986) Observed and computed lateral circulation patterns in a partly mixed estuary. *Estuarine, Coastal and Shelf Science* 22: 439–57. https://doi.org/10.1016/0272-7714(86)90067-3

Uncles, R.J., Clark, J.R., Bedington, M. and Torres, R. (2020) On sediment dispersal in the Whitsand Bay Marine Conservation Zone: neighbour to a closed dredge-spoil disposal site. In J. Humphreys and R. Clark (eds) *Marine Protected Areas: Science, Policy and Management*, pp. 599–629. Elsevier Inc., Oxford. https://doi.org/10.1016/B978-0-08-102698-4.00031-9

Van De Graaff, J. and Van Overeem, J. (1979) Evaluation of sediment transport formulae in coastal engineering practice. *Coastal Engineering* 3: 1–32. https://doi.org/10.1016/0378-3839(79)90002-4

Van Niekerk, L., Taljaard, S., Adams, J.B., Lamberth, S.J., Huizinga, P., Turpie, J.K. and Wooldridge, T.H. (2019) An environmental flow determination method for integrating multiple-scale echyhydrologyical and complex ecosystem processes in estuaries. *Science of The Total Environment* 656: 482–94. https://doi.org/10.1016/j.scitotenv.2018.11.276

Van Rijn, L.C. (1997) Sediment transport and budget of the central coastal zone of Holland. *Coastal Engineering* 32: 61–90. https://doi.org/10.1016/S0378-3839(97)00021-5

Vane, C.H., Kim, A.W., Emmings, J.F., Turner, G.H., Moss-Hayes, V., Lort, J.A. and Williams, P.J. (2020) Grain size and organic carbon controls polyaromatic hydrocarbons (PAH), mercury (Hg) and toxicity of surface sediments in the River Conwy Estuary, Wales, UK. *Marine Pollution Bulletin* 158: 111412. https://doi.org/10.1016/j.marpolbul.2020.111412

Voss, A. (1979) Waves, currents, tides-problems and prospects. *Energy* 4: 823–31. https://doi.org/10.1016/0360-5442(79)90014-8

Wang, Y., Chen, J., Cai, H., Yu, Q. and Zhou, Z. (2021) Predicting water turbidity in a macro-tidal coastal bay using machine learning approaches. *Estuarine, Coastal and Shelf Science* 252: 107276. https://doi.org/10.1016/j.ecss.2021.107276

Warren, I.R. and Bach, H.K. (1992) MIKE 21: a modelling system for estuaries, coastal waters and seas. *Environmental Software* 7: 229–40. https://doi.org/10.1016/0266-9838(92)90006-P

Wolanski, E., King, B. and Galloway, D. (1995) Dynamics of the turbidity maximum in the Fly River estuary, Papua New Guinea. *Estuarine, Coastal and Shelf Science* 40: 321–37. https://doi.org/10.1016/S0272-7714(05)80013-7

Wolanski, E., Boorman, L.A., Chicharo, L., Langlois-Saliou, E., Lara, R., Plater, A.J., Uncles, R.J. and Zalewski, M. (2004) Ecohydrology as a new tool for sustainable management of estuaries and coastal waters. *Wetlands Ecology and Management* 12: 235–76. https://doi.org/10.1007/s11273-005-4752-4

Wolanski, E. and McLusky, D., 2011. Elsevier reference module in earth systems and environmental sciences: treatise on estuarine and coastal science. Elsevier Inc., Oxford. https://www.sciencedirect.com/referencework/9780080878850/treatise-on-estuarine-and-coastal-science

Xiao, Z.Y., Wang, X.H., Song, D., Jalón-Rojas, I. and Harrison, D. (2020) Numerical modelling of suspended-sediment transport in a geographically complex microtidal estuary: Sydney Harbour Estuary, NSW. *Estuarine, Coastal and Shelf Science* 236: 106605. https://doi.org/10.1016/j.ecss.2020.106605

Young, M.H. and Murray, S.P. (1988) Vertical slice numerical model for density-driven coastal currents. *Mathematical and Computer Modelling* 11: 96–100. https://doi.org/10.1016/0895-7177(88)90461-X

Zhang, H., Madsen, O.S., Sannasiraj, S.A. and Chan, S. (2004) Hydrodynamic model with wave–current interaction in coastal regions. *Estuarine, Coastal and Shelf Science* 61: 317–24. https://doi.org/10.1016/j.ecss.2004.06.002

Zigic, S., King, B. and Lemckert, C. (2005) Modelling the two-dimensional flow between an estuary and lake connected by a bi-directional hydraulic structure. *Estuarine, Coastal and Shelf Science* 63: 33–41. https://doi.org/10.1016/j.ecss.2004.11.001

CHAPTER 6

Marine Plastics: Emerging Challenges and Priorities for Estuaries and Coasts

ANTHONY W. GALLAGHER, MALCOLM D. HUDSON, OLIVER T. BROOKS and JESSICA L. STEAD

Abstract

Plastics are a highly successful range of materials, the development of which has outstripped any concomitant system of governance aimed at managing them. The effects of this are keenly felt in the oceans, with plastic leakage into the global marine environment estimated to be in the order of 10 million tonnes per year. Without significant changes in behaviour and practice, this will only increase. The impacts of marine plastics are wide ranging and, in the case of entanglement, ingestion and smothering of marine wildlife, are highly visible, gaining widespread global attention and raising significant concerns. Pathways to marine pollution are multifarious, but an estimated 80% of all plastic found in ocean originates from land-based sources. In addressing this concern, there are significant challenges, both scientific and political. While marine plastics can be highly visible and lead to mortality, their consequences when passed up the food chain are more difficult to determine. Work on a range of methodological approaches, as well as material flow and hot spot analysis, is ongoing to provide a better evidence base for policy development, while a lack of data and understanding of long-term liability point to the need for a precautionary approach. While existing mechanisms and agreements could be better coordinated and strengthened, the solutions require deep-rooted behavioural change as well as a joining up of efforts to better reflect the transboundary and transdisciplinary nature of the problem. Without effective intervention, we will continue to see an increasing concentration of synthetic polymers in coastal and estuarine environments, threatening ecological thresholds that have yet to be determined.

Keywords: Marine plastics pollution, methods, pathways, hot spots, policy integration

Correspondence: anthony.gallagher@evolvedresearch.co.uk

Introduction

Plastics are a highly successful range of synthetic or semi-synthetic polymers that have developed at a pace far outstripping global economic growth rates over recent decades. Their characteristics and versatility have led them to be used in an increasing range of applications, outpacing the development of any system of governance aimed at managing them. The impact of this is keenly felt in the oceans, with plastic leakage into the marine environment estimated to be between 4.8 and 12.7 m tonnes of annual inputs (Andrady

Anthony W. Gallagher, Malcolm D. Hudson, Oliver T. Brooks and Jessica L. Stead, 'Marine Plastics: Emerging Challenges and Priorities for Estuaries and Coasts' in: *Challenges in Estuarine and Coastal Science*. Pelagic Publishing (2022). © Anthony W. Gallagher, Malcolm D. Hudson, Oliver T. Brooks and Jessica L. Stead. DOI: 10.53061/CIKC3737

2011; Lebreton, Greer and Borrero 2012; Jambeck et al. 2015), as shown in Fig. 6.1. With cumulative plastic consumption at 8.3 bn tonnes and projected growth in plastic production estimated to reach 34 bn tonnes by 2050 (Geyer, Jambeck and Law 2017), this flow of plastics into the global oceans will only worsen without a significant change in behaviour.

The impacts of marine plastics are wide ranging and in the case of entanglement, ingestion and smothering of marine wildlife, are highly visible, gaining widespread global attention and raising significant concerns as to their impact. However, they also act as vectors for species and disease, and represent a pathway for ecotoxicological impacts. The consequence of these negative impacts on marine and coastal ecosystem services affecting fisheries, aquaculture, recreation and wellbeing have been estimated to cost up to £1.9 tn per year (Beaumont et al. 2019), with concerns, based on limited evidence, over the potential impact on human health (Vethaak and Legler 2021).

In addressing this global concern, the challenges are both scientific and political. To date, despite a surge of scientific enquiry, there is still insufficient information on the fate of plastic litter and microplastic particles, to make a robust assessment of the risks. While there have been numerous studies into the distribution and ubiquity of plastic materials, many of the methods used are still in their infancy, and often applied inconsistently, making comparison between studies difficult. The increasing diversity of plastics and their use across a range of products such as tyres and antifoulant paints, for example, as well as their impact as nanoparticles, means that assessing the risks to ecosystems and human health is also challenging. Although it is clear that ingestion by individuals can be significant and lead to mortality, the consequences of that when passed up the food chain are more difficult to determine. Work on the development of a range of methodological approaches, as well as material flow and hot spot analysis, is ongoing to provide a better understanding for policy development.

The challenges for effective policy are significant. While the problem is being discussed and initiatives are under way at both national and global levels, the solutions are not easy

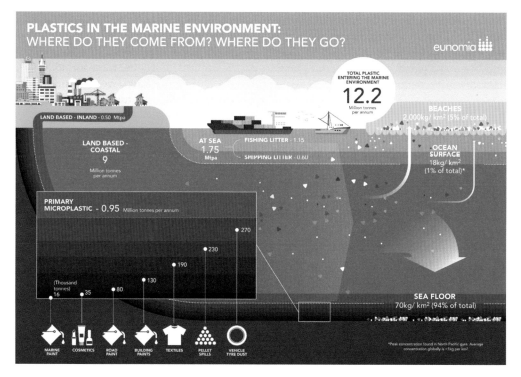

Figure 6.1 Plastics in the marine environment (from Sherrington 2016)

and require deep-rooted behavioural change as well as a joining up of efforts across all levels of governance to better reflect the transboundary and transdisciplinary nature of the problem. Without effective intervention, we will continue to see an increasing concentration of synthetic polymers in coastal and estuarine environments threatening, or potentially threatening, ecological thresholds that have yet to be determined.

This chapter reviews the key challenges facing estuaries and costs, both scientific and political, though first it will review the nature of the marine plastics problem with respect to polymers, their attributes, behaviour and impact.

The marine plastic problem facing estuarine and coastal environments

There are now many hundreds of polymers and mixtures of polymers in commercial production, with the market being particularly dominated by polyethylene (PE – as both high density, HDPE, and low density, LDPE), polypropylene (PP), polyvinyl chloride (PVC), polyurethane (PUR), polystyrene (PS) and polyethylene terephthalate (PET). These six polymers make up about 80% of plastics production and form a large proportion of most marine litter. Typically, pure polymers, or mixes of polymers, are combined with a range of additives including various plasticisers, colourants, and stabilisers to produce compounds that enhance their properties and enable the production of an increasing range of materials. These can be shaped or moulded into specific products, leading to an increasing number of plastics offering technical solutions to practical problems.

These properties are designed to be fit for purpose, with high volume 'commodity plastics' such as PE, PP, PS, PVC and polymethyl methacrylate (PMMA) being used in applications where exceptional material properties are not required. This is typically the case with packaging, pots and containers, and also twine and ropes. Where properties need to be enhanced for specific applications, however, 'engineering grade plastics' can be used, with examples of these being polycarbonate (PC), polyamides (PA) and Acrylonitrile butadiene styrene (ABS). The properties of plastics that are so successful, however, also determine its behaviour when left uncontrolled in marine environments. For example, the buoyancy of a plastic is dependent upon the specific gravity (SG) or relative density of the material, with the SG being the ratio of a material's density to that of a reference point, typically water, as shown in Table 6.1. Plastics that are more buoyant will of course continue to exhibit this behaviour when uncontrolled, being therefore more likely to remain on the surface for longer.

Since high-volume production polymers (e.g. PE, PP) are buoyant, and even heavier polymers float just below the surface layers, oceanographic conditions and coastal and estuarine processes are important in determining the distribution of plastics, with oceanic and coastal currents, as well as surface-driven winds directly affecting spatial concentrations of MPP. This not only creates large-scale pollution features such as the Pacific Garbage Patch or its equivalent in the Bay of Bengal, associated with a 'dead zone' of anoxic conditions, but also transports plastic pollution from one location to another, often creating significant problems for particular coastlines or marine features and ecosystems.

Size is an important feature of plastic pollution of all sorts, affecting its likely source, transportation and impact. Plastic pollutants can be classified into the following size groups, with all sizes of plastics having the potential to cause harm:

- macroplastics (>5 mm diameter);
- mesoplastics (≤5 to >1 mm);
- microplastics (≤1 mm to >0.1 μm);
- nanoplastics (≤0.1 μm).

Table 6.1 The specific gravity of common polymers

Polymer	Application	Specific Gravity (SG)	Behaviour
Polystyrene (EPS)	Insulation/cool boxes; floats and buoys, including fish aggregating devices	0.02–0.64	
Polypropylene (PP)	Nets (mostly gillnet and trawl nets), rope, mesh, textiles, packaging, bottles, food containers	0.90–0.92	Float
Polyethylene (PE)	Nets (mostly trawl nets, purse seine); longlines; aquaculture: rope, cage, floats, tubes, disks, pipes, containers, packaging, films, insulation covers	0.91–0.95	
Seawater		1.03	
Polystyrene (PS)	Drinking cups	1.05	
Polyamide/Nylon (PA, PA6)	Nets (mostly gillnet and seine), textiles, automotive components, packaging	1.13–1.15	
Polyvinyl chloride (PVC)	Pipes, tubing, containers, canvas	1.16–1.30	Sink
Polyurethane (PUR)	Insulation, foams, elastomer components, sealants, textiles	1.20	
Polyethylene terephthalate (PET)	Bottles, strapping, insulation textiles, food containers, packaging, glass fibres, tapes	1.34–1.39	
Glass fibre reinforced plastic	Boats	>1.35	

All littering can be seen as the product of both accidental and deliberate discard and ineffective waste management. The presence of the smaller fractions in the marine environment can result from the degradation of in situ larger plastic objects, either through photo-oxidation and chemical action or physical breakdown through wave action. The longer that marine plastics therefore remain on the surface, the greater the degree of breakdown, with PS being particularly susceptible to fragmentation. However, plastics also shed microplastics irrespective of where they are in the water column, leading to ingestion by filter feeders and fish. Exact degradation rates under real marine conditions are as yet unknown. Marine ropes can rapidly produce microplastics within the first few months of use (Welden and Cowie 2017), though fibres have been found to be shedding off from the surface of the trawl fragments that have been in the marine environment for more than 30 years, while the material itself remains sturdy and robust. Marine life can also act as a source of microplastics to the marine and estuarine environments with non-native sphaeromatid isopods in Australia having been documented as attacking polystyrene floats, leading to the release of millions of particles into the marine environment (Davidson 2012). Such floats are commonly used in coastal and estuarine environments for a variety of purposes, including in aquaculture.

In addition to the breakdown of larger items, it should be stated that meso- and microplastics are also directly released into the marine environment from a variety of other sources, particularly direct discharges from land-based sources such as industrial processes, domestic uses and wastewater treatment plants. These discharges are typically particulate or fibrous in form, with a significant proportion of the fibres resulting from sources such as facial cleaners, cigarette butts and clothes. Of the particulates, a proportion, known as microbeads, are used as components in products such as toothpastes and cleaning products, while particulates also include the raw material, virgin pellets of plastic production that are melted down to form other plastic products and packaging. These are prominent in the marine environment, with large numbers found in rivers, estuaries and coastal environments, and many lost to the marine environment through shipping and transportation. Both particulates and fibres are small enough to pass through wastewater treatment plants, allowing for large quantities to enter watercourses, but it is the fibres that are most often cited as being the more significant component (Andrady 2011). By way of explanation, experimentation has showed that a single piece of clothing could shed 1,900 pieces of polymer fibre per wash, and that these fibres tend to be present in higher numbers during the winter months as more washes are done during this period (Browne et al. 2011).

Given the relative pathways of microplastics into the marine environment, it is perhaps not surprising that studies have shown concentrations to be higher in more populated areas, owing to the greater presence of industrial and domestic sources. However, fishing gear is also a significant source of global marine debris, with the Food and Agriculture Organization of the United Nations (FAO, 2016) estimating that abandoned, discarded and lost fishing gear accounts for 10% of global marine plastic pollution, and a recent study on the 'Great Pacific Garbage Patch' plastic accumulation zone in the North Pacific Ocean also estimating that fishing nets alone represent 46% of the 79,000 tons of plastic observed within the 1.6 million km^2 region surveyed (Lebreton et al. 2018).

Scientific challenges of marine plastics for estuarine and coastal environments

There are numerous scientific challenges associated with marine plastics, not least quantifying and characterising the pathways to pollution and their inherent threat. Addressing these will provide a more solid evidence base from which to develop a better system of governance.

Quantifying and mapping pathways

Understanding the pathways to pollution requires not only an understanding of the key sources but also of the stores, processes and flows of plastics through from primary manufacture, design and production to packaging, distribution and transportation through suppliers, wholesalers and retailers to end users. Quantifying these stages enables material flows to be mapped and hot spots to be identified.

Pathways to marine pollution are multifarious, with maritime activities such as fisheries and shipping seen as important sources. However, an estimated 80% of all plastic pollution found in the marine environment originates from land-based sources (Horton et al. 2017) via highly populated catchments, with many of the great transboundary river systems, estuaries and deltas seen as global hotspots for marine plastics. Estuaries have therefore been identified as a primary transfer zone of plastics between continental and marine waters (Alligant, Tassin and Gasperi 2018). Lebreton et al. (2017) estimated an annual output of plastic debris by rivers into coastal seas of 1.15–2.47 m tonnes. However, estuaries may not only be a transfer zone, but also a sink for plastic debris. They are known as 'sinks' and 'filters' for other suspended particles, such as sediment (Schubel and Carter 1984; Holmes, Turner and Thompson 2014), and have features such as salinity fronts that have been observed to retain plastic debris (Acha et al. 2003). However, the limited consideration of estuarine processes within many sampling studies means that further determination of this is difficult. The limited comparability of these studies owing to the varied methods utilised is also problematic.

When reviewing factors that may influence the transport of plastic debris in estuaries, both riverine and marine processes need to be considered (Dris et al. 2020), with estuaries being particularly dynamic and complicated to interpret. These factors include source inputs, tidal cycles, salinity gradients, the maximum turbidity front and river flow (ibid.). Some of these factors have been investigated, and for individual systems some conclusions can be drawn. While limited consideration is given to the influence of the tidal cycle – out of 18 studies reviewed by Dris et al. (2020), seven did not specify tidal cycle information about when samples were taken – it seems that when reviewing estuarine water samples, meteorological conditions may have more of an influence on microplastic abundance (Dris et al. 2020). A number of studies report seasonal differences in microplastic abundance or distribution in estuaries, with tropical regions showing pronounced rainy and dry seasons (Barletta, Lima and Costa 2019). It is unclear whether macroplastics show the same trends as microplastics in this, as there are fewer studies, though it is likely that they do.

Within estuaries, and across estuarine types, source factors will vary. In well-mixed estuaries with strong tidal mixing and short residence times, plastic debris will be flushed out into coastal seas relatively quickly. However, in estuaries with longer residence times, such as estuaries in Western Australia, which may be only connected to the sea for short periods of time if at all (e.g. Stokes Inlet), plastic debris will be retained within these systems for much longer periods. Understanding and including information on the key characteristics about estuaries within marine plastics research will aid the comparability of studies and a better interpretation of results.

Broad global conclusions about the fate of plastic debris in estuaries cannot and should not be drawn currently. Estuaries can vary widely in terms of hydrodynamics and morphodynamics even over small geographic distances, and the role of estuaries therefore as source, sink or slow conduit of plastic debris may vary within a particular estuary over different timescales, from short (daily tidal cycles) to long term (decades). Better understanding the interaction between marine plastics and estuarine processes is therefore conditional on research design and the selection of an appropriate methodology.

Methodological divergence and convergence

The emergence of marine plastics as a global issue has stimulated enormous growth in the amount of research into marine plastics from a wide range of perspectives, often focusing on the fate of these materials once they are lost in the environment. As the scientific community has sought to tackle these challenges, numerous issues have arisen relating to the selection and use of the methods available.

The assessment of macroplastics in coastal systems is sometimes seen as straightforward (Pinheiro et al. 2021), although the characteristics of plastics mean that materials will behave differently dependent upon their properties such as SG, or their density, morphology and debris age. Since monitoring is resource-intensive, the assessment of litter has often been carried out using citizen science methods. This has the added advantage of engaging the public in both science and the problem of marine plastics, with citizen science beach cleans such as the long-term programme organised by the UK Marine Conservation Society providing valuable indications of the amounts and nature of material accumulating on shoreline areas. However, these approaches very often target more accessible locations only and may have issues with methodology (Nelms et al. 2020). In the UK, Thames 21 captured detailed spatial temporal information on the accumulation of litter on the foreshore, but lacked the resources to monitor the material accumulating within the river or to assess smaller items such as microplastics. Evidence suggests this could be significant, with Morritt et al. (2014) reporting 8490 plastic items trapped by fyke nets at seven locations in the Thames over a three-month period, over 20% of which were from sanitary products probably originating from wastewater treatment plants, an association that also shown in other studies (Pinheiro et al. 2021). Such systematic studies are rare but suggest a slowly moving flux seawards within industrialised estuaries, which may be much more severe in cities with less-developed infrastructure and waste management. For example, in the lower Paranaguá estuary (Brazil), 74.8% of marine debris on the shoreline comprised larger plastic fragments (Krelling and Turra 2019). This study also revealed how meteorological and oceanographic events such as heavy rain can significantly affect the amounts of plastic litter recorded, and highlighted fisheries alongside wastewater as significant sources.

Assessing microplastics in coastal environments presents particular challenges such as the prevention and effective monitoring of contamination. Microplastics are ubiquitous in the environment, so contamination threatens all microplastic research efforts. Contamination can occur while sampling in the field, or later while processing or analysing in the laboratory. Clothing made of synthetic fibres sheds microfibres during normal wear, so contamination could occur while environmental samples are being collected or from clothing worn by investigators or others working in the same space. While clothing as a source of contamination can be addressed by wearing natural fibre garments, synthetic polymer contamination via airborne contamination is also a major issue and remains a challenge (Dris et al. 2017). Ambient air (indoors or outdoors) is a source of contamination with microplastics present both indoors and in remote outdoor locations (Bergmann et al. 2019). Ideally, experimental analysis should be carried out in forensic-standard clean rooms with filtered air and positive pressure, or at least in laminar flow cabinets, but these are not standard in many research facilities set up for examining environmental samples, and the costs are considerable. Contamination can also occur from reagents, which should usually be filtered, and from glassware or equipment, which where possible should be acid-washed subject to high temperature treatment to destroy any polymer materials (Prata et al. 2021).

These issues are not insurmountable, but many published studies may be flawed owing to inadequate control (Prata et al. 2021), with the literature being inconsistent (to date) in reporting the precautions. For example, Thiele et al. (2021) reviewed studies of microplastics in coastal and marine fish, and found that 55% were inadequate owing to either a lack of or limited contamination control; along with a focus on larger microplastics (>1 mm); and lack of established techniques (e.g. vibrational spectroscopy) to confirm polymer types without which misidentification or under/over counting was considered likely, and the assessment therefore limited.

While contamination issues remain an ongoing challenge, researchers have struggled to develop consistent methods and approaches for assessing microplastics in the marine environment (Prata et al. 2018). This makes comparison between studies difficult, as the methods and units reported do not follow a standard approach. Seawater can be sampled and filtered in situ or in the laboratory, with different filter or sieve pore sizes reflecting more convenience or availability than scientific rigour. As plastics are not distributed uniformly in the water column, there is also the question of from where samples should be taken. While SG may play a part, it is likely that microplastics are initially concentrated in the surface microlayer, held by surface tension and buoyancy (Anderson et al. 2018), although denser or weathered/fouled materials will sink below the surface, remaining suspended for an indeterminate amount of time depending on their properties and environmental conditions. When samples are being analysed, there are also numerous published methods for assessing the same or similar media. For example, the assessment of microplastics in shellfish has been reported using enzymes, alkalis, acids and density separation (Thiele, Hudson and Russell 2019). Each of these methods vary in their efficiency and efficacy for recovering microplastics. Some (e.g. acid) degrade certain polymers, while others may be ineffective in releasing the microplastics from their media, meaning they cannot be extracted or located.

There is also a lack of standardised approaches for characterisation of the extracted microparticles. Many early studies relied on light microscopy, and while this does offer opportunities for counting, characterisation and measurement it is likely to generate false positive identification. Ideally, vibrational spectroscopy (e.g. Raman laser or Fourier Transfer Infra-Red) should be used, as these approaches give accurate counts of different polymer types though without directly allowing inference of mass. However, these methods are costly and can themselves produce under-reporting or false positives. One way forward may be the use of Gas Chromatography–Mass Spectroscopy and related technology, as these have the potential to reveal the different proportions of contaminants present even in very low concentrations, and to infer masses, but not the numbers or sizes of particles (Haave et al. 2021).

Determining what methods are best for sampling and analysis in estuarine and coastal environments raises some significant challenges in delivering a scientifically rigorous evidence base.

Risk assessment

As with the selection of appropriate methods for sampling and analysis, ecotoxicological risk characterisation of marine plastics is vital to informing relevant policy and legislative action (Burns and Boxall 2018). Risk assessment relies on the availability of high-quality data pertaining to environmental concentrations of plastics and the ecological effects of exposure (Gottschalk, Sun and Nowack 2013; Burns and Boxall 2018). Without accurate characterisation of risk, the present and future implications of marine plastic pollution are uncertain, and hence are likely to limit the efficacy of regulatory action.

Despite marine plastics being widely studied, their behaviour within the marine environment is not well understood. This is compounded by the perpetual breakdown of plastic fragments into microplastics, which are ubiquitous and highly bioavailable, inducing largely unknown and unquantified impacts (Andrady 2011; Eriksen et al. 2014). Therefore, mapping particle transport and sinks lacks accuracy (Cózar et al. 2014; Pedrotti et al. 2016) and inherently poses challenges in understanding which marine communities are exposed to which types of plastic debris. Environmental sampling inadequacies amplify this challenge, as the smallest size classes are often not captured using common techniques such as plankton net tows (van Sebille et al. 2015; Anderson et al. 2018), despite these particle sizes being recorded as the most common (Cózar et al. 2014). As a result, present measured environmental concentrations (MECs) could underestimate the true occurrence of microplastics by an order of magnitude (Lindeque et al. 2020). This in turn limits the applicability of current ecotoxicological risk assessments, which to date, based on projections up to the year 2100, found no significant risk at current MECs (Everaert et al. 2018). However, this may change as techniques improve and environmental microplastic sampling methods develop to retrieve smaller size classes (Burns and Boxall 2018).

The ecological impacts of macroplastic particles, including entanglement, suffocation and starvation (Andrady 2011), are relatively well studied compared with micro- and nanoplastics. However, there is mounting evidence that indicates toxicity of these smaller fragments, including implications for reproduction (Zhang et al. 2019), growth and direct mortality (Connors et al. 2017). The persistence and fragmentation of plastics mean they are ubiquitous in the marine environment: on shorelines, in the deep sea, free-floating and within a plethora of marine organisms (Cózar et al. 2014). Moreover, even if plastic introduction into the marine environment were to cease, pre-existing fragments will be broken down mechanically and through photodegradation, continuing the input of nano- and microplastics to the environment (Pedrotti et al. 2016). Yet laboratory and controlled field studies of microplastic ecotoxicology are widely criticised for exposing organisms to unrealistically high concentrations of plastics that are often uncommon morphologies and polymer types in the natural environment (Browne et al. 2011; Burns and Boxall 2018). This limits the applicability of these studies to the natural environment and consequently reduces the efficacy of the risk assessment process (Burns and Boxall 2018). As sampling protocols develop to retrieve the smallest plastic size classes, historically excluded from sampling efforts (Lindeque et al. 2020), the environmental relevance of current ecotoxicology assays may improve, facilitating improved risk assessment. However, as with environmental sampling, development of a standardised and robust protocol is crucial (Adam, Yang and Nowack 2019).

Policy challenges

While the scientific community has worked to reflect the emergence of the marine plastics pollution problem, the need for an integrated and coherent policy response at global, regional and national scales is clearly evident. Key global mechanisms aimed at managing and protecting the marine environment such as the UN Convention on the Law of the Sea, the United Nations Environment Programme's Regional Seas Programme (RSP) and MARPOL 73/78, which relates to the prevention of marine and atmospheric pollution from ships, have been in place for many years, with all three making direct reference to marine plastic pollution. However, on an operational level a lack of implementation and enforcement often restricts their effectiveness. Moreover, while efforts are ongoing to strengthen policy mechanisms, they are also not coordinated.

Over recent years, public policy responses have trended upward (Karasik et al. 2020), with 28 international policies on plastic pollution agreed since 2000, though these have been generally non-binding. Discussions continue, with initiatives such as the Global Partnership on Marine Litter launched, and a series of United Nations Environment Assembly resolutions passed, stressing the importance of addressing the marine plastics issue. However, an analysis of international policies highlights key gaps in the response at this level. To date, no global, binding, specific and measurable targets have been agreed to reduce plastic pollution, and while the discussions provide an opportunity to address the issue more directly through regional strategies, protocols and plans, there is still a need for action.

The current global and regional policy framework could be strengthened by encouraging states that have not yet ratified certain agreements to do so, promoting compliance with respective instruments, and implementing stringent practices to enforce them. The most appropriate mechanisms for achieving this might be the RSPs such as the South Asian Seas Programme (SASP) in the South Asia Region or associated RSPs such as OSPAR for the North-East Atlantic. In South Asia, for example, all the coastal nations, Maldives aside, are ranked among the top 20 most polluting nations ranked by the volume of mismanaged plastic waste, while modelled estimates of floating microplastics suggest that the Bay of Bengal Large Marine Ecosystem has the highest oceanic plastic concentration (Jambeck et al. 2015). As a response the SASP, facilitated by the South Asia Co-operative Environment Programme (SACEP) has facilitated the development of the Regional Marine Litter Action Plan, which has highlighted a series of gaps in current governance, as shown in Box 6.1, indicating the scale of the challenges involved.

The South Asian Seas Action Plan calls on members to, *inter alia*, improve coordination; enhance interagency cooperation; undertake regular monitoring and enforcement; develop and update a marine litter database; review product modification; and enhance waste management. All represent pertinent and valid actions to enable change. However, to date there is little evidence that any of these actions are being implemented.

The number of policy developments at national level is also increasing, but again fragmented and reducing marine plastics will require more than policy development. Bangladesh was the first country to ban single-use plastic bags, in 2002, and several countries, including the USA, Canada, New Zealand, France, Sweden, South Korea and the UK, have introduced a legislative ban on rinse-off cosmetic products containing plastic microbeads. Yet the issue of plastic pollution still pervades in coastal and estuarine environments. The policy challenge is not only to develop new, better, more coherent policies, but also to promote societal and behavioural change and to support that by providing the investment needed to address all aspects of the marine plastic problem, including a more circular economy, and enhanced waste management capacity and infrastructure.

Box 6.1   South Asian Seas Programme Regional Marine Litter Action Plan, 2018

Major gaps and challenges

- Lack of marine litter data in the SAS Region
- Poor institutional system for the management of marine litter
- Non-availability of a legal framework for marine litter management
- Poor and insufficient enforcement of international conventions, agreements, laws, regulations and treaties
- Limited implementation of direct development activities for marine management
- Lack of research and surveys on marine litter
- Weak formulation and enforcement of the regulatory framework
- Lack of marine litter production and consumption policy and strategies
- Lack of education and awareness for marine litter management
- Lack of marketing and economic instruments for marine litter management

The integration and coordination of policies on plastic pollution is a related issue. Policy is fragmented not only because plastic pollution is a transdisciplinary and transboundary problem but because complex issues surround the integration of governance at global, regional and national level. Discussions, initiatives, policy development, implementation and enforcement are still nascent, meaning the challenges to address the problem of marine plastics in coastal and estuarine environments are significant.

Conclusion

Since the turn of the millennium, marine plastic has emerged as a significant global issue and a highly visible reflection of anthropogenic impact. With an estimated 80% of all marine plastic pollution originating from land-based sources via highly populated catchments, it is not surprising that many of the great transboundary river systems, estuaries and deltas are seen as global hotspots for marine plastics. The challenges of marine plastic pollution in coastal and estuarine environments are both scientific and political, and so developing, implementing and enforcing a system of governance that addresses all aspects of a problem is therefore both transdisciplinary and transboundary in nature.

Scientifically, the subject has become increasingly well documented, with a raft of related research papers quantifying and characterising plastics in different species, ecosystems and environments. These have reflected emerging areas of interest but have also demonstrated a lack of standardisation in their design, with study boundaries and the selection of methods and analysis limiting comparability and restricting the research outcomes, particularly for highly complex and dynamic environments such as estuaries. There is a lack of understanding, therefore, with regard to marine plastic impact pathways including, for example, the extent of nanoparticle exposure, which could pose a significant threat to human health or when tipping points may be reached. There is thus a need for these limitations to be addressed.

Irrespective of the scientific challenges however, the need for a coherent, connected and effective system of governance is also clear. A lack of data and understanding of long-term liability would point to the need for a precautionary approach. However, currently policies aimed at addressing the issue are fragmented at all scales of governance, globally, regionally and nationally. On a global scale, while existing mechanisms and agreements could be better coordinated and strengthened, it is still likely that there will be a gap – and that to really address the problem there is a need for an integrated policy response with agreed, binding, specific and measurable targets. This is more likely to accrue from the development of a new global instrument or agreement on marine litter and plastics.

References

Acha, E.M. et al. (2003) The role of the Río de la Plata bottom salinity front in accumulating debris. *Marine Pollution Bulletin* 46 (2): 197–202. https://doi.org/10.1016/S0025-326X(02)00356-9

Adam, V., Yang, T. and Nowack, B. (2019) Toward an ecotoxicological risk assessment of microplastics: Comparison of available hazard and exposure data in freshwaters. *Environmental Toxicology and Chemistry* 38 (2): 436–47. https://doi.org/10.1002/etc.4323

Alligant, S., Tassin, B. and Gasperi, J. (2018) Microplastic contamination in the Seine River estuary. In *Sixth International Conference on Estuaries and Coasts (ICEC-2018)*, Caen, France. https://hal-enpc.archives-ouvertes.fr/hal-01873428

Anderson, Z.T., Cundy, A.B., Croudace, I.W., Warwick, P.E., Celis-Hernandez, O. and Stead, J.L. (2018) A rapid method for assessing the accumulation of microplastics in the sea surface microlayer (SML) of estuarine systems. *Scientific Reports* 8: 9428. https://doi.org/10.1038/s41598-018-27612-w

Andrady, L.A. (2011) Microplastics in the marine environment. *Marine Pollution Bulletin* 62: 1596–1605. https://doi.org/10.1016/j.marpolbul.2011.05.030

Barletta, M., Lima, A.R.A. and Costa, M.F. (2019) Distribution, sources and consequences of nutrients, persistent organic pollutants, metals and microplastics in South American estuaries. *Science of The Total Environment* 651: 1199–1218. https://doi.org/10.1016/j.scitotenv.2018.09.276

Beaumont, N.J., Aanesen, M., Austen, M.C., Börger, T., Clark, J.R., Cole, M., Hooper, T., Lindeque, P.K., Pascoe, C., Wyles, K.J. (2019) Global ecological, social and economic impacts of marine plastic. *Marine Pollution Bulletin* 142: 189–95. https://doi.org/10.1016/j.marpolbul.2019.03.022

Bergmann, M., Mützel, S., Primpke, S., Tekman, M.B., Trachsel, J. and Gerdts, G. (2019) White and wonderful? Microplastics prevail in snow from the Alps to the Arctic. *Science Advances* 14 August 2019: eaax1157. https://doi.org/10.1126/sciadv.aax1157

Browne, M., Crump, P., Niven, S.J., Teuten, E., Tonkin, A., Galloway, T. and Thompson, R. (2011) Accumulation of microplastic on shorelines worldwide: sources and sinks. *Environmental Science and Technology* 45 (21): 9175–9. https://doi.org/10.1021/es201811s

Burns, E.E. and Boxall, A.B.A. (2018) Microplastics in the aquatic environment: evidence for or against adverse impacts and major knowledge gaps. *Environmental Toxicology and Chemistry* 37 (11): 2776–96. https://doi.org/10.1002/etc.4268

Cózar, A., Echevarría, F., González-Gordillo, J.I., Irigoien, X., Úbeda, B., Hernández-León, S., Palma, Á.T., Navarro, S., García-de-Lomas, J., Ruiz, A., Fernández-de-Puelles, M.L. and Duarte, C.M. (2014) Plastic debris in the open ocean. *Proceedings of the National Academy of Sciences* 111 (28): 10239–44. https://doi.org/10.1073/pnas.1314705111

Davidson, T.M. (2012) Boring crustaceans damage polystyrene floats under docks polluting marine waters with microplastic. *Marine Pollution Bulletin* 64: 1821–8. https://doi.org/10.1016/j.marpolbul.2012.06.005

Dris, R., Gasperi, J., Mirande, C., Mandin, C., Guerrouache, M., Langlois, V. and Tassin, B. (2017) A first overview of textile fibers, including microplastics, in indoor and outdoor environments. *Environmental Pollution* 221: 453–8. https://doi.org/10.1016/j.envpol.2016.12.013

Dris, R. et al. (2020) Plastic debris fowing from rivers to oceans: the role of the estuaries as a complex and poorly understood key interface. in T. Rocha-Santos, M. Costa and C. Mouneyrac (eds) *Handbook of Microplastics in the Environment*, pp. 1–28. Cham: Springer International Publishing. https://doi.org/10.1007/978-3-030-10618-8_3-1

Eriksen, M., Lebreton, L.C.M., Carson, H.S., Thiel, M., Moore, C.J., Borerro, J.C., Galgani, F., Ryan, P.G. and Reisser, J. (2014) Plastic pollution in the world's oceans: more than 5 trillion plastic pieces weighing over 250,000 tons afloat at sea. *PLoS ONE* 9 (12): 1–15. https://doi.org/10.1371/journal.pone.0111913

Everaert, G. et al. (2018) Risk assessment of microplastics in the ocean: modelling approach and first conclusions. *Environmental Pollution* 242 (Part B): 1930–8. https://doi.org/10.1016/j.envpol.2018.07.069

FAO (Food and Agriculture Organization of the United Nations) (Eric Gilman, Francis Chopin, Petri Suuronen and Blaise Kuemlangan) (2016). Abandoned, lost and discarded gillnets and trammel nets: methods to estimate ghost fishing mortality, and the status of regional monitoring and management. FAO Fisheries and Aquaculture Technical Paper No. 600. Rome, Italy.

Geyer, R., Jambeck, J.R. and Law, K.L. (2017) Production, use and fate of all plastics worldwide. *Science Advances*. https://doi.org/10.1126/sciadv.1700782

Gottschalk, F., Sun, T.Y. and Nowack, B. (2013) Environmental concentrations of engineered nanomaterials: review of modeling and analytical studies. *Environmental Pollution* 181: 287–300. https://doi.org/10.1016/j.envpol.2013.06.003

Haave, M. et al. (2021) Documentation of microplastics in tissues of wild coastal animals. *Frontiers in Environmental Science* 9. https://doi.org/10.3389/fenvs.2021.575058

Horton, A. et al. (2017) Microplastics in freshwater and terrestrial environments: Evaluating the current understanding to identify the knowledge gaps and future research priorities. *Science of The Total Environment* 586: 127–41. https://doi.org/10.1016/j.scitotenv.2017.01.190

Jambeck, J.R., Geyer, R., Wilcox, C., Siegler, T.R., Perryman, M., Andrady, A., Narayan, R. and Lavender Law, K. (2015) Plastic waste inputs from land into the ocean. *Science* 347 (6223): 768–771. https://doi.org/10.1126/science.1260352

Karasik, R., Vegh, T., Diana, Z., Bering, J., Caldas, J., Pickle, A., Rittschof, D. and Virdin, J. (2020) 20 years of government responses to the global plastic pollution problem: the Plastics Policy Inventory. NI X 20–05. Duke University, Durham, NC.

Koelmans, A.A., Kooi, M., Law, K. and van Sebille, E. (2017) All is not lost: deriving a top-down mass budget of plastic at sea. *Environmental Research Letters* 12: 11. https://doi.org/10.1088/1748-9326/aa9500

Krelling, A.P. and Turra, A. (2019) Influence of oceanographic and meteorological events on the quantity and quality of marine debris along an estuarine gradient. *Marine Pollution Bulletin* 139: 282–98. https://doi.org/10.1016/j.marpolbul.2018.12.049

Lebreton, L.C.M. et al. (2017) River plastic emissions to the world's oceans. *Nature Communications* 8 (1): 15611. https://doi.org/10.1038/ncomms15611

Lebreton, L. et al. (2018) Evidence that the Great Pacific garbage patch is rapidly accumulating plastic. *Science Reports* 8: 4666. https://doi.org/10.1038/s41598-018-22939-w

Lebreton, L.C., Greer, S.D. and Borrero, J.C. (2012) Numerical modelling of floating plastic debris in the world's oceans. *Marine Pollution Bulletin* 64: 653–61. https://doi.org/10.1016/j.marpolbul.2011.10.027

Lindeque, P.K. et al. (2020) Are we underestimating microplastic abundance in the marine environment? A comparison of microplastic capture with nets of different mesh-size. *Environmental Pollution* 265 (Part A). https://doi.org/10.1016/j.envpol.2020.114721

Morritt, D., Paris, V.S., Pearce, D., Crimmen, O.A. and Clark, P.F. (2014) Plastic in the Thames: a river runs through it. *Marine Pollution Bulletin* 78 (1–2): 196–200. https://doi.org/10.1016/j.marpolbul.2013.10.035

Nelms, S.E., Eyles, L., Godley, B.J., Richardson, P.B., Selley, H., Solandt, J.-L. and Witt, M.J. (2020) Investigating the distribution and regional occurrence of anthropogenic litter in English marine protected areas using 25 years of citizen-science beach clean

data. *Environmental Pollution* 263 (Part B): 114365. https://doi.org/10.1016/j.envpol.2020.114365

Pedrotti, M.L., Petit, S., Elineau, A., Bruzaud, S., Crebassa, J.-C., Dumontet, B., Martí, E., Gorsky, G. and Cózar, A. (2016) Changes in the floating plastic pollution of the Mediterranean Sea in relation to the distance to land. *PLoS ONE* 11 (8): 1–14. https://doi.org/10.1371/journal.pone.0161581

Pinheiro, L.M., Agostini, V.O., Lima, A.R.A., Ward, R.D. and Pinho, G.L.L. (2021) The fate of plastic litter within estuarine compartments: An overview of current knowledge for the transboundary issue to guide future assessments. *Environmental Pollution* 279: 116908. https://doi.org/10.1016/j.envpol.2021.116908

Prata, J.C., da Costa, J.P., Duarte, A.C. and Rocha-Santos, T. (2019) Methods for sampling and detection of microplastics in water and sediment: a critical review. *Trends in Analytical Chemistry* 110 (January): 150–9. https://doi.org/10.1016/j.trac.2018.10.029

Prata, J.C., Reis, V., da Costa, J.P., Mouneyrac, C., Duarte, A.C. and Rocha-Santos, T. (2021) Contamination issues as a challenge in quality control and quality assurance in microplastics analytics. *Journal of Hazardous Materials* 403: 123660. https://doi.org/10.1016/j.jhazmat.2020.123660

SACEP (South Asia Co-operative Environment Programme) (2018) Towards litter free Indian Ocean, summary of the Regional Marine Litter Action Plan for South Asian Seas Region. Colombo, SACEP.

Schubel, J.R. and Carter, H.H. (1984) The estuary as a filter for fine-grained suspended sediment. In *The Estuary as a Filter*, pp. 81–105. Elsevier. https://doi.org/10.1016/B978-0-12-405070-9.50011-6

Sherrington, C. (2016) *Plastics in the Marine Environment*. Bristol, Eunomia.

Thiele, C.J., Hudson, M. and Russell, A. (2019). Evaluation of existing methods to extract microplastics from bivalve tissue: adapted KOH digestion protocol improves filtration at single-digit pore size. *Marine Pollution Bulletin* 142: 384–93. https://doi.org/10.1016/j.marpolbul.2019.03.003

Thiele, C.J., Hudson, M., Russell, A.E., Saluveer, M. and Sidaoui-Haddad, G. (2021). Microplastics in fish and fishmeal: an emerging environmental challenge? *Scientific Reports* 11 (1): 2045. https://doi.org/10.1038/s41598-021-81499-8

Van Sabille, E. et al. (2015) A global inventory of small floating plastic debris. *Environmental Research Letters* 10. https://doi.org/10.1088/1748-9326/10/12/124006

Vethaak, A.D. and Legler, J. (2021) Microplastics and human health. *Science* 371: 672–4. https://doi.org/10.1126/science.abe5041

Welden, A.W. and Cowie, P.R. (2017) Degradation of common polymer ropes in a sublittoral marine environment. *Marine Pollution Bulletin* 118 (1–2): 248–53. https://doi.org/10.1016/j.marpolbul.2017.02.072

Estuarine Tidal Freshwater Zones in a Changing Climate: Meeting the Challenge of Saline Incursion and Estuarine Squeeze

SALLY LITTLE, JONATHAN P. LEWIS, HELEN PIETKIEWICZ and KRYSIA MAZIK

Abstract

The environmental effects of climate change (e.g. global sea-level rise, increasing temperatures and changing precipitation patterns) and other human activities in the catchment (e.g. over-abstraction of freshwater, changes to river flow and land-use change) are myriad; however, to date, the ecological impacts of increasing saline incursion on freshwater coastal areas (through surface waters) remain poorly understood. This is particularly the case for estuarine tidal freshwater zones. The tidal freshwater zone is an aquatic transition zone that forms between the non-tidal freshwater river and tidal brackish waters in meso- and macrotidal estuaries with high river flow. Projected saline incursion into these estuarine freshwater zones is likely to result in habitat reduction or loss (termed here 'estuarine squeeze'), changes in ecosystem structure and functioning, and loss of key ecosystem services and societal goods and benefits. Here we highlight the importance of tidal freshwater zones and review the potential impacts of climate change, rising sea levels and catchment changes as drivers of saline incursion into these systems. We identify future challenges and research priorities that need to be addressed in order to inform and guide estuarine ecosystem-based management decisions globally. We stress that to appropriately implement the holistic ecosystem approach to estuarine management and policy, the critically important tidal freshwater section must be included.

Keywords: tidal freshwater, saline incursion, estuarine squeeze, climate change, estuary

Correspondence: sally.little@ntu.ac.uk

Introduction

Estuaries face unprecedented pressures from the environmental effects of climate change (e.g. increasing temperatures, changing precipitation patterns and global sea-level rise) and other human activities in the sea, channel and catchment (e.g. over-abstraction of freshwater, changes to river flow and land-use change) (Kennish 2002; Robins et al. 2016; Oppenheimer et al. 2019; Wolanski et al. 2019). Researching, monitoring and mitigating the

Sally Little, Jonathan P. Lewis, Helen Pietkiewicz and Krysia Mazik, 'Estuarine Tidal Freshwater Zones in a Changing Climate: Meeting the Challenge of Saline Incursion and Estuarine Squeeze' in: *Challenges in Estuarine and Coastal Science*. Pelagic Publishing (2022). © Sally Little, Jonathan P. Lewis, Helen Pietkiewicz and Krysia Mazik. DOI: 10.53061/VCCR7195 impacts of these pressures on ecosystem structure and functioning is relatively advanced in the brackish upper, middle and lower estuary, where the majority of estuarine research over the decades has been focused. In contrast, far less is known about the impacts of these pressures on the uppermost estuarine zones, which are subject to tidal action, but are beyond the limit of saline incursion (Odum 1988; Schuhardt et al. 1993). These tidal freshwater zones (TFZs) (also known as tidal freshwater areas and tidal freshwater reaches; McLusky 1994) are particularly prevalent in mesotidal (1–4 m) and macrotidal (>4 m) estuaries that are subject to high river flow (e.g. Schuchardt and Schirmer, 1991; Muylaert and Sabbe, 1999; Little, 2012).

The TFZ is part of the estuary, rather than the river, as it falls within the tidal limits (Dionne (1963), in Fairbridge (1980)) and is defined here as the freshwater (or limnetic <0.5) section of an estuary, which occurs between the upstream extent of saline incursion and the limit of tidal rise (delimiting the downstream and upstream TFZ boundaries respectively), as classified by McLusky (1994) (Fig 7.1). This zone is, however, often omitted from fluvial studies owing to its tidal influence, and from estuarine studies owing to the lack of salt and presence of freshwater fauna (e.g. Odum 1988; Schuhardt et al. 1993; Attrill, Rundle and Thomas 1996; Rundle, Attrill and Arshad 1998; Sousa, Guilhermino and Antunes 2005), even though in some estuaries the TFZ can constitute a substantial proportion of the estuarine ecosystem. For example, the TFZ of the Columbia River Estuary, USA, is 233 km long, three-quarters of the estuary's total length (Simenstad et al. 2011).

The specific environmental conditions and ecology of the TFZ, and its role in linking terrestrial, river and estuarine systems, may mean that it plays an important role in the structure and functioning of the estuarine ecosystem as a whole (Schuhardt et al. 1993; Williams and Williams 1998; Lehman 2007; Xu et al. 2021). This includes biogeochemical cycling, food-web function and the provision of trophic subsidies to the brackish estuary, but a lack of research into this zone means that our understanding is limited (Muylaert et al. 2005). It is clear, however, that the TFZ is at impending risk of loss, particularly owing to climate-driven increases in sea level and reductions in river flow, which will act to salinise these freshwater zones through saline incursion (Prandle and Lane 2015; Little,

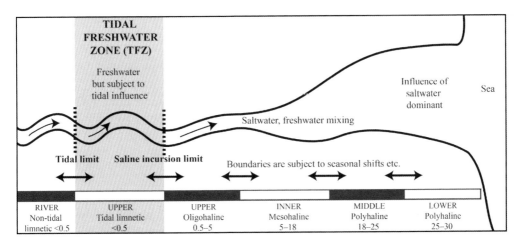

Figure 7.1 An example of an unbounded temperate estuary, showing the entire salinity transition from the sea to the river, with special focus on the tidal freshwater zone (shaded in grey), i.e. the uppermost tidal freshwater (salinity <0.5) section of the estuary located between the tidal limit and the limit of saline incursion. This estuary is divided into zones as classified by McLusky (1994), based on the definition by Dionne (1963, in Fairbridge 1980) and the salinity zones of the Venice System (1959). Schematic modified from Park (1999).

Wood and Elliott 2017; Rodrigues, Fortunato and Freire 2019; Jones et al. 2020). The need for TFZ research is therefore pressing because if, as postulated, these zones play a key role in the functioning of the estuarine ecosystem, then their loss owing to saline incursion and estuarine squeeze could be detrimental for the ecosystem services and societal goods and benefits they provide (e.g. Tully et al. 2019).

This chapter highlights the importance of TFZs as unique zones within estuaries, which support distinctive and productive communities and contribute to the functioning of the wider estuarine ecosystem (Williams and Williams 1998; Seys, Vincx and Meire 1999; Muylaert et al. 2005; McLachlan, Haghkerdar and Greig 2019). We discuss major challenges facing TFZs, with a focus on the potential impacts and consequences of climate change, rising sea levels and catchment changes as drivers of freshwater salinisation and estuarine squeeze (Prandle and Lane 2015; Little, Wood and Elliott 2017; Jones et al. 2020). Our focus here is on temperate estuaries (owing to similarities of environmental conditions), but it is important to recognise that these zones occur in estuaries all around the world.

The tidal freshwater zone

Temperate TFZs typically look more like rivers than estuaries (Fig. 7.1), with narrower steeper-sided channels, narrow intertidal zones and often fringing reedbeds of *Phragmites australis* and *Phalaris arundinacea* (Pihl et al. 2002). It is therefore unsurprising that TFZs have classically been regarded as riverine stretches, as they are also freshwater habitats, with some freshwater fauna and largely river water chemistry (Jones et al. 2020). The TFZ is, however, subject to a number of specific physicochemical processes that distinguish these zones from both rivers and the brackish sections of estuaries (e.g. Schuhardt et al. 1993; Ysebaert et al. 1998; Muylaert, Sabbe and Vyverman 2000; Elliot and McLusky 2002; McLachlan, Haghkerdar and Greig 2019; Jones et al. 2020). These include tidal physics (oscillating water levels and bidirectional currents), long water residence times, high turbidity and low dissolved oxygen content. They are also hotspots of biogeochemical cycling, particularly for the processing, recycling and retention of key nutrients (C, N, P and Si) delivered from the terrestrial catchment (Knights et al. 2017; Xu et al. 2021) in addition to the complex biogeochemical reactions that occur at the downstream boundary of this zone where freshwater and saltwater meet (the freshwater–seawater interface (Morris et al. 1978). The substrate commonly ranges from muds (i.e. fine silts) through to coarse gravels, though low sediment supply or sediment scouring can result in areas of hard substrate (Pihl et al. 2002; Chen et al. 2005; Flemming 2011). These conditions make it a unique and particularly harsh environment in which to live, and TFZ inhabitants must be able to cope with, or at least have strategies for dealing with, these conditions, which creates a distinct and specialised flora and fauna (Odum 1988; Williams and Williams 1998; Seys, Vincx and Meire 1999; Little, Wood and Elliott 2017; McLachlan, Haghkerdar and Greig 2019).

The TFZ is generally considered to be net heterotrophic, with ecosystem metabolism being predominately fuelled by the breakdown of allochthonous organic matter (particulate and dissolved organic matter; POM and DOM) sourced from the catchment and transported via river flow (Van den Meersche et al. 2009). The zones are, however, also areas of high autochthonous productivity with the TFZ regularly recording higher levels of primary productivity than the adjoining river and brackish estuary (Schuchardt and Schirmer 1991; Muylaert et al. 2005). These zones support a diverse range of primary producers including phytoplankton (particularly diatoms, Chlorophyceae and cyanobacteria), emergent and submerged macrophytes (and associated epiphytic algae), macroalgae and benthic microalgae and riparian terrestrial vegetation (Grimaldo et al. 2009). In some TFZs, over the productive period (spring to autumn) multiple chlorophyll-a peaks are recorded, with the first usually occurring in spring/early summer, and additional peak(s) later in summer and autumn (Schuchardt and Schirmer 1991; Muylaert et al. 2005). Pelagic diatoms are frequently the most abundant phytoplankton in these areas (e.g. Schuchardt and Schirmer 1991; Carbonnel et al. 2009) particularly during spring, where they rapidly utilise dissolved silica and nutrients delivered by the river via high winter flows (Rocha et al. 2002; Carbonnel et al. 2009).

High allochthonous POM and DOM input and autochthonous productivity mean large quantities of organic matter are available to decomposers, zooplankton, detritus-feeding invertebrates and other estuarine consumers. Detritus dominates the available POM in the TFZ, with phytoplankton believed to generally comprise <10% of the total POM available to consumers (Deegan and Garritt 1997; Hughes et al. 2000; Chanton and Lewis 2002; Sobczak et al. 2005). However, phytoplankton have a higher nutritional value than detritus, and have been shown to be an important source of POM for the pelagic food web and upper trophic levels (Hoffman, Bronk and Olney 2008). The importance and relative contributions of allochthonous and autochthonous organic matter to the food web will depend on the physical characteristics of the TFZ (e.g. depth, turbidity, tidal energy, river flow and residence time) and seasonality (e.g. the dieback of emergent vegetation in autumn and winter, spring/summer phytoplankton blooms). During low river flow periods, longer residence times means that autochthonous phytoplankton biomass can accumulate (Hoffman and Bronk 2006). Conversely, during high flow periods, shorter residence times can mean autochthonous populations are unable to establish, though allochthonous OM to the TFZ can increase, augmenting autochthonous organic matter supplies (Hoffman and Bronk 2006).

Pelagic and littoral/benthic food webs in the TFZ are likely interdependent, which may enhance their resilience and help buffer environmental variability (Young et al. 2021). Zooplankton, for example, feed on emergent and submerged aquatic vegetation when high river flows reduce phytoplankton biomass (Young et al. 2021). In shallow water habitats, fish often display more generalist feeding behaviours, and as such are reliant on prey that consume a variety of primary producers in the TFZ (Young et al. 2021) and this can vary based on life stage. For instance, Flounder *Platichthys flesus* migrate into the TFZ during larval development where they feed on zooplankton; however, during juvenile development they predate on the benthic macrofauna (Williams and Williams 1998; Dias et al. 2020). Zooplankton in the TFZ are generally dominated by rotifers and crustacea (particularly copepods and cladocerans). Rotifers can reach extremely high abundances (e.g. Muylaert et al. 2000b; Park and Marshall 2000; Tackz et al. 2004), particularly in spring, when they often peak, being the first zooplankters to have a grazing impact on the spring phytoplankton bloom, while copepods and cladocerans often peak later in summer and/ or autumn (Muylaert et al. 2000b; Connelly et al. 2020).

Benthic macroinvertebrates are an integral part of the TFZ food web, being one of the most important primary consumers and fulfilling key ecosystem functions (e.g. detrital decomposition, sediment mixing, nutrient cycling and energy flow as predators/prey; Covich, Palmer and Crowl 1999). The TFZ macrobenthos community is distinct, consisting of both brackish- and freshwater-derived species that are tolerant of the specific environmental conditions (Little, Wood and Elliott 2017). These species are a mix of hydrological generalists, species that could potentially inhabit adjacent river and brackish areas (but may be outcompeted) and taxa not found in other freshwater habitats locally (Sousa, Antunes and Guilhermino 2007; Little, Wood and Elliott 2017). Species richness and diversity is lower than the adjoining non-tidal river, and some taxa such as oligochaetes and chironomids (e.g. *Limnodrilus hoffmeisteri*, *Tubifex tubifex*), occur in high abundance and biomass (Williams and Williams 1998; Dias et al. 2020). Brackish- and freshwater-derived amphipods (e.g. *Gammarus pulex*, *G. zaddachi*, *Corophium multisetosum*), isopods (e.g. *Asellus aquaticus*, *Cyathura carinata*) and gastropod and bivalve molluscs are also common. Many of the latter are non-indigenous (some now naturalised) and can dominate the macrobenthos biomass (i.e. Asian Clam *Corbicula fluminea*, New Zealand Mudsnail *Potamopyrgus antipodarum*, Zebra Mussel *Dreissena polymorpha*; Strayer and Smith 2001; Sousa, Antunes and Guilhermino 2007). Where suitable sediment and dissolved oxygen conditions allow, freshwater insect nymphs of the orders Plecoptera, Ephemeroptera, Trichoptera and Megaloptera also occur along with coleopterans, triclads and other dipterans (Williams and Williams 1998; Little, Wood and Elliott 2017). The macrobenthos form the main food item for many bird and fish species, at various life stages, and are an important trophic link to adjacent terrestrial and aquatic zones. For example, during juvenile development in the TFZ, the mean wet weight of Flounder in the Aber estuary increased more than 100 times (from 5 to 540 mg) over a 30–5 week period on a diet of freshwater chironomid larvae (Williams and Williams 1998).

The TFZ is important for many economically, recreationally and ecologically important fish species, providing spawning grounds, nursery habitats, foraging resources and pathways for diadromous (catadromous or anadromous) migrations (Pihl et al. 2002; Kraus and Secor 2005;). While many of these species might not be permanent residents, perhaps just passing through (e.g. McIvor and Odum 1988; Schuhardt et al. 1993; Elliott et al. 2007; Le Pichon et al. 2017), this habitat might be of importance to up to 50% of fish species that utilise estuaries at some stage in their lifecycle (Pihl et al. 2002). These include threatened species such as the catadromous Eel *Anguilla anguilla*, the anadromous Salmon *Salmo salar* and River Lamprey *Lampetra fluviatilis* (Masters et al. 2006; Van Liefferinge et al. 2012; Wilson, Giltrap and Kelly 2016; Le Pichon et al. 2017). In European estuaries, it has been estimated that on average 28 fish species utilise the TFZ for feeding, 10 for spawning, 8 for diadromous migrations and 19 as nursery grounds (Pihl et al. 2002). These consist of marine, estuarine, diadromous (anadromous, catadromous, semi-catadromous and amphidromous) and freshwater species (Potter et al. 2015), with Flounder and Sand Goby *Pomatoschistus minutus* often the most commonly recorded (Williams and Williams 1998; Maes, Stevens and Ollevier 2005; Breine et al. 2009; Wilson, Giltrap and Kelly 2016). This community composition is distinctly different from the adjoining river and brackish estuary, despite some inevitable overlap (e.g. Odum, Rozas and McIvor 1987; Vetemaa et al. 2006; Breine et al. 2009; Wilson, Giltrap and Kelly 2016).

Externally, the TFZ provides an important trophic link to adjacent terrestrial and aquatic zones, providing trophic subsidies in the form of organisms, organic matter and nutrients, and as such is a critical facilitator of biological connectivity in estuaries. Organic matter exported from the TFZ to the brackish portion of the estuary is consumed and incorporated into the food web (Dias et al. 2016). For example, in the Minho estuary (north-west Iberian Peninsula) in August 2011, POM (largely comprised of phytoplankton) from the TFZ subsidised up to 80% of consumers' biomass in the lower estuary (Dias et al. 2016). In the Aber estuary, an estimated 62.6 kg of freshwater invertebrates were exported into the brackish estuary for consumers per annum via downstream drift, compared with just 2.5 kg carried upstream by incoming tides (reverse drift) (Williams and Williams 1998). Aquatic insect emergence from the TFZ (predominantly chironomids) also provides a critical nutritional subsidy to riparian consumers (i.e. spiders, bats and birds) with emergence rates comparable to that of non-tidal rivers (albeit at the lower end of the range; Kautza and Sullivan 2016; Zapata and Sullivan 2018).

Future threats: drivers of saline incursion into the TFZ

As the nexus between catchment, estuary and coast, the TFZ is subject to a myriad of pressures (Neumann et al. 2015; Robins et al. 2016; Oppenheimer et al. 2019). These pressures do not occur in isolation; they are linked and interact, leading to cumulative, synergistic and antagonistic effects on the TFZ over a range of spatial and temporal scales

(Scavia et al. 2002; Lotze et al. 2006; Little et al. 2017). These include exogenic pressures such as climate change (including sea-level rise, temperature and precipitation changes and extreme weather events), and endogenic pressures such as pollution and human modification of the catchment and channel (Elliott 2011). Here, we focus on saline incursion and the resultant estuarine squeeze, which is driven by reductions in river flow and rising sea levels (Fig 7.2) (Little, Wood and Elliott 2017; Ensign and Noe 2018). This is an important, recurring threat, which is difficult to isolate, being both a consequence of climate change and other human pressures on the channel and catchment (Fig. 7.3) (Ross et al. 2015; Little, Wood and Elliott 2017).

Exogenic pressures: climate-driven changes

Global sea-level rise

Global sea levels are predicted to rise between 0.29 and 1.10 m by 2100 (Oppenheimer et al. 2019); however, a rise of 2 m by 2100 cannot be ruled out owing to uncertainties over the contribution of the Greenland and Antarctic ice sheets (Oppenheimer et al. 2019). As nearshore waters deepen, the volume of tidal water entering the estuary channel will increase, increasing the depth and decreasing frictional drag of the tidal wave in outer estuarine channels (Talke and Jay 2020). Converging banks will act to propagate the tidal wave upstream, increasing tidal amplitude, current speeds and salinity in inner and upper estuarine channels, and driving saltwater into TFZs and formerly non-tidal rivers (Prandle and Lane 2015; Ensign and Noe 2018).

The most significant effects will occur where land level is sinking (e.g. through glacio-isostatic adjustment and geological subsidence; Shennan and Horton 2002), thereby contributing to increases in sea levels relative to the land (relative sea-level rise; RSLR) (Nicholls and Klein 2005). Sea-level rise will also increase the intensity of extreme sea-level events (Church et al. 2013; Tully et al. 2019; Talke and Jay 2020). Church et al. (2013) noted that a 0.5 m rise in mean sea level will result in a dramatic increase in the frequency of high-water extremes by an order of magnitude, or more in some regions, increasing the frequency and intensity of saline incursion pulse events.

The physical responses of temperate estuaries to sea-level rise will vary considerably owing to differences in estuary morphology, hydrology and sedimentary characteristics, and non-uniform local factors such as uplift, subsidence, river flow, storm surge frequency and magnitude, tidal factors and sedimentation rates (Prandle and Lane 2015; Khojasteh et al. 2021). Anthropogenic modification (i.e. channelisation) will exacerbate natural responses to sea level rise in intensively modified estuaries (Talke and Jay 2020). Despite these differences, it is generally accepted that where uninhibited, estuaries will naturally shift landward in response to sea-level rise (Ensign and Noe 2018).

River flow

The upstream extent of saline incursion in estuaries is, in part, dependent upon downstream river flow, opposing the upstream force of the tide (Prandle 2009; Little, Wood and Elliott 2017; Rodrigues, Fortunato and Freire 2019). River flow is primarily determined by catchment run-off, which is in turn determined by climate (i.e. precipitation and temperature) and the characteristics of and human activities within the catchment (e.g. land-use, water storage and abstraction) (O'Briain 2019).

In north-west Europe, rivers generally have a sustained flow, but with seasonal peaks (e.g. in winter/spring) following heavy precipitation and occasional snow melt events (Whitehead et al. 2009). Future climate projections for Europe include warmer, wetter winters and hotter, drier summers with more frequent high intensity extreme weather events (Arnell 2003; Christierson et al. 2012; Vautard et al. 2014). These are expected to result in seasonal shifts in river flow, including more frequent high and low flow periods and reductions in groundwater recharge (Murphy et al. 2009). In the UK, river catchment models project winter increases (by up to 25%) and summer decreases in flow (by 40–80%), dependent on catchment location, geology, land-use, soils and model uncertainty (Christierson et al. 2012; Fowler and Wilby 2010; Prudhomme et al. 2012). Decreasing trends will be exacerbated by human pressures (Vörösmarty et al. 2000; Johnson et al. 2009) and it is projected that increased abstraction (for domestic, agricultural and urban use), owing to higher temperatures and reduced reservoir levels, may considerably exceed any future effects of climate change on river flow regimes (Vörösmarty et al. 2010).

Endogenic pressures: catchment land-use and human modification of channels

Intense urban, industrial and agricultural development on upper estuarine floodplains has resulted in TFZ channels that have been heavily modified for navigation and flood defence, profoundly impacting estuarine morphodynamics (Kennish 2002; Zonneveld and Barendgret 2009). Channels have been artificially deepened and narrowed, with steep-sided flood banks removing connection to the flood plain and restricting the extent of intertidal areas (Talke and Jay 2020). Removal of these flood storage areas and reducing the cross-sectional area of the channel increases tidal flow velocity, funnelling and propagating the tidal wave upstream, acting to increase saline incursion (Schuhardt et al. 1993). Here we use the term incursion rather than intrusion to distinguish between the salinisation of coastal surface and groundwater systems (the latter concerned with freshwater aquifer salinisation).

In Europe, the majority of estuaries are bounded at their upper limits by in-channel engineering structures (e.g. barrages, weirs, dams, sluices and gates) to prevent tidal flooding by regulating and/or restricting flow (Little, Wood and Elliott 2017; van Puijenbroek et al. 2019). These structures fragment river and estuarine channels, restricting movement of fauna between zones and habitats and changing the environmental conditions (Kukulka and Jay 2003; Simenstad et al. 2011; van Puijenbroek et al. 2019). Van Puijenbroek et al. (2019) found that out of 33 large European rivers, only two were still free flowing to the sea, with a loss of anadromous fish species coinciding with the decrease in accessibility.

Saline incursion and estuarine squeeze

Where the upper estuary is bounded (e.g. by dams or weirs), future SLR and river flow scenarios risk 'squeezing out' the TFZ (Little, Wood and Elliott 2017). RSLR and reductions in river flow will increase the proportion of estuarine haline zones (oligo-, meso- and poly-) and drive saltwater into the TFZ (Little 2012). This will be exacerbated by sustained or pulsed salinity increases driven by the combination of droughts and extreme high-water events (i.e. storm surges) and human modifications to channel morphology (i.e. channelisation) (Talke and Jay 2020). In an unmodified natural system, the estuarine transition (of which the TFZ is part) would simply shift upstream into the currently non-tidal river in response to SLR (Ensign and Noe 2018), but in bounded estuaries the upstream incursion of the tide is blocked. The TFZ will therefore be squeezed out by the upstream incursion of saltwater, resulting in the degradation and reduction or loss of these tidal freshwater habitats. We have termed this 'estuarine squeeze' (Fig. 7.2).

Prandle and Lane (2015) postulated that 1 m SLR will likely have a significant effect on half of all UK estuaries, with saline incursion increasing by more than 25% in estuaries with depths <10 m. Changing river flows of 25% (either increase or decrease) had significant effects on both vertical mixing and saline incursion extent. Little, Wood and Elliott (2017)

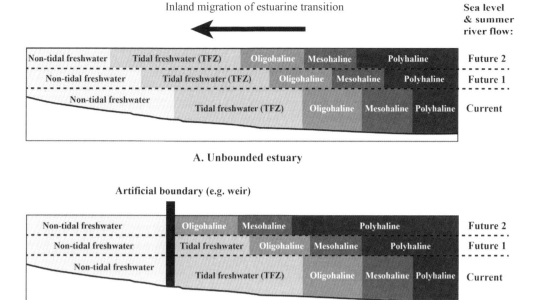

Figure 7.2 Estuarine squeeze schematic: hypothetical spatial change of estuarine salinity zones (based on the Venice System) in response to increased sea level and decreased summer river flow. In (a), an unbounded system, the estuarine transition naturally migrates inland through time (Futures 1 and 2), but where inland migration is prevented by a physical barrier, as in (b), squeezing of the upper estuarine tidal freshwater zone occurs (Future 1), potentially leading to complete loss of the zone (Future 2).

observed that just small differences between summer and winter flows (0.51 and 0.71 $m^3 s^{-1}$), resulted in large differences in saline incursion extents in the Adur and Ouse estuaries in Sussex, UK (3.7 km and 1.6 km respectively). Applying projected modelled reductions of 32% (as per Romanowicz et al. 2006) to the summer river flow of the River Adur, resulted in over five times the increase in saline incursion extent predicted for RSLR alone (from 0.32 km to 2 km for a 67 cm RSL). In the Tagus Estuary in Portugal, the impact of drought on river flow had a greater impact than sea level on saline incursion, with increased salinity into the TFZ correlated most strongly with duration of drought conditions (Rodrigues, Fortunato and Freire 2019).

The impacts of estuarine squeeze may therefore be seen more rapidly than would be expected for SLR alone, as many river basins in Europe already suffer from unsustainable extraction regimes and experience some degree of water stress (i.e. the proportion of water withdrawal with respect to total renewable resources) during summer months (Rodda 2006; Herrera-Pantoja and Hiscock 2008). Severe water stress in river basins in Europe is forecasted to increase under future climate scenarios, along with an increasing number of people living in water-stressed areas (Lehner et al. 2001; Schröter et al. 2005).

Response: potential ecological consequences of increasing saline incursion and estuarine squeeze

Salinity is considered the dominant environmental variable determining floral and faunal communities in estuaries, with species distributions along the estuarine transition primarily determined by salinity tolerance (De Jonge 1974; Attrill 2002; Telesh and Khlebovich 2010).

Large-scale shifts in estuarine salinity structure, particularly under low flow (drought) conditions, are likely to have serious deleterious effects upon TFZ flora and fauna across all trophic levels, altering estuarine food webs through resource loss, competition for resources (inter and intra-specific) and changing predator and prey interactions (Attrill, Rundle and Thomas 1996; Muylaert and Sabbe 1999; Martinho et al. 2007; Bessa et al. 2010; Kasai et al. 2010).

Where unbounded, the flora and fauna of the estuarine transition may naturally shift upstream following the extension of their associated salinity zone (Fig. 7.3). Benthic macroinvertebrates, for instance, are considered particularly sensitive and rapid indicators of salinity change, and significant saline incursion events into the TFZ (e.g. owing to droughts) have resulted in the downstream loss of freshwater fauna and upstream shift of mobile marine and brackish species (Andrews 1977; Attrill, Rundle and Thomas 1996; Bessa et al. 2010). However, while the estuarine transition will shift upstream with increasing saline incursion (Little 2012; Ensign and Noe 2018; Rodrigues, Fortunato and Freire 2019), it does not necessarily follow that there will be an associated shift in the distribution of the benthos (as per salinity tolerances *sensu* the Remane Curve; Whitfield et al. 2012). This is owing to any physiological mechanisms and/or capacity for adaptation that may allow freshwater species to survive saline incursion (Tills et al. 2010) and the response of all species to factors (biotic and abiotic) other than salinity that limit their distributions. For example in the Adur estuary in Sussex, UK, the transition from sandy mud with gravel to fine-grained silty mud in the mid-estuary was a barrier to the upstream extension of many marine and brackish species, despite the upstream extension of suitable salinity conditions (Little, Wood and Elliott 2017). Increasing saline incursion in this system would be represented by a landward extension of a low diversity/high abundance/high biomass community (typically associated with the oligohaline zone), populated by opportunistic euryhaline brackish-water species, rather than a landward shift in the set structure of the

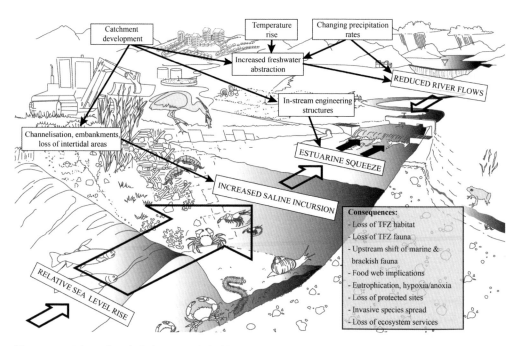

Figure 7.3 A broad-scale linkage model of how endogenic managed and exogenic (climate-driven) unmanaged pressures act together to drive saline incursion and estuarine squeeze of the tidal freshwater zone in a bounded estuary and some of the key consequences.

Remane Curve (Little 2012). In contrast, in some systems it could be expected that as the TFZ shifts upstream into non-tidal rivers, changing substrate (into fluvial sands and gravels; e.g. Williams and Williams 1998) and decreasing depth may increase the richness and diversity of these zones as habitat conditions suit additional freshwater species.

Extreme variations in saline incursion extents between summer and winter (with more frequent saline pulse events) in addition to high flow flash-flood events (Robins et al. 2016), will also act to modify the communities of the upper estuary and TFZ. This may make the distributions of mobile estuarine benthos more variable and unable to progress beyond early benthic succession as favourable conditions extend and contract (Santos et al. 1996; Ysebaert et al. 2005), although the much less variable interstitial salinities of the underlying sediment may act to buffer infauna from rapid changes in surface water salinity (Chapman 1981). However, increased freshwater abstraction in the catchment may negate or reduce any climate-driven increase in winter river flows, reducing any significant seasonal differences in saline incursion extents (Vörösmarty et al. 2000; Vörösmarty et al. 2010; Lester et al. 2011).

Distribution changes of key TFZ species may alter estuarine, riverine and terrestrial food webs through changing trophodynamics. Where estuarine amphipods (e.g. *Gammarus* sp.) constitute a large proportion of fish species' diets, shifts in the distributions and abundances of these could result in a local decline in fish populations (e.g. the reported decline in juvenile flounder in the Thames estuary since the mid-1980s; Thomas 1998). A mismatch between the location of emerging freshwater insect larvae and their terrestrial riparian consumers will occur as the TFZ shifts upstream. For obligate freshwater fish species (e.g. Chub *Squalius cephalus*, Tench *Tinca tinca*, Rudd *Scardinius erythrophthalmus*, Roach *Rutilus rutilus*), predicted increases in saline incursion might result in the restriction of their current downstream distributions at high tide or during drought conditions, dependent upon salinity tolerance, capacity for adaptation and additional environmental factors that have been shown to influence habitat use (Roessig et al. 2004; Love, Gill and Newhard 2008; Baptista et al. 2010; Dolbeth et al. 2010; Gillson 2011; Purcell, Klerks and Leberg 2010).

Low river flows and increased salinities and temperatures may also provide ideal microhabitats for the propagation of non-native species through increased larval retention and development time (Cohen et al. 1995; Herborg et al. 2005). For instance, the low flows and increased saline incursion in the Thames estuary, as a result of the 1989–92 drought, were linked to the large upstream increase of the invasive Chinese Mitten Crab *Erocheir sinesis* between 1992 and 1996 (Herborg et al. 2005). The invasive potential of non-natives also improves substantially following system disturbance, which may result from increased variability of saline incursion in the upper estuary, particularly when in combination with reduced water quality (Crooks, Chang and Ruiz 2011). The TFZ is clearly susceptible to invasion, with several non-native (some now naturalised) species dominating the macrobenthos biomass in some systems (i.e. *Corbicula fluminea*, *Potamopyrgus antipodarum*, *Dreissena polymorpha*; Strayer and Smith 2001; Cole and Caraco 2006; Sousa, Antunes and Guilhermino 2007). Invasive species may have a profound effect on estuarine ecology through predation and competition with native species, significantly altering food webs and trophic structure (Hanson and Sytsma 2008; Dittel and Epifanio 2009).

Where estuaries are bounded and estuarine squeeze occurs, it is predicted that as salinity increases the more sensitive macroinvertebrate freshwater species will be lost, while more tolerant species remain, albeit living in stressed suboptimal conditions (Muller and Mendl 1979; Williams and Williams 1998; Piscart et al. 2005; Blinn and Ruiter 2006; Williams 2009). These species will be joined by the upstream extension of mobile brackish and marine species and potentially non-native species (Crooks, Chang and Ruiz 2011). This may result in novel assemblages and reorganised communities of organisms in the upper estuary that have not previously interacted (Rius et al. 2014; Wolanski and Elliott 2015). These new community configurations may have a significant impact on functional biodiversity (e.g. Cole and Caraco 2006) and may play a key role in the provision of ecosystem services into the future. For example, macrobenthos in the TFZ play an important functional role in the breakdown of allochthonous organic matter (OM), some of which is transported to the lower estuary and utilised by consumers. Studies have shown that where there is a mismatch between detrital litter type, salinity regime and the detritivore community, decomposition rates are reduced (Franzitta et al. 2015). Allochthonous detritus from the catchment (which contributes a substantial proportion of the OM to the TFZ) decomposes much more rapidly when broken down in freshwater by detritivores in the TFZ community than in high salinity waters by marine detritivores (Franzitta et al. 2015). Loss of the TFZ community through estuarine squeeze could therefore reduce allochthonous OM decomposition in the upper estuary and alter nutrient cycling.

Where the TFZ is completely squeezed out, the loss of freshwater flora and fauna below the in-stream engineering structure will create a sharp division between upstream non-tidal freshwater and downstream brackish or marine habitats (depending on the degree of squeeze). Complete removal of the TFZ will result in the complete loss of the functions it provides, which will detrimentally impact both the brackish estuary and adjacent river and terrestrial zones (Ensign and Noe 2018).

The significance of the impact of estuarine squeeze on the structure and functioning of estuaries will be dependent on the spatial extent of the current TFZ and the rate of squeeze. Rate will depend on the bathymetry of the estuary (depth, cross-sectional area and bed gradient), catchment and channel modification, RSLR and reductions in river flow. Some estuaries are already saline up to their anthropogenic boundary (e.g. Tees, Mersey, Tyne estuaries; Wright and Worrall 2001), whereas the TFZ of others are being reduced in size, with shallower estuaries in low lying areas at most risk (Little, Wood and Elliott 2017). In some estuaries, future saline incursion may reach many kilometres inland (e.g. Grabemann et al. 2001; Bhuiyan and Dutta 2012; Hong and Shen 2012; Rice, Hong and Shen 2012; Hoitink and Jay 2016). But even where absolute distances are small (<500 m), salinity stress through bank breach or tidal flooding could result in the loss of protected coastal freshwater habitats and species (e.g. Little, Wood and Elliott 2017).

Concluding remarks: future challenges and priorities

We have highlighted the potential importance of the TFZ in estuarine ecosystem functioning and the potential ecological consequences of one of the most pressing future threats to this zone: increasing saline incursion and estuarine squeeze driven by accelerating SLR, climate change and other human impacts on the channel and catchment. Where estuaries are unbounded, the loss of TFZ ecosystem functions owing to salinisation at the downstream boundary may well be offset by the upstream tidal extension into the currently non-tidal river (Ensign and Noe 2018). However, where an estuary is bounded at its tidal limit, as in most of north-west Europe, these functions will be lost through estuarine squeeze. The future loss of these zones may require mitigation through a new tidal freshwater conservation agenda, focused on restoration through appropriate channel management and reconnecting and creating (e.g. through managed realignment) tidal freshwater floodplains (e.g. Beauchard et al. 2013; Temmerman et al. 2013). Such an approach would align with the drive for nature-based solutions for the post-2020 nature, climate and sustainable development agendas (Cohen-Shacham et al. 2016; Hobbie and Grimm 2020).

It is clear that we are only at the beginning of fully understanding the importance of the TFZ in terms of its ecological functionality and significance to the wider coastal zone (i.e. adjoining estuarine, riverine and terrestrial habitats), such as through the provision of trophic subsidies of nutrients, organisms and OM, and the ability to support high fish predator populations and overwintering birds. This lack of knowledge needs to be addressed, and quickly, or we risk losing an important part of the estuary before we can fully determine its extent and function. As such, the TFZ must be recognised as an integral part of the estuary, contributing to its functioning and ecosystem service value (~US$28,917 per ha per year for coastal systems; de Groot et al. 2012). We cannot fully realise the benefits of the integrated holistic ecosystem approach to management (CBD 2004) while still routinely ignoring the TFZ in estuarine research and management agendas. Only by understanding the role the TFZ plays in the integrated coastal network can we fully assess the impact of the loss of this zone owing to estuarine squeeze. The TFZ should be a leading part of the estuarine research agenda post-2022, and this should centre on detailed studies of the structure and functioning of this zone, improving metrics for assessment (e.g. Wilson, Dunne and Giltrap 2017) and appropriate management strategies that would identify zones at risk of degradation and loss, thus focusing conservation attention and, if necessary, intervention.

References

Andrews, M.J. (1977) Observations on the fauna of the Metropolitan River Thames during the drought of 1976. *London Naturalist* 56: 44–56.

Arnell, N.W. (2003) Relative effects of multi-decadal climatic variability and changes in the mean and variability of climate due to global warming: future streamflows in Britain. *Journal of Hydrology* 270: 195–213. https://doi.org/10.1016/S0022-1694(02)00288-3

Attrill, M.J. (2002) A testable linear model for diversity trends in estuaries. *Journal of Animal Ecology* 71: 262–9. https://doi.org/10.1046/j.1365-2656.2002.00593.x

Attrill, M.J., Rundle, S.D. and Thomas, R.M. (1996) The influence of drought-induced low freshwater flow on an upper-estuarine macroinvertebrate community. *Water Research* 30: 261–8. https://doi.org/10.1016/0043-1354(95)00186-7

Baptista, J., Martinho, F., Dolbeth, M., Viegas, I., Cabral, H. and Pardal, M. (2010) Effects of freshwater flow on the fish assemblage of the Mondego estuary (Portugal): comparison between drought and non-drought years. *Marine and Freshwater Research* 61: 490–501. https://doi.org/10.1071/MF09174

Beauchard, O., Jacobs, S., Ysebaert, T. and Meire, P. (2013) Sediment macroinvertebrate community functioning in impacted and newly-created tidal freshwater habitats. *Estuarine, Coastal and Shelf Science* 120: 21–32. https://doi.org/10.1016/j.ecss.2013.01.013

Berezina, N.A. (2003) Tolerance of freshwater invertebrates to changes in water salinity. *Russian Journal of Ecology* 34: 261–6. https://doi.org/10.1023/A:1024597832095

Bessa, F., Baeta, A., Martinho, F., Marques, S. and Pardal, M.Â. (2010) Seasonal and temporal variations in population dynamics of the *Carcinus maenas* (L.): the effect of an extreme drought event in a southern European estuary. *Journal of the Marine Biological Association of the United Kingdom* 90: 867–76. https://doi.org/10.1017/S0025315409991421

Bhuiyan, M.J.A.N. and Dutta, D. (2012) Assessing impacts of sea level rise on river salinity in the Gorai river network, Bangladesh. *Estuarine, Coastal and Shelf Science* 96: 219–27. https://doi.org/10.1016/j.ecss.2011.11.005

Blinn, D.W. and Ruiter, D.E. (2006) Tolerance values of stream caddisflies (Trichoptera) in the lower Colorado River basin, USA. *The Southwestern Naturalist* 51: 326–37. https://doi.org/10.1894/0038-4909(2006)51[326:TVOSCT]2.0.CO;2

Breine, J., Stevens, M., Maes, J., Van den Bergh, E. and Elliott, M. (2009) Tidal marshes as habitat for juvenile fish in the Zeeschelde estuary (Belgium). In: Breine, J. Fish assemblages as ecological indicator in estuaries: the Zeeschelde (Belgium). PhD theses of the Research Institute for Nature and Forest, T. 2009.1: pp. 109–29.

Carbonnel, V., Lionard, M., Muylaert, K. and Chou, L. (2009) Dynamics of dissolved and biogenic silica in the freshwater reaches of a macrotidal estuary (The Scheldt, Belgium). *Biogeochemistry* 96: 49–72. https://doi.org/10.1007/s10533-009-9344-6

CBD (2004) Decision VII/11 Ecosystem approach. Decision adopted by the Conference of Parties to the Convention on Biological Diversity at its seventh meeting, 9–20 and 27 February 2004, Kuala Lumpur, Malaysia.

Chadwick, M.A. and Feminella, J.W. (2001) Influence of salinity and temperature on the growth and production of a freshwater mayfly in the Lower Mobile River, Alabama. *Limnology and Oceanography* 46: 532–42. https://doi.org/10.4319/lo.2001.46.3.0532

Chadwick, M.A., Hunter, H., Feminella, J.W. and Henry, R.P. (2002) Salt and water balance in *Hexagenia limbata* (Ephemeroptera: Ephemeridae) when exposed to brackish water. *Florida*

Entomologist 85: 650–1. https://doi.org/10.1653/0015 -4040(2002)085[0650:SAWBIH]2.0.CO;2

Chanton, J. and Lewis, F.G. (2002) Examination of coupling between primary and secondary production in a river-dominated estuary: Apalachicola Bay, Florida, USA. *Limnology and Oceanography* 47: 683–97. https://doi.org/10.4319/lo.2002.47.3.0683

Chapman, P. (1981) Seasonal changes in the depth distributions of interstitial salinities in the Fraser River estuary, British Columbia. *Estuaries and Coasts* 4: 226–8. https://doi.org/10.2307/1351480

Chen, M.S., Wartel, S., Van Eck, B. and Van Maldegem, D. (2005) Suspended matter in the Scheldt estuary. *Hydrobiologia* 540: 79–104. https://doi.org/10.1007/ s10750-004-7122-y

Christierson, B.V., Vidal, J.-P. and Wade, S.D. (2012) *Journal of Hydrology* 424–5: 48–67. https://doi.org/ 10.1016/j.jhydrol.2011.12.020

Church, J.A., Clark, P.U., Cazenave, A., Gregory, J.M., Jevrejeva, S., Levermann, A., Merrifield, M.A., Milne, G.A., Nerem, R.S., Nunn, P.D., Payne, A.J., Pfeffer, W.T., Stammer, D., Unnikrishnan, A.S., 2013. Sea Level Change, in: Stocker, T.F., Qin, D., Plattner, G.-K., Tignor, M., Allen, S.K., Boschung, J., Nauels, A., Xia, Y., Bex, V. and Midgley, P.M. (eds) *Climate Change 2013: The Physical Science Basis. Contribution of Working Group 1 to the Fifth Assessment Report of the Intergovernmental Panel on Climate Change*. Cambridge, United Kingdom and New York: Cambridge University Press.

Cohen-Shacham, E., Walters, G., Janzen, C. and Maginnis, S. (2016) Nature-based solutions to address global societal challenges, IUCN, Gland, Switzerland. https://doi.org/10.2305/IUCN.CH. 2016.13.en

Cohen, A.N., Carlton, J.T. and Fountain, M.C. (1995) Introduction, dispersal and potential impacts of the green crab *Carcinus maenas* in San Francisco Bay, California. *Marine Biology* 122: 225–37.

Cole, J.J. and Caraco, N.F. (2006) Primary production and its regulation in the tidal-freshwater Hudson River. In: Levinton, J. and Waldman, J. (eds), *The Hudson River Estuary*, pp. 307–21. New York: Cambridge University Press. https://doi.org/ 10.1017/CBO9780511550539.011

Connelly, K.A., Rollwagen-Bollens, G. and Bollens, S.M. (2020) Seasonal and longitudinal variability of zooplankton assemblages along a river-dominated estuarine gradient. *Estuarine, Coastal and Shelf Science* 245: 106980. https://doi.org/10.1016/j.ecss. 2020.106980

Covich, A.P., Palmer, M.A. and Crowl, T.A. (1999) The role of benthic invertebrate species in freshwater ecosystems: zoobenthic species influence energy flows and nutrient cycling. *BioScience* 49: 119–27. https://doi.org/10.2307/1313537

Crooks, J.A., Chang, A.L. and Ruiz, G.M. (2011) Aquatic pollution increases the relative success of invasive species. *Biological Invasions* 13: 165–76. https://doi.org/10.1007/s10530-010-9799-3

de Groot, R., Brander, L., van der Ploeg, S., Costanza, R., Bernard, F., Braat, L., Christie, M., Crossman, N., Ghermandi, A., Hein, L., Hussain, S., Kumar, P., McVittie, A., Portela, R., Rodriguez, L.C., ten Brink, P. and van Beukering, P. (2012) Global estimates of the value of ecosystems and their services in monetary units. *Ecosystem Services* 1 (1): 50–61. https://doi.org/10.1016/j.ecoser.2012.07.005

De Jonge, V. (1974) Classification of brackish coastal inland waters. *Aquatic Ecology* 8: 29–39. https://doi. org/10.1007/BF02254903

Deegan, L.A. and Garritt, R.H. (1997) Evidence for spatial variability in estuarine food webs. *Marine Ecology Progress Series* 147: 31–47. https://doi.org/ 10.3354/meps147031

Dias, E., Barrosa, A.G., Hoffman, J.C., Antunes, C. and Moraise, P. (2020) Habitat use and food sources of European flounder larvae (*Platichthys flesus*, L. 1758) across the Minho River estuary salinity gradient (NW Iberian Peninsula). *Regional Studies in Marine Science* 34: 101196. https://doi.org/10. 1016/j.rsma.2020.101196

Dias, E., Morais, P., Cotter, A.M., Antunes, C. and Hoffman, J.C. (2016) Estuarine consumers utilize marine, estuarine and terrestrial organic matter and provide connectivity among these food webs. *Marine Ecology Progress Series* 554: 21–34. https:// doi.org/10.3354/meps11794

Dionne, J.C. (1963) Towards a more adequate definition of the St. Lawrence estuary. *Zeitschrift für Geomorphologie* 7: 36–44.

Dittel, A.I. and Epifanio, C.E. (2009) Invasion biology of the Chinese mitten crab Eriochier sinensis: a brief review. *Journal of Experimental Marine Biology and Ecology* 374: 79–92. https://doi.org/10.1016/j. jembe.2009.04.012

Dolbeth, M., Martinho, F., Freitas, V., Costa-Dias, S., Campos, J. and Pardal, M.A. (2010) Multiyear comparisons of fish recruitment, growth and production in two drought-affected Iberian estuaries. *Marine and Freshwater Research* 61: 1399–1415. https://doi.org/10.1071/MF10002

Elliott, M. and McLusky, D.S. (2002) The need for definitions in understanding estuaries. *Estuarine, Coastal and Shelf Science* 55: 815–27. https://doi.org/ 10.1006/ecss.2002.1031

Elliott, M. (2011) Marine science and management means tackling exogenic unmanaged pressures and endogenic managed pressures – a numbered guide. *Marine Pollution Bulletin* 62: 651–5. https:// doi.org/10.1016/j.marpolbul.2010.11.033

Elliott, M., Whitfield, A.K., Potter, I.C., Blaber, S.J.M., Cyrus, D.P., Nordlie, F.G. and Harrison, T.D. (2007) The guild approach to categorizing estuarine fish assemblages: a global review. *Fish and Fisheries* 8: 241–68. https://doi.org/10.1111/j.1467-2679.2007. 00253.x

Ensign, S.H. and Noe, G.B. (2018) Tidal extension and sea-level rise: recommendations for a research agenda. *Frontiers in Ecology and Environment* 16: 37–43. https://doi.org/10.1002/fee.1745

Fairbridge, R.W. (1980) The estuary: its definition and geodynamic cycle. In E. Olausson and I. Cato (eds) *Chemistry and Biochemistry of Estuaries*, pp. 1–35. New York: John Wiley and Sons.

Flemming, B.W. (2011) Geology, morphology and sedimentology of estuaries and coasts. In B.W. Flemming and J.D. Hansom (eds) *Treatise on Estuaries and Coasts, Volume 3, Estuarine and Coastal Geology and Morphology*, pp. 7–38. Amsterdam: Elsevier. https://doi.org/10.1016/B978-0-12-374711-2.00302-8

Fowler, H.J. and Wilby, R.L. (2010) Detecting changes in seasonal precipitation extremes using regional climate model projections: Implications for managing fluvial flood risk. *Water Resources Research* 46: W03525. https://doi.org/10.1029/2008WR007636

Franzitta, G., Hanley, M.E., Airoldi, L., Baggini, C., Bilton, D.T., Rundle, S.D. and Thompson, R.C. (2015) Home advantage? Decomposition across the freshwater–estuarine transition zone varies with litter origin and local salinity. *Marine Environmental Research* 110: 1–7. https://doi.org/10.1016/j.marenvres.2015.07.012

Gillson, J. (2011) Freshwater flow and fisheries production in estuarine and coastal systems: where a drop of rain is not lost. *Reviews in Fisheries Science* 19: 168–86. https://doi.org/10.1080/10641262.2011.560690

Grabemann, H., Grabemann, I., Herbers, D., Muller, A. (2001) Effects of a specific climate scenario on the hydrography and transport of conservative substances in the Weser estuary Germany: a case study. *Climate Research* 18: 77–87. https://doi.org/10.3354/cr018077

Grimaldo, L.F., Stewart, A.R. and Kimmerer, W. (2009) Dietary segregation of pelagic and littoral fish assemblages in a highly modified tidal freshwater estuary. *Marine and Coastal Fisheries: Dynamics, Management, and Ecosystem Science* 1: 200–17. https://doi.org/10.1577/C08-013.1

Hanson, E. and Sytsma, M. (2008) The potential for mitten crab *Eriocheir sinensis* H. Milne Edwards, 1853 (Crustacea: Brachyura) invasion of Pacific northwest and Alaskan estuaries. *Biological Invasions* 10: 603–14. https://doi.org/10.1007/s10530-007-9156-3

Herborg, L.M., Rushton, S.P., Clare, A.S. and Bentley, M.G. (2005) The invasion of the Chinese mitten crab (*Eriocheir sinensis*) in the United Kingdom and its comparison to continental Europe. *Biological Invasions* 7: 959–68. https://doi.org/10.1007/s10530-004-2999-y

Herrera-Pantoja, M. and Hiscock, K.M. (2008) The effects of climate change on potential groundwater recharge in Great Britain. *Hydrological Processes* 22: 73–86. https://doi.org/10.1002/hyp.6620

Hobbie, S.E. and Grimm, N.B. (2020) Nature-based approaches to managing climate change impacts in cities. *Philosophical Transactions of the Royal Society B* 375: 20190124. https://doi.org/10.1098/rstb.2019.0124

Hoffman, J.C. and Bronk, D.A. (2006) Inter-annual variation in stable carbon and nitrogen isotope biogeochemistry of the Mattaponi River, Virginia. *Limnology and Oceanography* 51: 2319–32. https://doi.org/10.4319/lo.2006.51.5.2319

Hoffman, J.C., Bronk, D.A. and Olney, J.E. (2008) Organic matter sources supporting lower food web production in the tidal freshwater portion of the York River estuary, Virginia. *Estuaries and Coasts* 31: 898–911. https://doi.org/10.1007/s12237-008-9073-4

Hoitink, A.J.F. and Jay, D.A. (2016) Tidal river dynamics: Implications for deltas. *Review of Geophysics* 54: 240–72. https://doi.org/10.1002/2015RG000507

Hong, B. and Shen, J. (2012) Responses of estuarine salinity and transport processes to potential future sea-level rise in the Chesapeake Bay. *Estuarine, Coastal and Shelf Science* 104–5: 33–45. https://doi.org/10.1016/j.ecss.2012.03.014

Hughes, J.E., Deegan, L.A., Peterson, B.J., Holmes, R.M. and Fry, B. (2000) Nitrogen flow through the food web in the oligohaline zone of a New England estuary. *Ecology* 81: 433–52. https://doi.org/10.1890/0012-9658(2000)081[0433:NFTTFW]2.0.CO;2

Johnson, A.C., Acreman, M.C., Dunbar, M.J., Feist, S.W., Giacomello, A.M., Gozlan, R.E., Hinsley, S.A., Ibbotson, A.T., Jarvie, H.P., Jones, J.I., Longshaw, M., Maberly, S.C., Marsh, T.J., Neal, C., Newman, J.R., Nunn, M.A., Pickup, R.W., Reynard, N.S., Sullivan, C.A., Sumpter, J.P. and Williams, R.J. (2009) The British river of the future: how climate change and human activity might affect two contrasting river ecosystems in England. *Science of the Total Environment* 407: 4787–98. https://doi.org/10.1016/j.scitotenv.2009.05.018

Jones, A.E., Hardison, A.K., Hodges, B.R., McClelland, J.W. and Moffett, K.B. (2020) Defining a riverine tidal freshwater zone and its spatiotemporal dynamics. *Water Resources Research* 56: e2019WR026619. https://doi.org/10.1029/2019WR026619

Kasai, A., Kurikawa, Y., Ueno, M., Robert, D. and Yamashita, Y. (2010) Salt-wedge intrusion of seawater and its implication for phytoplankton dynamics in the Yura Estuary, Japan. *Estuarine, Coastal and Shelf Science* 86: 408–14. https://doi.org/10.1016/j.ecss.2009.06.001

Kautza, A. and Sullivan, S.M.P. (2016) The energetic contributions of aquatic primary producers to terrestrial food webs in a midsize river system. *Ecology* 97: 694–705. https://doi.org/10.1890/15-1095

Kennish, M.J. (2002) Environmental threats and environmental future of estuaries. *Environmental Conservation* 29: 78–107. https://doi.org/10.1017/S0376892902000061

Khojasteh, D., Glamore, W., Heimhuber, V. and Felder, S. (2021) Sea level rise impacts on estuarine dynamics: a review. *Science of the Total Environment*

780: 146470. https://doi.org/10.1016/j.scitotenv.2021.146470

Knights, D., Sawyer, A.H., Barnes, R.T., Musial, C.T. and Bray, S. (2017) Tidal controls on riverbed denitrification along a tidal freshwater zone. *Water Resources Research* 53: 799–816. https://doi.org/10.1002/2016WR019405

Kraus, R.T. and Secor, D.H. (2005) Application of the nursery-role hypothesis to an estuarine fish. *Marine Ecology Progress Series* 291: 301–5. https://doi.org/10.3354/meps291301

Kukulka, T. and Jay, D.A. (2003) Impacts of Columbia River discharge on salmonid habitat: 1. A nonstationary fluvial tide model. *Journal of Geophysical Research* 108: 3293. https://doi.org/10.1029/2002JC001382

Le Pichon, C., Coustillas, J., Zahm, A., Bunel, M., Gazeau-Nadin, C. and Rochard, E. (2017) Summer use of the tidal freshwaters of the River Seine by three estuarine fish: coupling telemetry and GIS spatial analysis. *Estuarine, Coastal and Shelf Science* 195: 83–96. https://doi.org/10.1016/j.ecss.2017.06.028

Lehman, P.W. (2007) The influence of phytoplankton community composition on primary productivity along the riverine to freshwater tidal continuum in the San Joaquin River, California. *Estuaries and Coasts* 30: 82–93. https://doi.org/10.1007/BF02782969

Lehner, B., Heinrichs, T., Döll, P. and Alcamo, J. (2001) *EuroWasser – Model-Based Assessment of European Water Resources and Hydrology in the Face of Global Change.* Kassel World Water Series 5, Center for Environmental Systems Research, University of Kassel, Kassel, Germany.

Lester, R.E., Webster, I.T., Fairweather, P.G. and Young, W.J. (2011) Linking water-resource models to ecosystem-response models to guide water-resource planning – an example from the Murray-Darling Basin, Australia. *Marine and Freshwater Research* 62: 279–89. https://doi.org/10.1071/MF09298

Little, S. (2012) The impact of increasing saline penetration upon estuarine and riverine benthic macroinvertebrates. PhD Thesis Geography and Environment, Loughborough University, Loughborough, p. 308.

Little, S., Wood, P.J. and Elliott, M. (2017) Quantifying salinity-induced changes on estuarine benthic fauna: the potential implications of climate change. *Estuarine, Coastal and Shelf Science* 198: 610–25. https://doi.org/10.1016/j.ecss.2016.07.020

Little, S., Spencer, K.L., Schuttelaars, H.M., Millward, G.E. and Elliott, M. (2017) Unbounded boundaries and shifting baselines: estuaries and coastal seas in a rapidly changing world. *Estuarine, Coastal and Shelf Science* 198 (Part B, 5): 311–19. https://doi.org/10.1016/j.ecss.2017.10.010

Lotze, H.K., Lenihan, H.S., Bourque, B.J., Bradbury, R.H., Cooke, R.G., Kay, M.C., Kidwell, S.M., Kirby, M.X., Peterson, C.H. and Jackson, J.B.C. (2006) Depletion, degradation, and recovery potential of estuaries and coastal seas. *Science* 312: 1806–9. https://doi.org/10.1126/science.1128035

Love, J., Gill, J. and Newhard, J. (2008) Saltwater intrusion impacts fish diversity and distribution in the Blackwater River drainage (Chesapeake Bay watershed). *Wetlands* 28: 967–74. https://doi.org/10.1672/07-238.1

Maes, J., Stevens, M. and Ollevier, F. (2005) The composition and community structure of the ichthyofauna of the upper Scheldt estuary: synthesis of a 10-year data collection (1991–2001). *Journal of Applied Ichthyology* 21: 86–93. https://doi.org/10.1111/j.1439-0426.2004.00628.x

Martinho, F., Leitão, R., Viegas, I., Dolbeth, M., Neto, J.M., Cabral, H.N. and Pardal, M.A. (2007) The influence of an extreme drought event in the fish community of a southern Europe temperate estuary. *Estuarine, Coastal and Shelf Science* 75: 537–46. https://doi.org/10.1016/j.ecss.2007.05.040

Masters, J.E.G., Jang, M.-H., Ha, K., Bird, P.D., Frear, P.A. and Lucas, M.C. (2006) The commercial exploitation of a protected anadromous species, the river lamprey (*Lampetra fluviatilis* (L.)), in the tidal River Ouse, north-east England. *Aquatic Conservation: Marine and Freshwater Ecosystems* 16: 77–92. https://doi.org/10.1002/aqc.686

McIvor, C.C. and Odum, W.E. (1988) Food, predation risk, and microhabitat selection in a marsh fish assemblage. *Ecology and Evolution* 69: 1341–51. https://doi.org/10.2307/1941632

McLachlan, J.R., Haghkerdar, J.M. and Greig, H.S. (2019) Strong zonation of benthic communities across a tidal freshwater height gradient. *Freshwater Biology* 64: 1284–94. https://doi.org/10.1111/fwb.13304

McLusky, D.S. (1994) Tidal freshwaters. In P.S. Maitland, P.J. Boon and D.S. McLusky (eds) *The Fresh Waters of Scotland: A National Resource of International Significance*, pp. 51–64. Chichester: John Wiley & Sons Ltd.

Morris, A.W., Mantoura, R.F.C., Bale, A.J. and Howland, R.J.M. (1978) Very low salinity regions of estuaries: important sites for chemical and biological interactions. *Nature* 274: 678–80. https://doi.org/10.1038/274678a0

Muller, K. and Mendl, H. (1979) Importance of a brackish water area for the stonefly colonization cycle in a coastal river. *Oikos* 33: 272–7. https://doi.org/10.2307/3544003

Murphy, J.M., Sexton, D.M., Jenkins, G.J., Booth, B.B., Brown, C.C., Clark, R.T., Collins, M., Harris, G.R., Kendon, E.J., Betts, R.A., Brown, S.J., Humphrey, K.A., McCarthy, M.P., McDonald, R.E., Stephens, A., Wallace, C., Warren, R., Wilby, R. and Wood, R.A. (2009) UK Climate Projections Science Report: Climate Change Projections. Met Office Hadley Centre, Exeter.

Muylaert, K. and Sabbe, K. (1999) Spring phytoplankton assemblages in and around the maximum turbidity zone of the estuaries of the

Elbe (Germany), the Schelde (Belgium/The Netherlands) and the Gironde (France). *Journal of Marine Systems* 22: 133–49. https://doi.org/10.1016/S0924-7963(99)00037-8

Muylaert, K., Sabbe, K. and Vyverman, W. (2000) Spatial and temporal dynamics of phytoplankton communities in a freshwater tidal estuary (Schelde, Belgium). *Estuarine Coastal and Shelf Science* 50: 673–87. https://doi.org/10.1006/ecss.2000.0590

Muylaert, K., Tackx, M. and Vyverman, W. (2005) Phytoplankton growth rates in the freshwater tidal reaches of the Schelde estuary (Belgium) estimated using a simple light-limited primary production model. *Hydrobiologia* 540: 127–40. https://doi.org/10.1007/s10750-004-7128-5

Muylaert, K., Van Mieghem, R., Sabbe, K., Tackx, M. and Vyverman, W. (2000) Dynamics and trophic roles of heterotrophic protists in the plankton of a freshwater tidal estuary. *Hydrobiologia* 432: 25–36. https://doi.org/10.1023/A:1004017018702

Neumann, B., Vafeidis, A.T., Zimmermann, J. and Nicholls, R.J. (2015) Future coastal population growth and exposure to sea-level rise and coastal flooding – a global assessment. *PLoS ONE* 10: e0118571. https://doi.org/10.1371/journal.pone.0118571

Nicholls, R.J. and Klein, R.J.T. (2005) Climate change and coastal management on Europe's coast. In J. Vermaat, W. Salomons, L. Bouwer and K. Turner (eds) *Managing European Coasts. Environmental Science*, pp. 199–226. Berlin, Heidelberg: Springer. https://doi.org/10.1007/3-540-27150-3_11

O'Briain, R. (2019) Climate change and European rivers: an eco-hydromorphological perspective. *Ecohydrology* 12: e2099. https://doi.org/10.1002/eco.2099

Odum, W.E. (1988) Comparative ecology of tidal freshwater and salt marshes. *Annual Revue of Ecology and Systematics* 19: 147–76. https://doi.org/10.1146/annurev.es.19.110188.001051

Odum, W.E., Rozas, L.P. and McIvor, C.C. (1987) A comparison of fish and vertebrate community composition in tidal freshwater and oligohaline marsh systems. In D.D. Hook, W.H. McKee Jr., H.K. Smith, J. Gregory, V.G. Burrell, M.R. DeVoe, R.E. Sojka, S. Gilbert, R. Banks, L.G. Stolzy, C. Brooks, T.D. Matthews and T.H. Shear (eds), *Ecology and Management of Wetlands*, pp. 112–32. London: Croom Helm.

Oppenheimer, M., Glavovic, B.C., Hinkel, J., van de Wal, R., Magnan, A.K., Abd-Elgawad, A., Cai, R., Cifuentes-Jara, M., DeConto, R.M., Ghosh, T., Hay, J., Isla, F., Marzeion, B., Meyssignac, B. and Sebesvari, Z. (2019) Sea level rise and implications for low-lying islands, coasts and communities. In H.-O. Pörtner, D.C. Roberts, V. Masson-Delmotte, P. Zhai, M. Tignor, E. Poloczanska, K. Mintenbeck, A. Alegría, M. Nicolai, A. Okem, J. Petzold, B. Rama and N.M. Weyer (eds) *IPCC Special Report on the Ocean and Cryosphere in a Changing Climate*. The Intergovernmental Panel on Climate Change.

Park, D. (1999) Waves, Tides and Shallow-Water Processes, 2nd edn. The Open University, Butterworth-Heinemann, Oxford.

Park, G.S. and Marshall, H.G. (2000) The trophic contributions of rotifers in tidal freshwater and estuarine habitats. *Estuarine, Coastal and Shelf Science* 51: 729–42. https://doi.org/10.1006/ecss.2000.0723

Pihl, L., Cattrijsse, A., Codling, I., Mathieson, S., McLusky, D.S. and Roberts, C. (2002) Habitat use by fishes in estuaries and other brackish areas. In M. Elliott and K. Hemingway (eds) *Fishes in Estuaries*, pp. 10–53. Oxford: Blackwell Science. https://doi.org/10.1002/9780470995228.ch2

Piscart, C., Lecerf, A., Usseglio-Polatera, P., Moreteau, J.-C. and Beisel, J.-N. (2005) Biodiversity patterns along a salinity gradient: the case of net-spinning caddisflies. *Biodiversity and Conservation* 14: 2235–49. https://doi.org/10.1007/s10531-004-4783-9

Potter, I.C., Tweedley, J.R., Elliott, M. and Whitfield, A.K. (2015) The ways in which fish use estuaries: a refinement and expansion of the guild approach. *Fish and Fisheries* 16: 230–9. https://doi.org/10.1111/faf.12050

Prandle, D. (2009) *Estuaries Dynamics, Mixing, Sedimentation and Morphology*. Cambridge: Cambridge University Press. https://doi.org/10.1017/CBO9780511576096

Prandle, D. and Lane, A. (2015) Sensitivity of estuaries to sea level rise: vulnerability indices. *Estuarine, Coastal and Shelf Science* 160: 60–8. https://doi.org/10.1016/j.ecss.2015.04.001

Prudhomme, C., Young, A., Watts, G., Haxton, T., Crooks, S., Williamson, J., Davies, H., Dadson, S. and Allen, S. (2012) The drying up of Britain? A national estimate of changes in seasonal river flows from 11 regional climate model simulations. *Hydrological Processes* 26: 1115–18. https://doi.org/10.1002/hyp.8434

Purcell, K.M., Klerks, P.L. and Leberg, P.L. (2010) Adaptation to sea level rise: does local adaptation influence the demography of coastal fish populations? *Journal of Fish Biology* 77: 1209–18. https://doi.org/10.1111/j.1095-8649.2010.02727.x

Rice, K., Hong, B. and Shen, J. (2012) Assessment of salinity intrusion in the James and Chickahominy Rivers as a result of simulated sea-level rise in Chesapeake Bay, East Coast, USA. *Journal of Environmental Management* 111: 61–9. https://doi.org/10.1016/j.jenvman.2012.06.036

Rius, M., Clusella-Trullas, S., McQuaid, C.D., Navarro, R.A., Griffiths, C.L., Matthee, C.A., von der Heyden, S. and Turon, X. (2014) Range expansions across ecoregions: interactions of climate change, physiology and genetic diversity. *Global Ecology and Biogeography* 23: 76–88. https://doi.org/10.1111/geb.12105

Robins, P.E., Skov, M.W., Lewis, M.J., Gimenez, L., Davies, A.G., Malham, S.K., Neill, S.P., McDonald, J.E., Whitton, T.A., Jackson, S.E. and Jago, C.F. (2016) Impact of climate change on UK estuaries: a review of past trends and potential projections.

Estuarine, Coastal and Shelf Science 169: 119–35. https://doi.org/10.1016/j.ecss.2015.12.016

Rocha, C. and Galvão, H.A.B. (2002) Role of transient silicon limitation in the development of cyanobacteria blooms in the Guadiana estuary, south-western Iberia. *Marine Ecology Progress Series* 228: 35–45. https://doi.org/10.3354/meps228035

Rodrigues, M., Fortunato, A.B. and Freire, P. (2019) Saltwater intrusion in the upper Tagus estuary during droughts. *Geosciences* 9: 400. https://doi.org/10.3390/geosciences9090400

Roessig, J.M., Woodley, C.M., Cech, J.J. and Hansen, L.J. (2004) Effects of global climate change on marine and estuarine fishes and fisheries. *Reviews in Fish Biology and Fisheries* 14: 251–75. https://doi.org/10.1007/s11160-004-6749-0

Romanowicz, R., Beven, K., Wade, S. and Vidal, J. (2006) Effects of climate change on river flows and groundwater recharge: a practical methodology. Interim report on rainfall-runoff modelling. UKWIR Report, London, CL/04.

Ross, A.C., Najjar, R.G., Li, M., Mann, M.E., Ford, S.E. and Katz, B. (2015) Sea level rise and other influences on decadal-scale salinity variability in a coastal plain estuary. *Estuarine, Coastal and Shelf Science* 157: 79–92. https://doi.org/10.1016/j.ecss.2015.01.022

Rundle, S.D., Attrill, M.J. and Arshad, A. (1998) Seasonality in macroinvertebrate community composition across a neglected ecological boundary, the freshwater–estuarine transition zone. *Aquatic Ecology* 32: 211–16. https://doi.org/10.1023/A:1009934828611

Santos, P.J., Castel, P.J. and Souza-Santos, L.P. (1996) Seasonal variability of meiofaunal abundance in the oligo-mesohaline area of the Gironde estuary, France. *Estuarine, Coastal and Shelf Science* 43: 549–63. https://doi.org/10.1006/ecss.1996.0087

Scavia, D., Field, J.C., Boesch, D.F., Buddemeier, R.W., Burkett, V., Cayan, D.R., Fogarty, M., Harwell, M.A., Howarth, R.W., Mason, C., Reed, D.J., Royer, T.C., Sallenger, A.H. and Titus, J.G. (2002) Climate change impacts on U.S. coastal and marine ecosystems. *Estuaries* 25: 149–64. https://doi.org/10.1007/BF02691304

Schuchardt, B. and Schirmer, M. (1991) Phytoplankton maxima in the tidal freshwater reaches of two coastal plain estuaries. *Estuarine, Coastal and Shelf Science* 32: 187–206. https://doi.org/10.1016/0272-7714(91)90014-3

Schröter, D., Cramer, W., Leemans, R., Prentice, I.C., Araújo, M.B., Arnell, N.W., Bondeau, A., Bugmann, H., Carter, T.R., Gracia, C.A., de la Vega-Leinert, A.C., Erhard, M., Ewert, F., Glendining, M., House, J.I., Kankaanpää, S., Klein, R.J.T., Lavorell, S., Linder, M., Metzger, M.J., Meyer, J., Mitchell, T.D., Reginster, I., Rounsevell, M., Sabaté, S., Sitch, S., Smith, B., Smith, J., Smith, P., Sykes, M.T., Thonicke, K., Thuiller, W., Tuck, G., Zaehle, S. and Zierl, B., 2005. Ecosystem service supply and vulnerability to global change in Europe. *Science* 310: 1333–1337.

Schuchardt, B.U.H. and Schirmer, M. (1993) The tidal freshwater reach of the Weser estuary: riverine or estuarine. *Netherlands Journal of Aquatic Ecology* 27: 215–26. https://doi.org/10.1007/BF02334785

Seys, J., Vincx, M. and Meire, P. (1999) Spatial distribution of oligochaetes (Clitellata) in the tidal freshwater and brackish parts of the Schelde estuary (Belgium). *Hydrobiologia* 406: 119–32. https://doi.org/10.1023/A:1003751512971

Shennan, I. and Horton, B. (2002) Holocene land- and sea-level changes in Great Britain. *Journal of Quaternary Science* 17: 511–26. https://doi.org/10.1002/jqs.710

Simenstad, C.A., Burke, J.L., O'Connor, J.E., Cannon, C., Heatwole, D.W., Ramirez, M.F., Waite, I.R., Counihan, T.D. and Jones, K.L. (2011) Columbia River estuary ecosystem classification – concept and application, U.S. Geological Survey Open-File Report 2011–1228. https://doi.org/10.3133/ofr20111228

Sobczak, W.V., Cloern, J.E., Jassby, A.D., Cole, B.E., Schraga, T.S. and Arnsberg, A. (2005) Detritus fuels ecosystem metabolism but not metazoan food webs in San Francisco estuary's freshwater delta. *Estuaries* 28: 124–37. https://doi.org/10.1007/BF02732759

Sousa, R., Antunes, C. and Guilhermino, L. (2007) Species composition and monthly variation of the Molluscan fauna in the freshwater subtidal area of the River Minho estuary. *Estuarine, Coastal and Shelf Science* 75: 90–100. https://doi.org/10.1016/j.ecss.2007.02.020

Sousa, R., Guilhermino, L. and Antunes, C. (2005) Molluscan fauna in the freshwater tidal area of the River Minho estuary, NW of Iberian Peninsula. *Annales De Limnologie-International Journal of Limnology* 41: 141–7. https://doi.org/10.1051/limn/2005009

Strayer, D.L. and Smith, L.C. (2001) The zoobenthos of the freshwater tidal Hudson River and its response to the zebra mussel (*Dreissena polymorpha*) invasion. *Archiv fur Hydrobiologie Supplementband* 139: 1–52.

System, V. (1959) Final resolution. The Venice System for the classification of marine waters according to salinity. 8–14 April 1958, Venice, Italy. In D. Ancona (ed.), *Symposium on the Classification of Brackish Waters, Archives Oceanography and Limnology* 11 (supplement): 243–8.

Tackz, M., De Pauw, N., Van Mieghem, R., Azémar, F., Hannouti, A., Van Damme, S., Fiers, F., Daro, N. and Meire, P. (2004O) Zooplankton in the Schelde estuary, Belgium and The Netherlands. Spatial and temporal patterns. *Journal of Plankton Research* 26: 133–41. https://doi.org/10.1093/plankt/fbh016

Talke, S.A. and Jay, D.A. (2020) Changing tides: the role of natural and anthropogenic factors. *Annual Review of Marine Science* 12: 121–51. https://doi.org/10.1146/annurev-marine-010419-010727

Telesh, I.V. and Khlebovich, V.V. (2010) Principal processes within the estuarine salinity gradient: a review. *Marine Pollution Bulletin* 61: 149–55. https://doi.org/10.1016/j.marpolbul.2010.02.008

Temmerman, S., Meire, P., Bouma, T.J., Herman, P.M.J., Ysebaert, T. and De Vriend, H.J. (2013) Ecosystem-based coastal defence in the face of global change. *Nature* 504: 79–83. https://doi.org/10.1038/nature12859

Thomas, R.M. (1998) Temporal changes in the movements and abundance of Thames Estuary fish populations. In M.J. Attrill (ed.) *A Rehabilitated Estuarine Ecosystem. The Environment and Ecology of the Thames Estuary*, pp. 115–40. Dordrecht: Kluwer Academic Publishers. https://doi.org/10.1007/978-1-4419-8708-2_7

Tills, O., Spicer, J.I. and Rundle, S.D. (2010) Salinity-induced heterokairy in an upper-estuarine population of the snail *Radix balthica* (Mollusca: Pulmonata). *Aquatic Biology* 9: 95–105. https://doi.org/10.3354/ab00231

Tully, K., Gedan, K., Epanchin-Niell, R., Strong, A., Bernhardt, E.S., Bendor, T., Mitchell, M., Kominoski, J., Jordan, T.E., Neubauer, S.C. and Weston, N.B. (2019) The invisible flood: the chemistry, ecology, and social implications of coastal saltwater intrusion. *BioScience* 69 (5): 368–78. https://doi.org/10.1093/biosci/biz027

Van den Meersche, K., Van Rijswijk, P., Soetaert, K. and Middelburg, J.J. (2009) Autochthonous and allochthonous contributions to mesozooplankton diet in a tidal river and estuary: integrating carbon isotope and fatty acid constraints. *Limnology and Oceanography* 54: 62–74. https://doi.org/10.4319/lo.2009.54.1.0062

Van Liefferinge, C., Dillen, A., Ide, C., Herrel, A., Belpaire, C., Mouton, A., de Deckere, E. and Meire, P. (2012) The role of a freshwater tidal area with controlled reduced tide as feeding habitat for European eel (*Anguilla anguilla* L.). *Journal of Applied Icthyology* 28: 572–81. https://doi.org/10.1111/j.1439-0426.2012.01963.x

van Puijenbroek, P.J.T.M., Buijse, A.D., Kraak, M.H.S. and Verdonschot, P.F.M. (2019) Species and river specific effects of river fragmentation on European anadromous fish species. *River Research and Application* 35: 68–77. https://doi.org/10.1002/rra.3386

Vautard, R., Gobiet, A., Sobolowski, S., Kjellström, E., Stegehuis, A., Watkiss, P., Mendlik, T., Landgren, O., Nikulin, G., Teichmann, C. and Jacob, D. (2014) The European climate under a 2°C global warming. *Environmental Research Letters* 9: 034006. https://doi.org/10.1088/1748-9326/9/3/034006

Vetemaa, M., Eschbaum, R., Verliin, A., Albert, A., Eero, M., Lillemägi, R., Pihlak, M. and Saat, T. (2006) Annual and seasonal dynamics of fish in the brackish-water Matsalu Bay, Estonia. *Ecology of Freshwater Fish* 15: 211–20. https://doi.org/10.1111/j.1600-0633.2006.00134.x

Vörösmarty, C.J., Green, P., Salisbury, J. and Lammers, R.B. (2000) Global water resources: vulnerability from climate change and population growth.

Science 289: 288284. https://doi.org/10.1126/science.289.5477.284

Vörösmarty, C.J., McIntyre, P.B., Gessner, M.O., Dudgeon, D., Prusevich, A., Green, P., Glidden, S., Bunn, S.E., Sullivan, C.A., Liermann, C.R. and Davies, P.M. (2010) Global threats to human water security and river biodiversity. *Nature* 467: 555–61. https://doi.org/10.1038/nature09440

Whitehead, P.G., Wilby, R.L., Battarbee, R.W., Kernan, M. and Wade, A.J. (2009) A review of the potential impacts of climate change on surface water quality. *Hydrological Sciences Journal* 54: 101–23. https://doi.org/10.1623/hysj.54.1.101

Whitfield, A.K., Elliott, M., Basset, A., Blaber, S.J.M. and West, R.J. (2012) Paradigms in estuarine ecology: a review of the Remane diagram with a suggested revised model for estuaries. *Estuarine, Coastal and Shelf Science* 97: 78–90. https://doi.org/10.1016/j.ecss.2011.11.026

Williams, D.D. (2009) Coping with saltwater: the conditions of aquatic insects in estuaries as determined by gut content analysis. *The Open Marine Biology Journal* 3: 21–7. https://doi.org/10.2174/1874450800903010021

Williams, D.D. and Hamm, T. (2002) Insect community organisation in estuaries: the role of the physical environment. *Ecography* 25: 372–84. https://doi.org/10.1034/j.1600-0587.2002.250314.x

Williams, D.D. and Williams, N.E. (1998) Seasonal variation, export dynamics and consumption of freshwater invertebrates in an estuarine environment. *Estuarine, Coastal and Shelf Science* 46: 393–410. https://doi.org/10.1006/ecss.1997.0280

Wilson, J.G., Dunne, N. and Giltrap, M. (2017) Assessing candidate metrics for the ecological quality of TFTW (tidal freshwaters in transitional waters) in Ireland using benthic invertebrates. *Ocean & Coastal Management* 143: 115–21. https://doi.org/10.1016/j.ocecoaman.2017.02.018

Wilson, J.G., Giltrap, M. and Kelly, F. (2016) Fish in tidal freshwater transitional waters in Ireland: recommendations for assessment, policy and management of ecological quality under the Water Framework Directive (WFD). *Biology and Environment: Proceedings of the Royal Irish Academy* 116B: 221–32. https://doi.org/10.3318/bioe.2016.28

Wolanski, E., Day, J.W., Elliott, M. and Ramachandran, R. (2019) *Coasts and Estuaries. The Future.* Amsterdam: Elsevier.

Wolanski, E. and Elliott, M. (2015) *Estuarine Ecohydrology: An Introduction.* Amsterdam: Elsevier. https://doi.org/10.1016/B978-0-444-63398-9.00001-5

Wright, J. and Worrall, F. (2001) The effects of river flow on water quality in estuarine impoundments. *Physics and Chemistry of the Earth, Part B: Hydrology, Oceans and Atmosphere* 26: 741–46. https://doi.org/10.1016/S1464-1909(01)00079-X

Xu, X., Wei, H., Barker, G., Holt, K., Julian, S., Light, T., Melton, S., Salamanca, A., Moffett, K.B., McClelland, J.W. and Hardison, A.K. (2021) Tidal freshwater zones as hotspots for biogeochemical cycling: sediment organic matter decomposition in the lower reaches of two south Texas rivers. *Estuaries and Coasts* 44: 722–33. https://doi.org/10.1007/s12237-020-00791-4

Young, M., Howe, E., O'Rear, T., Berridge, K. and Moyle, P. (2021) Food web fuel differs across habitats and seasons of a tidal freshwater estuary. *Estuaries and Coasts* 44: 286–301. https://doi.org/10.1007/s12237-020-00791-4

Ysebaert, T., Fettweis, M., Meire, P. and Sas, M. (2005) Benthic variability in intertidal soft-sediments in the mesohaline part of the Schelde estuary. *Hydrobiologia* 540: 197–216. https://doi.org/10.1007/s10750-004-7144-5

Ysebaert, T., Meire, P., Coosen, J. and Essink, K. (1998) Zonation of intertidal macrobenthos in the estuaries of Schelde and Ems. *Aquatic Ecology* 32: 53–71. https://doi.org/10.1023/A:1009912103505

Zapata, M.J. and Sullivan, S.M.P. (2018) Spatial and seasonal variability of emergent aquatic insects and nearshore spiders in a subtropical estuary. *Marine and Freshwater Research* 70: 541–53. https://doi.org/10.1071/MF18130

Zonneveld, I.S., Barendgret, A., 2009. Human activities in European tidal freshwater wetlands, in: Barendregt, A., Whigham, D.F. and Baldwin, A.H. (eds), *Tidal Freshwater Wetlands*. Leiden: Backhuys.

CHAPTER 8

Defining Habitat Losses due to Coastal Squeeze

NIGEL PONTEE, JAMES A. TEMPEST, KENNETH PYE and SIMON J. BLOTT

Abstract

The loss of habitats caused by the presence of coastal defences and rising sea levels is commonly referred to as 'coastal squeeze'. Coastal squeeze losses have driven the creation of significant numbers of compensatory habitat creation schemes in England and Wales. However, there have been inconsistencies in the definition of coastal squeeze and the methods used to assess it. An agreed definition and a standard assessment methodology are needed to correctly determine the causes of habitat loss and to ensure correct management actions and compensatory habitats targets. A revised definition of coastal squeeze is presented, focusing on whether the natural landward movement of habitats under rising sea levels is slowed or prevented by anthropogenic structures or management actions. The work described in this chapter shows how coastal squeeze can affect a range of coastal habitat types and be caused by a range of structures and management actions. A new assessment method is described, emphasising the importance of understanding the multiple causes of coastal habitat loss and outlining the datasets available to do this. A summary of the main findings from the trial of the methodology at four sites is presented. The work shows that although the role of coastal squeeze as a cause of past habitat losses may have been overstated in some instances, it is important to carefully consider whether coastal squeeze could become more widespread in the future under a scenario of more rapid sea-level rise.

Keywords: sea-level rise, management actions, coastal squeeze

Correspondence: nigel.pontee@jacobs.com

Introduction

Coastal habitats, such as saltmarshes and beaches, provide spaces for plants and animals as well as numerous other valuable ecosystem services including flood and erosion protection, carbon sequestration, pollution absorption and recreation opportunities (Jones et al. 2011; Constanza et al. 2014). In the UK, large areas of coastal habitats have been lost due to progressive phases of land claim since Romano-British times (Davidson et al. 1991). Although losses due to large-scale land claim have been much lower over the last 40 years, significant concerns remain over the indirect losses of coastal habitats caused by the presence of coastal defences and sea-level rise (SLR; Fig. 8.1). In the UK, the term 'coastal squeeze' is commonly used to describe this process (Doody 2004).

Nigel Pontee, James A. Tempest, Kenneth Pye and Simon J. Blott, 'Defining Habitat Losses due to Coastal Squeeze' in: *Challenges in Estuarine and Coastal Science.* Pelagic Publishing (2022). © Nigel Pontee, James A. Tempest, Kenneth Pye and Simon J. Blott. DOI: 10.53061/BQPM4918

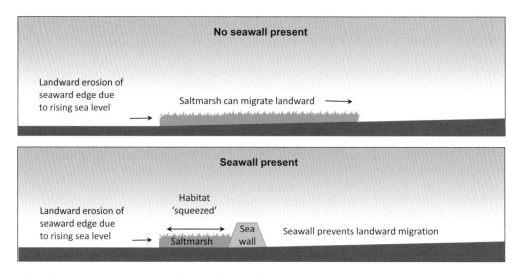

Figure 8.1 Example of a commonly used diagram illustrating the concept of coastal squeeze

The importance of managing, enhancing and creating coastal priority habitats is set out in a number of national biodiversity strategies and climate change national adaptation schemes (Defra 2018; HM Government 2018). Coastal habitats are also protected by national and international designations. To date, one of the most important pieces of legislation protecting coastal habitats has been the EU Habitats and Birds Directive. This has provided a legal requirement to protect designated conservation sites, known as Natura 2000 sites, from deterioration since 1992. The Habitats Directive was adopted in England at this time and implemented in 1994 (Miles and Richardson 2018). The reduction in extent of habitats, owing to coastal squeeze, is one possible cause of deterioration, which can lead to the requirement to create compensatory habitat.

In England and Wales, coastal squeeze losses have been a key component in determining compensatory habitat targets, for intertidal habitats leading to significant numbers of compensatory habitat creation schemes. Further schemes are planned in areas believed to be suffering from coastal squeeze. The provision of compensatory habitat is costly; individual schemes can cost in excess of £20m, and the total cost of delivering the 11,500 ha identified by Shoreline Management Plan (SMP) policies until 2060 (Adaptation Sub-Committee Progress Report 2013) would be at least £575m (Pontee 2017). This estimate was based on a modest cost of £50,000 per hectare, although the costs for schemes can be substantially higher depending on site conditions. There is therefore a strong economic impetus to improve our understanding of coastal squeeze impacts and to consistently define, measure and recognise the uncertainty of habitat losses due to coastal squeeze. Understanding the causes of coastal habitat loss is also important in determining the most appropriate management responses.

This chapter is based on a recent Environment Agency-led research project entitled 'What is Coastal Squeeze?' (Pontee et al. 2020). The project was developed to address a number of inconsistencies and limitations in previous definitions and assessment methodologies that have been identified over the last ten years. The project builds upon in a series of projects for the Environment Agency and several technical papers (Pontee 2011, 2013a, 2013b, 2017). It was recognised from the outset that the project would be of significant interest to a wide range of stakeholders involved in the management of coastal habitats. The project was therefore designed and run with collaboration in mind. This was essential in order to develop a definition that would apply to a range of coastal settings in the UK and be accepted by a range of organisations. The project was developed in discussion with the Environment Agency, Natural England and Natural Resources Wales. Once initiated, the project was steered by a Project Board of 13 members from the Environment Agency, Defra, Natural England, Scottish Environment Protection Agency, Marine Management Organisation, Natural Resources Wales and Welsh Government. Numerous progress calls/discussion meetings were held with the Project Board in order to reach consensus on various aspects of the work. A stakeholder workshop was held in the early stages of the project to gain input from across industry and organisations on the scope and requirements of a definition and assessment method. Attendees at this event included members of the Project Board, consultancy companies, Network Rail, county councils and universities.

The chapter starts with a review of the factors influencing coastal habitat extents and the previous definitions of coastal squeeze. The following sections present the new definition of coastal squeeze and the assessment method that have been developed. A summary of the some of the main findings from trial applications of the method is then presented. The chapter ends by explaining how the work will be used to inform future coastal management in England and Wales.

Factors influencing the extent of coastal habitat

Central to the definition of coastal squeeze is that human structures can impede the natural transgression of habitats in response to SLR. Such structures include seawalls, revetments, flood embankments, railway/road embankments and quay walls. SLR has the potential to bring about changes to habitats in a number of ways including (a) increased wave attack, leading to erosion of seawards edges of habitat; (b) increased inundation of habitats, leading to changes in habitat zonation (including extent, position and type) and (c) overtopping/breaching/landward movement of dunes/barrier beaches/barrier islands. There are, however, many other factors that can influence the type and extent of coastal habitat (Table 8.1) that do not represent coastal squeeze. These factors are natural, artificial or a combination of both. It is important to identify and understand these various factors so that coastal squeeze can be correctly identified.

The role of different factors, plus the type and scale of impacts on coastal habitats, are likely to vary according to geographic location and specific site conditions. In the context of the UK, several previous publications have shown that SLR has been a minor factor leading to loss of saltmarsh and intertidal flats in the past, compared with other factors such as fluctuations in wind/wave climate (e.g. Pye 2000; van der Wal and Pye 2004). In many instances, habitat change is likely to result from multiple causes. In these circumstances, it may be difficult to identify the main cause of change and a range of further, more detailed investigations may be required. Additionally, it is important to consider whether the changes are part of a progressive long-term trend or a shorter cycle. For example, sandy beaches might vary significantly in width in response to differences in wave conditions between the summer and winter. Therefore, losses of width that occur in the winter months might be reversed in the summer months, resulting in no net change when measured over the whole year. Similar patterns can occur over periods of years in response to periods of increased/decreased storminess. Identifying a progressive long-term trend such as coastal squeeze therefore needs to consider an appropriate time span.

Previous definitions of coastal squeeze

The origin of the term 'coastal squeeze' in the UK was documented by Doody (2004) who cited it as having initially arisen from observations of the loss of saltmarsh and mudflat in the Wash, owing to land claim. In the late 1980s and early 1990s, the term 'coastal

Table 8.1 Factors that can give rise to a change in extent of coastal habitat but do not constitute coastal squeeze

Factor	Potential influence on habitats	
Human intervention	**Flood and erosion related** • Barrages and barriers • Cliff remediation • Cliff stabilisation • Culverts • Groynes • Jetties, piers or breakwaters (shore connected) • Managed realignment/No active intervention • Offshore breakwater • Flood and erosion related	**Other activities** • Beach mining • Changes in grazing regime • Changes in land-use • Channel training works • Dams • Dredging • Industrial activities and traffic near habitat • Introduction of invasive species • Other changes in agricultural practices • Railway/road embankments/quay walls • Reclamations • Recreation and beach management • Water abstraction
Climate change	• Changes to annual rainfall patterns • Increase in CO_2 level[1] • Increase in drought • Increase in extreme rainfall • Increase in freshwater peak flows • Increase in temperature[1] • Short-term cyclical changes in wind direction • Short-term cyclical increases in SL[2] • Short-term cyclical increases in wave climate or change in direction • Short-term cyclical increases in storm surge frequency or magnitude • Short-term cyclical increases in wind speed	
Other	• Changing positions of banks and channels in estuaries • Changes to soil chemistry (e.g. waterlogging, redox potential, salinity) that are not due to SLR • Fungal pathogens • Herbivore activity • Introduction of invasive species • Wrack damage	

1　It is noted that CO_2 and temperature contribute to SLR but are indirect factors in coastal squeeze.
2　It is implicit in the definition of coastal squeeze that it is driven by a net rise in SLR over the long term. Therefore, short-term variations in sea level (e.g. owing to the lunar nodal tidal cycle) or other driving forces such as wind-wave climate or storm surges (e.g. decadal increases or decreases) do not in themselves cause coastal squeeze.

squeeze' was being used as part of a conservation argument against further saltmarsh reclamation in the Wash. The last application for embanking and land claim in the Wash at Gedney Drove End was dismissed following a public inquiry in 1981. The concept of 'squeezing' the intertidal zone by advancing the line of sea defence while the low water mark remained static, or did not move seaward by a corresponding amount, emerged in the years leading up this Inquiry, and was referred to explicitly by Doody (1987). The potential for SLR to encourage landward movement of the low water mark, contributing to coastal squeeze in the Wash and elsewhere, was also first recognised around this time. Quite separately, concerns about saltmarsh erosion in Essex and Kent, regardless of cause, were first being addressed by local engineers and HR Wallingford before being taken up the Nature Conservancy Council and Institute of Terrestrial Ecology (Boorman 1987; Boorman, Goss-Custard and McGorty 1989; Burd 1989). In the international context, one of the earliest references to the term 'squeeze' comes from Titus (1991) in the USA. This publication notes that wetlands losses might occur in the future if (a) rates of SLR exceed rates of vertical sediment accretion and (b) dikes or bulkheads (vertical walls) used to protect development restrict the natural ability of habitats to transgress landwards under rising sea levels.

In order to explore the usage of the term 'coastal squeeze', Pontee et al. (2020) reviewed 33 documents including coastal flood risk management strategies, SMPs, habitat creation programmes, journal papers and UK policy guidance notes. There are many similarities in terms of the elements included in the definitions (e.g. SLR, sea defences, intertidal habitats) as well as those that are commonly excluded (e.g. reclamation losses, high land, internal erosion). There are, however, some significant differences (Table 8.2).

Significant differences between definitions include:

- The processes driving the landward transgression of habitats. For example, Doody (2013) referred to rising sea levels and 'other factors such as increased storminess'. English Nature (2003a, b) also referred to coastal processes and storminess involved in the coastal squeeze process. In comparison, most other definitions referred only to SLR.
- The treatment of habitat quality. The majority of definitions, including those used within strategies and SMPs, did not include this, although English Nature et al. (2003) and Natural England (2015) did.
- The delineation of upper and lower limits of the intertidal zone: for example, highest astronomical tide (HAT), mean high water springs (MHWS), high water, back of beach, visible limit of habitat from aerial photographs. Although the boundary between mudflat and saltmarsh was universally taken to be mean high water neap tides, the upper limit of marsh was taken variously as MHWS or HAT.
- The type of habitats included. Most studies include intertidal mudflat and saltmarsh as habitats. Some studies also include sand flats, rocks and boulders. A number of studies covered transitional habitats at the top of marshes (transitional saltmarsh, transitional grassland, grazing marsh). One study (North West SMP; Halcrow 2010, 2012b, 2012c) included sand dunes, while another (the Tees; Black and Veatch 2007a, b) included shallow coastal waters. The Poole and Exe strategies (Atkins and Halcrow 2012a, b, c) also examined other terrestrial habitats.
- The type of structures included. Although most studies referred to flood and coastal erosion risk management (FCERM) defences, a range of terms were used, including seawall, flood defences, fixed sea defences. Some studies also referred to human-made defences, artificial structures, structures (generally) and inflexible structures (for example, roads, dykes, urbanisations, other facilities).

Outside the UK, the term 'coastal squeeze' has been applied to wetland habitats such as mangroves and saltmarshes (e.g., European Commission 2009; Torio and Chmura 2013), but also sandy beaches (e.g., Lester and Matella 2013). A number of authors have recognised that urbanisation and infrastructure can restrict the landward migration of coastal habitats

Table 8.2 Common inclusions, exclusions and differences in previous definitions of coastal squeeze

Inclusions	Exclusions	Differences
• Sea-level rise (SLR) • Defences preventing landward movement of habitats • Resulting losses in area being termed coastal squeeze • Intertidal habitats • Saltmarsh and mudflat loss in estuary environments • Reference to tidal levels to delineate habitats • Loss of designated habitats – usually Special Protection Areas/Special Areas of Conservation (SPA/SAC).	• Direct losses due to reclamation • Losses owing to naturally rising land • Changes in wind-wave climate and other coastal processes (for example, sediment supply) • Changes in habitat quality • Internal erosion of saltmarsh • Other impacts of defences • Other anthropogenic structures	• Processes driving the landward transgression • Treatment of habitat quality • Delineation of upper and lower limits of the intertidal zone • Type of habitats • Type of structures

(e.g. Climate Council of Australia 2009; Aukes 2017). Additionally, several authors have commented on the importance of vertical sedimentation in determining resilience of wetland habitats to SLR (e.g. Titus 1991; New Zealand Ministry for the Environment 2001).

Revised definition

A new definition of coastal squeeze has been developed for use within a UK context (Pontee et al. 2020):

> Coastal squeeze is the loss of natural habitats or deterioration of their quality arising from anthropogenic structures, or actions, preventing the landward transgression of those habitats that would otherwise naturally occur in response to sea-level rise in conjunction with other coastal processes. Coastal squeeze affects habitat on the seaward side of existing structures.

This definition was agreed by those organisations on the Project Board (see Introduction). Since the definition is intended to be used across the UK by a range of organisations and individuals, it is important that it is as unambiguous as possible. With this in mind the various elements of the definition have been carefully defined (Table 8.3). Coastal squeeze as defined excludes:

- Historical land claim landwards of currently existing structures.
- Other impacts of hard defences, such as reductions in sediment supply caused by protecting eroding sediment sources or interrupting longshore transport pathways. It is, however, recognised that decreases in sediment supply can potentially make downdrift habitats more susceptible to coastal squeeze. The role of these factors, and the implications for any mitigatory actions need to be appraised on a site-by-site basis.
- Impacts of other human activity/structures on habitats, such as alteration of estuary channel morphology owing to dredging, training walls, piers, or impacts on habitat quality owing to management practices or pollution.
- Other natural or human causes of habitat loss unrelated to creating barriers to landward transgression – for example, the lateral movement of channels, which may be unrelated to SLR and, while this would erode seaward edges of habitats, would not create landward transgression even under unconstrained conditions.
- Habitat loss against natural steeply rising land (that is, sloping coastal hinterlands) – such losses may need to be considered as a baseline scenario ('without defences') against which to judge coastal squeeze losses. It should be noted that some areas of rising land formed from unconsolidated sediments may erode relatively quickly in the future to provide accommodation space for habitats.

It is recommended that the above impacts should be assessed, described and accounted separately from coastal squeeze losses, even though the remedial measures may be linked or packaged with those taken to address coastal squeeze.

A review of Annex I (Habitats Directive 92/43/EEC, 1992 and Conservation of Habitats & Species Regulations, 2017) and Section 41 (NERC Act, 2006 for England) or 7 (Environment Act, 2016 for Wales) priority coastal/intertidal habitats suggests that the following habitats could be subject to coastal squeeze:

- boulder beaches
- shingle beaches and barriers
- intertidal seagrass beds
- intertidal reedbeds
- intertidal rock platforms

- mud and sandflats
- saline lagoons located in front of structures
- saltmarsh
- sand beaches
- sand dunes

All these habitats meet the following criteria: (a) the habitat, in an unconstrained scenario, is capable of transgressing landward in response to 'SLR and other coastal processes' – this means that the physical and/or biological components of the habitats are capable of being mobilised; (b) the habitats have a measurable area that could potentially be reduced in response to landward transgression being prevented; (c) there are relevant structure/s and/ or management actions that could prevent the landward transgression of the habitat. Such actions apply most readily to shingle beaches and sand dunes in the form of reprofiling activities or the planting of stabilising vegetation.

The relationship between these simple habitat names and the original habitat names/ types listed in Annex 1/Section 41 of Conservation of Habitats & Species Regulations (2017), NERC Act (2006) and Environment Act (2016) is given in Pontee et al. (2020). It should be noted that there may be some site-specific variations around the UK, and habitat types therefore need to be identified at a local level.

Table 8.3 Clarifications to the terms/items with the revised definition of coastal squeeze

Term/item	Clarification
Anthropogenic (human-created) structures	Includes features that act as barriers to the inland progression of marine waters and habitats such as flood and coastal erosion structures, quay walls and road/ railway embankments.
Anthropogenic actions	Includes activities that artificially prevent the landward transgression of habitats.
Natural habitats	Includes all relevant Annex I coastal/intertidal habitats found in the UK as defined in policy and legislation (including NERC s41 priority habitat or Environment Act Section 7 for Wales).
Habitat loss	Is considered in terms of planform area of the habitats and includes changes arising from frontal retreat (e.g. of a saltmarsh edge) as well as internal erosion (e.g. expansion of creeks within marshes).
Coastal processes relevant to identifying coastal squeeze	Includes those processes that, under natural unconstrained conditions, can lead to the landward migration of habitats under a scenario of SLR – such as waves for shingle beaches, winds for aeolian dunes and tidal inundation for saltmarshes.
Intertidal islands	The assessment of coastal squeeze in estuaries should consider whether the extent of any intertidal islands is affected by flood defences on the islands themselves or within the wider estuary. This consideration should also take into account the role of natural changes in channel position over time, which can influence the size and location of intertidal islands.
SLR	The net trend in relative sea level resulting from global eustatic variations (changes in ocean volume) and regional or local isostatic change (changes in land level). SLR excludes changes in water levels due to human interventions, for example, dredging, land claim, creation of flood storage/managed realignment areas. If these changes are relevant to an area, they should be assessed separately.
Habitat quality	Assessing coastal squeeze should consider whether there is deterioration in habitat quality or changes in vegetative species composition that may be occurring as a result of human structures/actions impeding the landward transgression of habitats. For example, in saltmarshes, SLR might lead to high marsh communities being replaced with lower marsh communities. These changes may occur ahead of, or at the same time as, areal losses.

Several aspects of the new definition are different from previous definitions or are worthy of special emphasis:

- The range of habitat types included. In the past, the term 'coastal squeeze' has most commonly been applied to saltmarshes. The new definition identifies ten main habitat types that could potentially be affected by coastal squeeze.
- The importance of landward transgression. The revised definition explicitly relates to habitat losses arising from the restriction of landward transgression and deliberately excludes reclamation losses, which were included in some previous definitions. The revised definition also highlights that habitat transgression requires the movement of both the upper and lower limits of habitats. The work demonstrates that the processes governing the landward extent of habitats differ between habitats. For habitats such as mudflats and saltmarshes the landward extent is controlled mainly by the limit of tidal inundation. For habitats such as beaches the landward extent is controlled by both tidal inundation and wave run-up. For sand dunes the landward extent is controlled by aeolian (wind) action. Understanding these processes is critical to determining whether observed habitat losses represent coastal squeeze. Processes that cause the loss of seaward limits of habitats but are not capable of causing the landward limits of the habitat to move inland (under a natural scenario with no defences) generally do not represent coastal squeeze. For example, the lateral movement of channels in an estuary could cause erosion of the edge of saltmarshes but would not cause the landward limit of saltmarsh to migrate inland. Hence saltmarsh loss would occur regardless of whether there were structures further landwards.
- Structures and activities included. In the past, the term 'coastal squeeze' was most commonly related to flood defence structures, typically earth embankments backing saltmarshes. The revised definition clarifies that a number of structures (Table 8.4) can prevent the landward migration of habitats including seawalls, revetments, flood embankments, railway/road embankments/quay walls and reclamations. The revised definition also includes management actions that could prevent the landward transgression of the habitat.
- Measurement of areal extent. Previous investigations of coastal squeeze in saltmarsh habitats have typically considered frontal retreat as a mechanism for habitat loss. The revised definition includes the internal erosion of saltmarshes through the expansion of creeks, since such losses could potentially be due to marshes failing to keep pace with sea level (i.e. drowning).
- Consideration of habitat quality. The revised definition includes habitat quality – something that most previous definitions do not include (except English Nature et al. 2003; Natural England 2015). Habitat quality is a complex subject, with various possible ways of measuring it and numerous factors contributing to changes in quality over time. In the context of coastal squeeze, we are interested in whether there is deterioration in habitat quality or changes in vegetative species composition that may be occurring as a result of human structures/actions impeding the landward transgression of habitats. For example, in saltmarshes, SLR might lead to high marsh communities being replaced with lower marsh communities. In the UK, concerns have been expressed that such changes may occur ahead of areal losses of habitat (Natural England pers. comm.) and thus could represent an advance warning of coastal squeeze.

Table 8.4 Examples of structures and management actions that could prevent the landward transgression of coastal habitats and result in coastal squeeze

Habitat	Example management action	Explanation	Example structure	Explanation
Intertidal seagrass beds	n/a	n/a	Earth embankments Seawalls/ revetments Sheet-piled quay walls Road/railway earth embankments	Prevention of tidal inundation can prevent the inland transgression of mudflats, seagrass, reedbeds and saltmarsh
Intertidal reedbeds	n/a	n/a		
Mud and sand flats	n/a	n/a		
Saltmarsh	Mowing Pesticide use Grazing			
Boulder beaches	n/a	n/a		Prevention of tidal inundation and wave overwashing prevents landward transgression of sand, gravel and boulder beaches
Shingle beaches and barriers	Reprofiling	Washover material bulldozed seawards to restore beach crest in more seaward position, thus preventing transgression		
Sand beaches	Sediment removal/ recycling Reprofiling	Washover or windblown material moved seawards and returned to beach, thus preventing transgression		
Sand dunes	Sediment removal/recycling (e.g. bulldozing sand off roads located on landward side of dunes) Vegetation (e.g. forestry)	Washover or windblown material moved seawards and returned to dunes, thus preventing transgression Sand trapping prevents wind mobilisation of sand), thus preventing transgression	Very large (high) seawalls Other new development requiring additional walls, raised earth embankments	Prevention of wind and wave overwashing prevents landward movement of sediment and transgression of dunes

Recommended assessment methodology

The assessment method that has been developed considers past and future losses separately (Figs 8.2 and 8.3). The main stages of each are similar and include an initial scoping stage that defines the study area (e.g. a length of defence/structure, a whole estuary), the habitats to be included and the period of interest. A subsequent screening stage, based on four questions, allows a rapid assessment of whether or not coastal squeeze is likely to be a potential cause for the habitat change. If the answer to any of these questions is negative, it is unlikely that there has been any coastal squeeze in the area. If any of the answers are positive, then it is possible that losses due to coastal squeeze have occurred and further assessment should be carried out. The method outlines how to quantify these changes, the

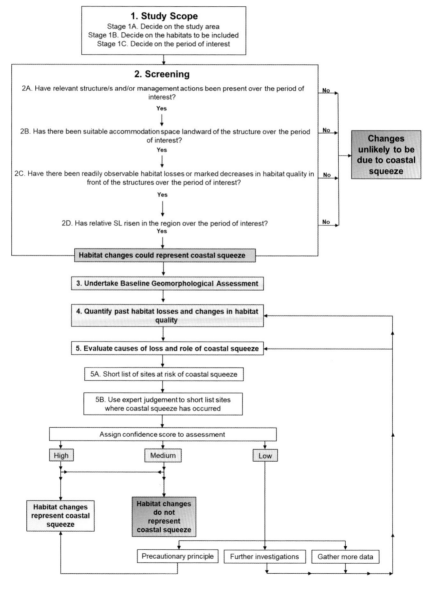

Figure 8.2 Flow diagram of the methodology for identifying past habitat losses attributable to coastal squeeze (from Pontee et al. 2020)

relevant data sources and causes of uncertainty that apply to each step of the method. The final stage of the assessment requires expert judgement to assess whether the observed/predicted changes represent coastal squeeze. The central question is whether the observed coastal habitat losses have been due to the prevention of the landward transgression.

The method also outlines how an assessment of confidence in the findings, 'high', 'medium' or 'low', should be made (Table 8.5). In some cases, it may be difficult to determine whether coastal squeeze has led, or is likely to lead, to losses of intertidal habitats; that is, confidence is low. In these circumstances, it is unlikely to be possible to demonstrate that coastal squeeze is having 'no adverse effect'. This may be owing to a lack of data or the existence of multiple potential factors that could be responsible for intertidal habitat loss. In these cases, two approaches can be taken:

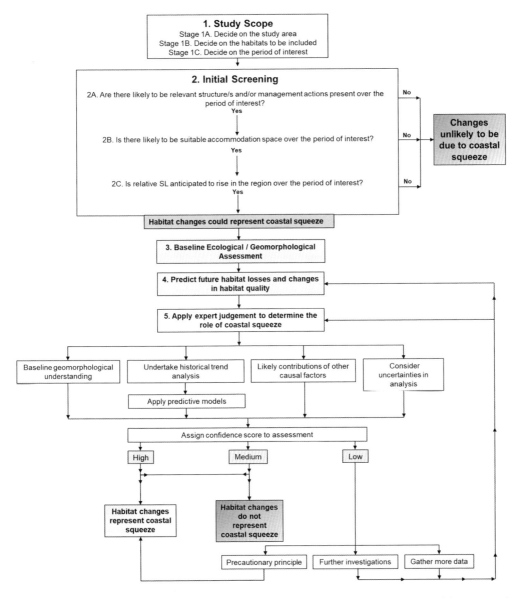

Figure 8.3 Flow diagram of the method for identifying future habitat losses attributable to coastal squeeze (Pontee et al. 2020)

Table 8.5 Criteria to assess confidence in coastal squeeze assessments

Confidence band	Criteria		
	Where habitat losses are likely to be coastal squeeze	Where there are no habitat losses	Where habitat losses are unlikely to be coastal squeeze
High	Clear evidence that: there is habitat loss or deterioration over time habitat loss or deterioration is due to SLR SLR exceeds sedimentation rates there are no other likely causes for habitat loss or deterioration	Clear evidence that: there is no habitat loss or deterioration over time SLR is less/equal to than sedimentation rates	Clear evidence that: there is habitat loss or deterioration over time SLR is less/equal to than sedimentation rates
	Supporting documentation: peer-reviewed publications, grey literature based on verifiable data analysis		
Medium	Clear evidence indicating habitat loss or deterioration over time Some evidence that habitat loss or deterioration is due to SLR Sedimentation rates believed to exceed SLR Some evidence that there are no other likely causes	Clear evidence that there is no habitat loss or deterioration over time Sedimentation rates believed to equal/exceed SLR	Clear evidence that there is habitat loss or deterioration over time Sedimentation rates believed to equal/exceed SLR Some evidence that habitat loss or deterioration is due to causes other than SLR
	Supporting documentation: non-peer reviewed publications (e.g. conference papers), grey literature (e.g. consultancy reports) based on data analysis		
Low	Some evidence for habitat loss or deterioration over time No clear evidence that SLR is the cause Sedimentation rates believed to exceed SLR	Some evidence for no change in habitat loss or deterioration over time Sedimentation rates believed to equal/exceed SLR	Some evidence for habitat loss or deterioration over time Sedimentation rates believed to equal/exceed SLR Multiple possible causes for habitat loss or deterioration
	Supporting documentation: absent or limited to suggestions for causes of habitat loss without supporting data analysis		

- Adopt the precautionary principle and assume that habitat losses or deterioration in quality results from coastal squeeze. The potential coastal squeeze losses should be reviewed in the future when further understanding or monitoring data may be available.
- Carry out additional studies to improve confidence in the findings. These studies might include more detailed morphological or ecological assessments to document the extent and cause of habitat loss or deterioration.

It should be noted that habitat losses not attributed to coastal squeeze may still require assessment and mitigation/compensation.

Case studies

Four case study areas were selected to develop and refine the method and to identify likely application issues elsewhere. The case study areas were chosen to reflect different geographical settings, different habitat assemblages and the varying amounts of data available (Fig. 8.4, Table 8.6). Although the case studies were not intended to provide comprehensive reviews of coastal squeeze at each site, they nevertheless identified several interesting aspects of coastal squeeze and the difficulties of assessing it in different settings.

The Lymington case study demonstrated that even under natural baseline conditions (without defences), habitat extent may decrease over time if steeply rising land results in there being insufficient room for landwards migration. Although some previous assessments have considered the presence of rising land elevations immediately behind present-day flood defences, the Lymington case study illustrates how highland further inland can limit

Figure 8.4 Location of four case studies used to test and refine the newly developed assessment method

Table 8.6 Characteristics of case study sites

Site	Setting	Habitats	Data availability
Aber Dysynni	Coast and estuary	Shingle beach/spit Mudflat, saltmarsh	Low data availability
Slaughden & Sudborne	Coast	Shingle beach	Moderate data availability
Lymington	Estuary	Mudflat, saltmarsh	Moderate data availability
Blackwater	Estuary	Mudflat, saltmarsh	High data availability

the potential for HAT transgression and thus the ability of tidally influenced landforms to migrate landwards. Such situations are likely to arise in many parts of the UK. These habitat losses should be considered as a form of 'natural' change rather than coastal squeeze.

The Blackwater case study (Fig. 8.5) demonstrated that although there have been losses of saltmarsh habitat in front of defences within the estuary, there is only limited evidence that SLR is the main cause of these losses. Much of the past and ongoing change in intertidal habitat extent can be attributed to wave action, local channel movements and renewed tidal influence in formerly reclaimed areas following breaching of defences, rather than marsh surfaces not being able to accrete vertically in line with SLR. There are currently areas of saltmarsh and mudflat accretion within the inner estuary, and net rates of saltmarsh loss, which are occurring mainly in the middle and outer estuary, are relatively low.

The Aber Dysynni and Slaughden/Sudbourne case studies revealed that the method of assessing coastal squeeze can be applied both to estuarine habitats such as marshes and to open coast habitats such as sand and shingle beaches, including areas where there is limited data availability. However, both case studies showed that assessing coastal squeeze on the open coast is subject to a number of difficulties. Unlike with tidally dominated landforms such as saltmarshes and mudflats (whose extent can be estimated by projecting tidal elevations inland), determining the degree to which a beach ridge system would have rolled landwards in the absence of structures or management actions is more problematic. Furthermore, assessment of the causes of changes in beach area must include the roles of changes in wave energy (some of which are unrelated to SLR) and sediment supply. These complexities may result in a low confidence in the final assessment of coastal squeeze and, therefore, indicate the need to either (a) adopt the precautionary principle or (b) carry out further studies to increase confidence in the findings.

A review of assessments of coastal squeeze undertaken previously in each setting showed that evidence for the loss of habitat being due to SLR against defences, as opposed to other causes, was often weak. The limitations of past studies (most commonly related to saltmarshes) include the failure to scientifically demonstrate that (a) habitat losses can be related to SLR and (b) habitat losses have not been due to other causes (e.g. increases in wind waves, lateral channel movements).

It is recommended that assessments of coastal squeeze need to be more scientifically rigorous in future, making use of additional data and developing a better understanding of past changes in habitat extent as well as the causes of these changes. This understanding is needed to make informed judgements about the likely future coastal squeeze losses. In some cases, additional datasets and/or analysis already exist to improve the scientific understanding. For the saltmarshes examined in this project, the required additional studies include bringing historical gains/loss studies up to date, gathering additional data on local rates of past SLR, considering habitat quality, examining the role of waves/other factors for vegetation loss and considering a range of future climate change scenarios for sediment supply and sea level. For the shingle beaches at Slaughden and Sudbourne, necessary

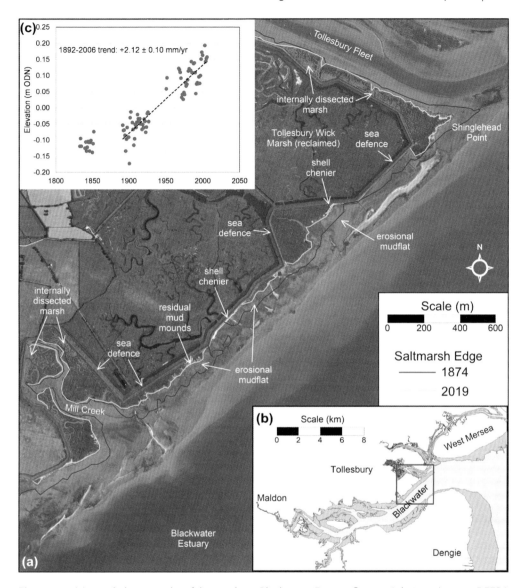

Figure 8.5 (a) Aerial photographs of the northern Blackwater Estuary flown 1 July 2018 (source: DEFRA Open Government Data), with the saltmarsh edge highlighted in yellow from aerial photographs flown 23 July 2019. The saltmarsh edge surveyed in 1874 (from 6 in Ordnance Survey maps) is highlighted in red; (b) location map of the Blackwater estuary; (c) mean sea-level record for the tide gauge at Sheerness (1833–2006, with a linear trend for the period 1892–2006, corrected to Ordnance Datum Newlyn

studies include gaining a better understanding the factors that influence alongshore and onshore/offshore movement of shingle, particularly in relation to the fate of shingle eroded during storms and its potential to move back onshore during fair weather periods. For the sand and shingle beaches of Aber Dysynni, further characterisation of several aspects is needed: sediment budget and transport pathways, wave and currents, and tidal levels. These factors influence beach volume and widths, and thus habitat extents.

An important implication from the case studies is that historical coastal squeeze losses in England and Wales, especially those of saltmarshes, might well be smaller than previous assessments have suggested. This is the case for a number of reasons:

- the habitat losses may have been caused by factors other than SLR and constriction against defences;
- increases in wave energy were overlooked as a cause of habitat loss in previous assessments; or
- the natural losses of habitats owing to steeply rising land may not have been fully accounted for.

Discussion and conclusions

The main aim of this project was to better understand the causes of coastal squeeze. This is particularly relevant where there is a legal obligation to compensate for the impacts of maintaining coastal flood management infrastructure or other infrastructure or management activities that could lead to coastal squeeze. There are, however, other policy and legislative drivers for positively managing, enhancing and creating coastal priority habitats, which are unrelated to coastal squeeze. It is nevertheless anticipated that the refined definition and assessment methodology will be helpful in improving our understanding of the likely rate and scale of impacts of accelerating SLR on coastal habitats in general, and promoting the need to periodically review the evidence available.

A revised definition of coastal squeeze has been developed that focuses on whether the natural landward movement of habitats under rising sea levels is slowed or prevented by artificial structures or management actions. The new definition differs from previous definitions by (a) including a wider range of habitats (including coastal wetlands, beaches, dunes and rock platforms), (b) including a wider range of structures (and also management actions) and (c) requiring the consideration of changes in habitat quality. The assessment method that has been developed emphasises the importance of understanding the multiple causes for coastal habitat losses to accurately attribute those caused by coastal squeeze. The method outlines the datasets available to make these assessments.

The new definition and method will refine how habitat losses are assessed and help inform compensatory habitat targets in the future. In some places, the new definition and assessment method may lead to a reduction in the amount of losses that are attributed to coastal squeeze and potentially the amount of compensatory habitat required. A 2–5% saving in this area would equate to at least £10–25m by 2060. By looking at the causes for coastal habitat loss/change more closely (e.g. SLR, changes in storminess, water quality), the new research will help coastal authorities consider how best to manage its coast/estuarine areas, clearly identifying how to best build the evidence base to support challenging decisions and improve coastal resilience.

The findings of the project are far reaching and relevant to the assessment of habitat losses at SMP, strategy and scheme levels. The work will have an impact on the management of FCERM structures such as flood embankments and walls, management actions such as shingle beach ridge reprofiling, and also non-FCERM structures such as road/railway embankments and quay walls. The work will affect all four nations of the UK:

- In England, the research will feed into a number of areas being led by the Environment Agency, including the (a) SMP refresh project and future SMP revisions, (b) Habitat Compensation Programme, (c) position statement on FCERM habitats work being produced by the Environment Agency/Natural England, (d) Natural England work to review and update protected site condition assessments, (e) biannual report to Defra on habitat compensation progress and (f) forthcoming saltmarsh restoration guide.
- In Scotland, the research will help to support local authorities who lead on implementing coast protection and flood defence actions.

- In Wales, this work will be used (a) within the SMP refresh project, (b) to inform new guidance on the assessment of coastal squeeze and (c) to support the Habitat Regulations Assessment work for coastal projects and plans.
- In Northern Ireland, the relevant authorities have had a watching brief on this research and are looking to use it to underpin future guidance on the management of coastal habitats.

Although the role of coastal squeeze as a cause of past habitat losses may have been overstated in some instances, it is important to carefully consider whether coastal squeeze could become more widespread in the future under a scenario of increased SLR. The behaviour of habitats in the coastal zone in response to SLR depends very much on the availability of sediment. With sufficient sediment supply, habitats may accrete vertically and maintain habitat extent or even prograde to increase their extent. However, where sediment supply is lower, habitats may erode and retreat landwards or even drown in situ. The combination of increasing rate of SLR coupled with the protection of extensive lengths of coast may limit the scope for future increases in sediment supply to help habitats keep pace with the change (Orford and Pethick 2006), leading to more habitats beginning to suffer coastal squeeze losses.

References

Adaptation Sub-Committee Progress Report (2013) Managing the land in a changing climate. Chapter 5: Regulating services – coastal habitats. Available at: https://www.theccc.org.uk/wp-content/uploads/2013/07/ASC-2013-Book-singles_2.pdf (Accessed: 24 March 2021).

Atkins and Halcrow (2012a) Habitat predictions and cause allocation in Poole Harbour. Poole & Wareham flood and coastal risk management strategy, 3rd draft report. Report prepared for Environment Agency by Atkins and Halcrow Group Ltd, August 2012.

Atkins and Halcrow (2012b) Potential habitat creation and enhancement opportunities in Poole Harbour. Poole and Wareham flood and coastal risk management strategy, 3rd draft report. Report prepared for Environment Agency by Atkins and Halcrow Group Ltd, October 2012.

Atkins and Halcrow (2012c) Habitat change predictions and cause allocation in the Exe Estuary. Exe estuary flood and coastal risk management strategy. 3rd draft report. Report prepared for Environment Agency by Atkins and Halcrow Group Ltd, June 2012.

Aukes, E.J. (2017) Framing coastal squeeze: understanding the integration of mega-nourishment schemes into the Dutch coastal management solutions repertoire: an interpretive analysis of coastal management processes. https://doi.org/10.3990/1.9789036543729

Black and Veatch (2007a) Tidal Tees flood risk management strategy: coastal squeeze study. Updated final report. Report prepared for Environment Agency by Black and Veatch Ltd, November 2007.

Black and Veatch (2007b) Appendix C: Calculations of Tees intertidal areas. Tidal Tees flood risk management strategy: coastal squeeze study. Updated final report. Report prepared for Environment Agency by Black and Veatch Ltd, November 2007.

Boorman, L.A. (1987) A survey of saltmarsh erosion along the Essex coast. Project 1064 (AW/ITE Contract F6–92–138). Institute of Terrestrial Ecology, Monks Wood.

Boorman, L.A., Goss-Custard, J.D. and McGorty, S. (1989) Climate change, rising sea level and the British coast. ITE Research Publication No. 1, HMSO, London.

Burd, F. (1989) The saltmarsh survey of Great Britain. Research & Survey in Nature Conservation No. 17, Nature Conservancy Council, Peterborough.

Climate Council of Australia (2009) Climate change risk to Australia's coast. A first pass national assessment. Department of Climate Change. Available at: https://www.environment.gov.au/climate-change/adaptation/publications/climate-change-risks-australias-coasts (Accessed: 24 March 2021).

Costanza, R., De Groot, R., Sutton, P., Van der Ploeg, S., Anderson, S.J., Kubiszewski, I., Farber, S. and Turner, R.K. (2014) Changes in the global value of ecosystem services. *Global Environmental Change* 26: 152–8. https://doi.org/10.1016/j.gloenvcha.2014.04.002

Davidson, N.C., Laffoley, D.A., Doody, J.P., Way, L.S., Gordon, J., Key, R., Drake, C.M., Pienkowski, M.W., Mitchell, R. and Duff, K.L. (1991) *Nature Conservation and Estuaries in Great Britain*. Nature Conservancy Council, Peterborough.

Defra (2018) The national adaptation programme and the third strategy for climate adaptation reporting. Making the country resilient to a changing climate, July 2018. Available at: https://assets.publishing.

service.gov.uk/government/uploads/system/uploads/attachment_data/file/727252/national-adaptation-programme-2018.pdf (Accessed: 24 March 2021).

Doody, J.P. (1987) The impact of 'reclamation' on the natural environment of the Wash. In J.P. Doody and B. Barnet (eds) *The Wash and its Environment*, pp. 165–72. Nature Conservancy Council. Research and Survey in Nature Conservation, No. 7. Peterborough, UK.

Doody, J.P. (2004) Coastal squeeze – an historical perspective. *Journal of Coastal Conservation* 10 (1): 129–38. https://doi.org/10.1652/1400-0350(2004)010[0129:CSAHP]2.0.CO;2

English Nature (2003a) England's best wildlife and geological sites: the condition of Sites of Special Scientific Interest in England in 2003. Report produced by English Nature.

English Nature (2003b) Unfavourable condition definitions for coasts and estuaries – inappropriate coastal management. Guidance note produced by English Nature, November 2003.

English Nature et al. (2003) Living with the Sea Life Project. Report by English Nature, Environment Agency, Natural Environment Research Council, Defra, EC.

European Commission (2009) The economics of climate change adaptation in EU coastal areas policy research corporation. Country Overview Assessment: Chapter 9 'Country overview and assessment – 8. Germany'. Germany. Report produced for Directorate-General for Maritime Affairs and Fisheries. Available at: https://ec.europa.eu/maritimeaffairs/sites/maritimeaffairs/files/docs/body/germany_climate_change_en.pdf (Accessed: 13 October 2020).

Environment Act (2016) Section 7: Biodiversity lists and duty to take steps to maintain and enhance biodiversity. Available at: https://www.legislation.gov.uk/anaw/2016/3/section/7 (Accessed: 24 March 2021).

Habitats Directive 92/43/EEC (1992) Council Directive 92/43/EEC of 21 May 1992 on the conservation of natural habitats and of wild fauna and flora. *Official Journal of the European Union* 206: 7–50.

Halcrow (2010) Appendix I: Regional assessment of coastal squeeze. Cell eleven tide and sediment transport study (CETaSS) Stage 2. Report prepared for the North West and North Wales Coastal Group by Halcrow Group Ltd, September 2010.

Halcrow (2012b) Annex C: Habitat losses and gains tables. North West England and North Wales shoreline management plan 2. Report prepared for the North West and North Wales Coastal Group by Halcrow Group Ltd, 2012.

Halcrow (2012c) North West England and North Wales, Shoreline management plan SMP2, Statement of environmental particulars. Report produced by Halcrow Group Ltd for the North West & North Wales Coastal Group, June 2012.

HM Government (2018) A green future: our 25 year plan to improve the environment. HM Government Report. Available at: https://assets.publishing.service.gov.uk/government/uploads/system/uploads/attachment_data/file/693158/25-year-environment-plan.pdf (Accessed: 24 March 2021).

Jones, M.L.M., Angus, S., Cooper, A., Doody, P., Everard, M., Garbutt, A., Gilchrist, P., Hansom, G., Nicholls, R. and Pye, K., Ravenscroft, N., Rees, S., Rhind, P. and Whitehouse, A. (2011) Chapter 11: Coastal margins. In *UK National Ecosystem Assessment: Understanding Nature's Value to Society*. Technical Report. UNEP-WCMC, Cambridge, pp. 2–47.

Lester, C. and Matella, M. (2013) Managing the coastal squeeze: resilience planning for shoreline residential development. *Stanford Environmental Law Journal* 36: 23–61.

Natural England (2015) Coastal management theme plan – developing a strategic and adaptive approach for flood and coastal erosion risk management for England's Natura 2000 sites 2015.

Natural Environmental and Rural Communities (NERC) Act (2006) Section 41: Biodiversity lists and action. Available at: https://www.legislation.gov.uk/ukpga/2006/16/section/41 (Accessed: 24 March 2021).

New Zealand Ministry for the Environment (2001) Planning for climate change effects on coastal margins. A report for the Ministry for the Environment as part of the New Zealand Climate Change Programme. September 2001.

Orford, J. and Pethick, J. (2006) Challenging assumptions of future coastal habitat development around the UK. *Earth Surface Processes and Landforms* 31: 1625–42. https://doi.org/10.1002/esp.1429

Pontee, N. (2011) Reappraising coastal squeeze: a case study from the NW England. *Maritime Engineering Journal* 164 (3): 127–38. https://doi.org/10.1680/maen.2011.164.3.127

Pontee, N. (2013a) Assessing coastal habitat loss: the role of coastal squeeze. In practice. *Bulletin of the Institute of Ecology and Environmental Management* 80 (June): 31–4.

Pontee, N. (2013b) Defining coastal squeeze: a discussion. *Ocean and Coastal Management* 84: 204–7. https://doi.org/10.1016/j.ocecoaman.2013.07.010

Pontee, N. (2017) Dispelling the myths surrounding coastal squeeze. Extended abstract for Flood and Coast 2017 Conference, Telford International Centre, 28–30 March.

Pontee, N., Tempest, J., Pye, K. and Blott, S. (2020) What is coastal squeeze? FRS17187. Report by Jacobs and Kenneth Pye Associates Ltd for the Environment Agency. Environment Agency, Bristol.

Pye, K. (2000) Saltmarsh erosion in south east England: mechanisms, causes and implications. In B.R. Sherwood, B.G. Gardiner and T. Harris (eds) *British Saltmarshes*, 360–96. London: Linnean Society.

Titus, J.G. (1991) Greenhouse-effect and coastal wetland policy – how Americans could abandon an area the size of Massachusetts at minimum cost. *Environmental Management* 15: 39–58. https://doi.org/10.1007/BF02393837

Torio, D.D. and Chmura, G.L. (2013) Assessing coastal squeeze of tidal wetlands. *Journal of Coastal Research* 29: 1049–61. https://doi.org/10.2112/JCOASTRES-D-12-00162.1

van der Wal, D. and Pye, K. (2004) Patterns, rates and possible causes of saltmarsh erosion in the Greater Thames area (UK). *Geomorphology* 61 (3–4): 373–91. https://doi.org/10.1016/j.geomorph.2004.02.005

CHAPTER 9

Carbon Storage in UK Intertidal Environments

WILLIAM E.N. AUSTIN, CRAIG SMEATON, PAULINA RURANSKA,
DAVID M. PATERSON, MARTIN W. SKOV, CAI J.T. LADD, LUCY McMAHON,
GLENN M. HAVELOCK, ROLAND GEHRELS, ROB MILLS, NATASHA L.M. BARLOW,
ANNETTE BURDEN, LAURENCE JONES and ANGUS GARBUTT

Abstract

We report on progress to assess carbon stocks in UK saltmarsh habitats, highlighting best practice in achieving national-scale assessments, including advances in field, laboratory and data methods. New understanding of coring disturbance highlights sediment compaction and its influence on carbon stock assessment; improvements in remote sensing methods are outlined and approaches to upscaling for carbon stock assessment are described. Here, we introduce the first UK-specific saltmarsh conversion for loss-on-ignition estimates of soil organic matter to soil organic carbon. The underlying drivers that determine the spatial distribution, magnitude and future vulnerability of these important natural capital assets are assessed, highlighting the significance of long-term sea-level drivers in shaping UK coastal environments and carbon stocks. The potential for management interventions that safeguard these long-term carbon stores through the protection, restoration and creation of saltmarsh habitats are also assessed. We highlight the emergent national policy opportunities for the inclusion of saltmarsh habitats in the UK greenhouse gas inventory, providing an important first step necessary to account for, protect and restore these long-term carbon stores, realising their potential for climate change mitigation.

Keywords: saltmarsh, carbon storage, sea-level change, managed realignment

Correspondence: wena@st-andrews.ac.uk

Introduction

Saltmarshes are one of the most common soft-sediment habitats of UK coastlines, being most expansive in the south-east and north-west (Boorman 2003; Burden et al. 2020) (Fig. 9.1). There is no definitive measure of the total saltmarsh extent in the UK at the present time, although a recent estimate of 44,102 ha (Burden et al. 2020) is consistent with a trend of marsh decline from 54,836 ha in 1945 to 46,631 ha in 2010 (Beaumont et al. 2014). Historically, losses have been driven by accelerated agricultural and industrial reclamation that commenced over 400 years ago (Hansom et al. 2001). This decline in UK saltmarsh

William E.N. Austin, Craig Smeaton, Paulina Ruranska, David M. Paterson, Martin W. Skov, Cai J.T. Ladd, Lucy McMahon, Glenn M. Havelock, Roland Gehrels, Rob Mills, Natasha L.M. Barlow, Annette Burden, Laurence Jones and Angus Garbutt, 'Carbon Storage in UK Intertidal Environments' in: *Challenges in Estuarine and Coastal Science*. Pelagic Publishing (2022). © William E.N. Austin et al. DOI: 10.53061/STPP2268

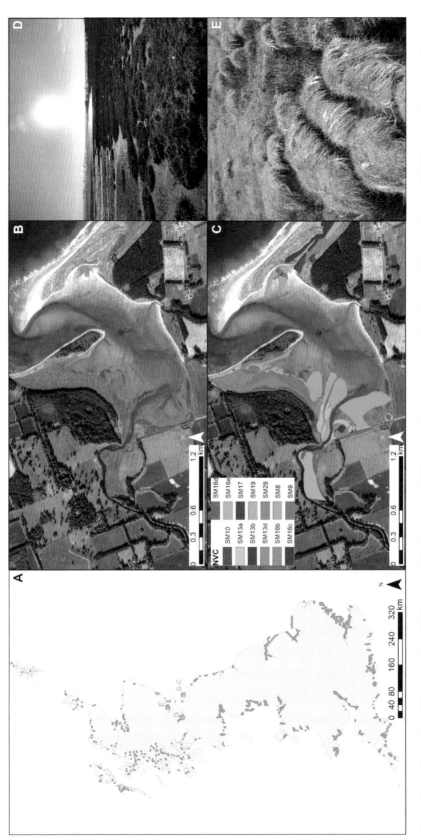

Figure 9.1 Saltmarsh at different scales. (A) UK saltmarsh habitat extent; (B) Tynemouth estuary aerial photo (www.digimap.edina.ac.uk); (C) Tynemouth estuary showing distribution of intertidal saltmarsh plant communities as per the National Vegetation Classification (NVC) (Haynes 2016); (D) Firth of Forth saltmarsh; (E) Tay estuary saltmarsh. Orange points indicate the location of panels B to E on panel A.

extent has slowed down since the 1980s (Morris et al. 2004), albeit losses continue in the south-east (Baily and Pearson 2001; Ladd et al. 2019). Marsh expansion has also occurred, particularly in west-coast marshes, stimulated by high sediment input and colonisation by the marsh-building plant Common Cordgrass *Spartina anglica* (syn. *Sporobolus anglicus*) (Ladd et al. 2019).

On balance, however, losses continue to exceed gains, including those from managed realignment (Phelan, Shaw and Baylis 2011, Boorman and Hazelden 2017) with a probable annual loss rate of <1% (Phelan, Shaw and Baylis 2011; Burden et al. 2020). There is considerable concern about future sea-level rise, as many marshes have human construction at the landward side, which blocks the inland migration of marshes. This process of coastal squeeze will, under projections of future sea-level rise, cause further saltmarsh habitat losses around the UK, and these losses will be driven, in part, by regional patterns of relative sea-level rise.

Since the nineteenth century, northern marshes have tended to gain in area, while those in the very south-east have eroded – a pattern explained by sediment supply, which is low in the south and high in the north (Ladd et al. 2019). Sediment supply is crucial to marsh erosion recovery, as recolonisation requires the seabed to exceed an elevation threshold where flooding frequency and hydrological disturbance is sufficiently low to permit plant colonisation and growth (Balke, Herman and Bouma 2014). Particles settling out of the water column include organic carbon (OC) that can represent a significant component of soil carbon stores (Rogers et al. 2019). Thus, sediment supply may be crucial to saltmarsh carbon storage through regulating recovery resilience and growth, by keeping marsh vertical accretion in pace with sea-level rise (Kirwan et al. 2016) and by directly injecting OC into accreting soils.

Marshes are naturally dynamic, and in local settings can undergo cyclical switching between expansion and erosion (Pringle 1995; Ladd 2021). Past shifts are evidenced by banded layers in sediment deposits, with multiple banding implying frequent switching in the depositional environment. The implications of marsh cyclicity for carbon stocks is unclear, although it seems likely that stocks will be positively correlated to stability (level of cyclicity), as disturbance causes carbon loss (Lang'at et al. 2014). Marsh stability varies geographically (Ladd 2021) and knowledge of regional variation in stability and sediment supply might help predict patterns of carbon stock and permanence across the UK. For example, sandy, west-coast marshes are thought to be more dynamic than east-coast marshes, although the implication to carbon stock is unknown. The nature of the constituent supply sediments, sandy versus muddy, also affects the carbon dynamics, with greater organic loading being associated with finer sediments (Sakamaki and Nishimura 2006).

Advances in field, laboratory and data methods

Current best practice for the estimation of blue carbon (including saltmarsh) OC stocks is summarised by Howard et al. (2014). These methods lay the foundation for the quantification of OC in intertidal environments. Advancements in sampling, mapping and laboratory analysis are constantly increasing the accuracy of saltmarsh C stock estimates across spatial and temporal scales.

Recent technological advances and the reduction in the costs of uncrewed aerial vehicles (UAV), for example, have revolutionised the mapping of coastal environments (Pham et al. 2019). UAVs are commonly employed as aerial photography platforms that, when used in conjunction with photogrammetry, produce high-resolution 2D and 3D maps widely used in coastal management (Gonçalves et al. 2015; Green et al. 2019; Pham et al. 2019). The applications for UAVs in coastal environments range beyond aerial photography, with their utilisation to characterise and quantify above-ground biomass (Doughty et al. 2019, 2021; Sani et al. 2019), topographic surveys utilising LiDAR (Pinto et al. 2020) and long-term monitoring of saltmarshes utilising the repeatability of surveys afforded by UAVs (Dai et al. 2020; Moore et al. 2021).

Soil core collection is another key to successful carbon stock assessment; currently, traditional paleo-environmental methods are largely employed, including gouge and Russian corers (Glew, Smol and Last 2002; Frew 2012) to collect material for analysis. More recently this has been supplemented with hammer coring techniques, where disposable plastic piping is hammered into the soil (Howard et al. 2014). The hammer coring method is low cost and can provide large quantities of sample material. However, recent research has highlighted that the hammer coring method can compact the soil, in turn altering the physical characteristics of the material, leading to significant errors in both OC stock and OC burial rates (Smeaton, Barlow and Austin 2020). Therefore, the tried and tested soil coring approaches (i.e. gouge and Russian corers) are recommended methods of sample collection for soil OC stock assessments.

The methods and techniques used to quantify OC in saltmarshes are well established and increasingly based upon the principles of elemental analysis, where the soil samples are combusted and the evolved gas measured to quantify the OC content of the material (Verardo, Froelich and McIntyre 1990; Nieuwenhuize, Maas and Middelburg 1994). However, the loss-on-ignition (LOI) method has been more widely used to estimate soil organic matter content and, indirectly, as a proxy for OC content (e.g. Craft et al. 1991), with recent evidence suggesting that the widely used conversion factor (LOI/OC = 1.724) may result in significant (20%) differences in some blue carbon OC stock estimates (Ouyang and Lee 2020). No such conversion factor has been available for UK saltmarsh soils until now, despite the wealth of historically available LOI data from these habitats and the growing evidence that global conversion factors may not always be reliable across a range of soil types (Pribyl 2010).

Using samples collected across multiple saltmarsh elevation transects as part of the UKRI-funded C-SIDE (carbon storage in intertidal environments) project (see Ruranska et al. 2020; Austin et al. 2021), we report a new UK saltmarsh OC to LOI relationship (equation 1; Fig. 9.2):

$$OC\% = 0.3693.LOI\% + 2.0888 \ (R^2 = 0.7792) \tag{1}$$

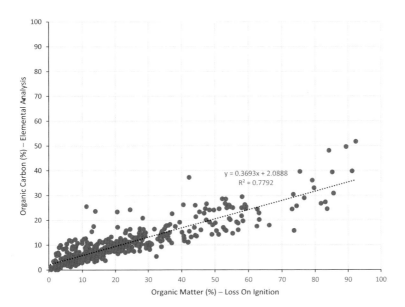

Figure 9.2 UK saltmarsh surface (10 cm) soil OC (%) and LOI data, highlighting the new UK saltmarsh soil OC–LOI relationship.

Within blue carbon research, there is also growing recognition that understanding the source of soil OC is a key consideration for the production of meaningful OC budgets and C accounting (Macreadie et al. 2019). Multiple biogeochemical approaches are available to estimate the quantity of OC originating from these autochthonous and allochthonous sources (Geraldi et al. 2019). The most commonly utilised approach uses light stable isotopes ($\delta^{13}C_{org}$ and $\delta^{15}N$) and elemental ratios (C/N, N/C) as tracers alongside mixing models (Thornton and McManus 1994) to estimate the input of OC from different sources (Cloern et al. 2002). More recently, these isotopic methods have been augmented with organic 'biomarkers', providing further insights into the source and composition of OC stored in saltmarshes (Bianchi and Canuel 2011; Vaughn et al. 2021). State-of-the-art analytical methods such as thermogravimetric analysis (Trevathan-Tackett et al. 2015), Fourier-transform infrared spectroscopy (Carnell et al. 2018) and nuclear magnetic resonance spectroscopy (Kelleway et al. 2017) are also being applied, providing greater insight into lability and molecular composition of the OC stored within saltmarsh soils.

After making point measurements of soil carbon, open-source tools can be used to extrapolate carbon contents across the entire length of a core (Hengl and MacMillan 2019); in combination with simple geometric and volumetric calculations, this allows saltmarsh soil OC stocks to be calculated (Howard et al. 2014). However, the application of state-of-the-art remote sensing and machine learning algorithms have been shown to produce superior marsh-scale stock estimates for several reasons: (a) the remotely sensed predictors of carbon, including elevation and vegetation characteristics (Normalised Difference Vegetation Index) are globally available, often free to use and have high spatial resolution (Pham et al. 2019); (b) a large variety of machine learning algorithms are now open source and relatively simple to apply (Pedregosa et al. 2011; Kuhn et al. 2020); (c) detailed maps can be produced showing how carbon is distributed in the landscape, and where the greatest levels of uncertainty in carbon predictions are located (Ladd et al., submitted). Additionally, if proxies for carbon stock can be derived from satellite data, image archives could be analysed to retrospectively track carbon stock change over time and to help inform how marsh carbon stocks may change in the future. New tools such as Google Earth Engine allow researchers to easily access satellite data for blue carbon research (Fitton et al. 2021).

Coastal research is attracting increasing interest from the general public in the natural environment. Citizen science engagement has allowed UK saltmarsh research to be scaled for the collection of samples and data nationwide, delivering new opportunities for national saltmarsh soil OC surveys. This and other such initiatives have been achieved by the growing interest of the public in the natural environment, coupled with projects that facilitate active participation (Martin, Christidis and Pecl 2016). Currently, there are numerous citizen science projects active in the saltmarsh and blue carbon sphere; these include the global Tea Composition H_2O project (www.bluecarbonlab.org/teacomposition-h2o/), which is placing teabags in coastal wetlands to better understand the degradation of fresh OC (Keuskamp et al. 2013; Mueller et al. 2018; Marley, Smeaton and Austin 2019). In the UK, the C-SIDE project has engaged with members of the public all over the country to collect surficial soil samples (top 10 cm) from saltmarshes (www.c-side.org/citizen-science-1); coupled with user-friendly tools (e.g. The Saltmarsh App, 2021), this has allowed the quantification of new national surficial saltmarsh soil OC stocks (Austin et al. 2021).

National-scale carbon stock assessments can be readily achieved, as outlined in the recent saltmarsh soil carbon stock assessment for Scotland (Austin et al. 2021) when (a) improved spatial estimates of present-day (and past for land-use change assessments) habitat extent is known; (b) information is available to constrain plant community (National Vegetation Classification, NVC) type and distributions across saltmarsh surfaces (e.g. Haynes 2016); (c) sufficient data exist to relate plant community (NVC) to underlying soil OC and soil density; (d) knowledge of the stratigraphy and soil OC content are available; and (e) appropriate upscaling statistical methods are used to incorporate uncertainty estimates.

Sea-level change around the UK – saltmarsh records of change

Relative sea-level (RSL) change is a consequence of changes in the volume and distribution of ocean mass, a wide variety of geophysical processes that operate on different spatial and temporal scales and a range of local and regional processes such as atmosphere/ocean dynamics and sediment compaction. In the UK, the primary pattern of present RSL change is driven by ongoing solid Earth response to the deglaciation of the British–Irish Ice Sheet (from *c.*20 ka BP), with land uplifting in Scotland and sinking in southern England (Lambeck, Smither and Johnston 1998; Peltier et al. 2002; Bradley et al. 2011; Shennan, Bradley and Edwards 2018). This results in a north–south regional gradient in RSL change in the UK, with faster rates of RSL rise in southern England and stable, or until recently even falling, RSL in north-west Scotland (Shennan, Bradley and Edwards 2018). The spatial pattern of the rate and magnitude of late Holocene RSL change (last *c.*2 ka) has been captured in high-resolution sea-level records reconstructed from saltmarsh sediments (e.g. Barlow et al. 2014; Long et al. 2014; Rushby et al. 2019).

At the start of the twenty-first century, global mean sea level was rising at about 3.2 mm/yr (Church et al. 2013). In the future, this rate will increase, and global mean sea level is projected to rise up to 1.1 m by 2100, dependent on the emission pathways (their representative concentration pathways, hereafter RCP) followed (Oppenheimer et al. 2019), largely owing to increases in thermal expansion of the ocean, and ice sheet and glacier melting. This gives a projected century-average sea-level rise rate of up to 11.0 mm/yr. The most up-to-date UK Climate Projections (UKCP18) deviate from these global mean projections owing to spatially variable solid Earth processes. Under RCP 8.5, sea level is projected to rise 0.53–1.15 m in London by 2100, and 0.30–0.90 m in Edinburgh (Palmer et al. 2018). Intertidal environments, and their associated blue carbon stores, are increasingly vulnerable to current and future projected high rates of RSL rise, with a potential reduction in coastal wetland owing to marsh submergence as RSL rise (RSLR) outpaces marsh platform accretion rates (Reed 1995; Allen 2000; Morris et al. 2002). Other studies (French 2006; Kirwan et al. 2016; Horton et al. 2018; Gonneea et al. 2019) suggest that the vulnerability of saltmarshes to future accelerated sea-level rise has been overestimated. Kirwan et al. (2016) and Gonneea et al. (2019) conclude that marsh survival is possible even at high rates of RSLR, especially in low saltmarsh zones and in areas where the marsh can migrate landwards. Here, vertical accretion rates may keep pace with RSLR, enabling the marsh to remain in stable dynamic equilibrium with sea level. Horton et al. (2018) used Holocene relative sea-level data to predict future marsh vulnerability, based on a dataset of over 780 Holocene sea-level data points from the UK. Their results show that when Holocene RSLR rates were similar to modern rates, few marshes actually retreated. However, these authors showed that marsh retreat in the Holocene has been far more common than marsh expansion during periods of rapid RSLR, with marshes nine times more likely to retreat than expand when RSLR rates are ≥7.1 mm/yr.

Saltmarshes have been recognised as globally significant in their capacity to store carbon (Mcleod et al. 2011; Duarte et al. 2013; Beaumont et al. 2014; Ouyang and Lee 2014; Macreadie et al. 2019). With sea level seen as a potentially dominant control, several studies have explored the relationship between RSLR and net OC accumulation in saltmarshes (Kirwan and Mudd 2012; FitzGerald and Hughes 2019; Gonneea et al. 2019; McTigue et al. 2019; Rogers et al. 2019; Haywood at al. 2020; Herbert, Windham-Myers and Kirwan 2021). One proposal is that instead of being vulnerable to increased RSLR, saltmarshes located in areas of accelerated sea-level rise are responding by burying even more carbon in their soils, owing to the creation of additional vertical accommodation space (space for particle settlement), but this is only sustainable if there is room for coastal wetlands to migrate inland (Rogers et al. 2019). However, inland migration is not always possible owing to man-made barriers, resulting in 'coastal squeeze' (Torio and Chmura, 2013). Other studies (Kirwan et al. 2016; Gonneea et al. 2019) suggest that the high rates of RSLR projected for the twenty-first century are likely to cause the replacement of high-marsh plant communities by low-marsh communities. With the high marsh unable to keep up with projected increases in RSLR, it will lose elevation in the tidal frame, resulting in the transgression of the more resilient low-marsh vegetation (that can sustain higher accretion rates) over the former high marsh. As the rate of RSLR is thought to be one of the dominant controls on saltmarsh accretion and net OC accumulation, the UK as a whole is considered an excellent location to test the hypothesis that the recent acceleration in relative sea-level rise has increased the capacity of salt marshes to store carbon, and to quantify the vulnerability of saltmarsh carbon stores to future sea-level change under current shoreline management plans (Austin et al. 2020).

Protection, restoration and creation of saltmarsh habitats

Coastal saltmarsh occupies the interface between land and sea, and its environment has some features reflected in both its flora and fauna. The vascular plants, essential for recognition of saltmarsh, are essentially terrestrial in origin and the habitat is best regarded as a highly modified terrestrial ecosystem, variously adapted to, or tolerant of, a semi-marine environment (Adam 1990). As a result, saltmarsh comprises a highly adapted and unique species assemblage protected by national and international regulation. Coastal saltmarsh is included in the EU Habitats Directive (EU 1992), of which four Annex 1 habitats are found wholly within this habitat (glassworts *Salicornia* and other annuals; cord-grass *Spartina* swards; Atlantic Salt Meadows; Mediterranean and thermo-Atlantic halophilous scrubs). The majority of saltmarsh is also designated under the Site of Special Scientific Interest framework (Rees et al. 2019), which gives protection under law from development and other damage.

Given the high-level protection saltmarshes are afforded, there have been considerable efforts to restore and create new habitat to make up for historical and ongoing losses through land reclamation or a changing climate; such initiatives increasingly recognise that carbon-rich sediments are trapped and sequestered in the process of saltmarsh restoration and creation.

There are a number of techniques that have been tested with various levels of success, ranging from small-scale interventions to wholesale realignment of the coastline. Transplanting turves of vegetation into unvegetated mudflat can quickly recreate stable saltmarsh communities, which in turn increase and stabilise sediment deposition rates (Maynard et al. 2011). However, saltmarshes are subject to internal erosion, particularly in the south and east of England, owing to changes in wind/wave climate (van der Wal and Pye 2004). Low-tech methods such as brushwood fences or coir logs have been used as 'leaky dams' to reduce erosion by buffering wave energy and helping trap sediment to maintain internal structural integrity. Moreover, the dams allow two-way flow of water, avoiding waterlogged sediments and at the same time reducing downward erosive forcing. The use of sediment fields, replicating Dutch-style systems, have also been trialled with less success, mainly owing to a higher tidal range in the UK and lower turbidity levels. The beneficial use of dredged material has also been used to replenish marshes depleted of sediment, but needs repeated application to maintain mudflat levels. Recharge with gravel and/or sand can be used to form artificial cheniers, which may protect eroding marshes from wave and current erosion. The creative use of 'firm' dredge spoil can combat undercutting and collapse of the saltmarsh edge and current-induced erosion (EA 2007). Regulated tidal exchange (RTE) is the controlled exchange of seawater to an area behind fixed sea defences, through engineered structure such as sluices, tide-gates or pipes, to create saline or brackish habitats (EA 2003). RTE can be an effective way of capturing and storing sediments with rates of up to 4 cm/yr in the early stages of site development (Marris et al. 2007).

Managed realignment

Managed realignment (MR), the deliberate breaching of coastal defences and subsequent tidal inundation to restore intertidal habitat, is the predominant method of saltmarsh restoration in the UK (Burden, Garbutt and Evans 2019). The first managed realignment scheme in the UK was carried out in 1991 at a site in south-east England as a demonstration for coastal defence (Doody 2013), and there are now 46 MR sites that have created saltmarsh habitat across Great Britain, identified by the 'Online managed realignment guide' (ABPmer 2021). These managed realigned schemes have been a core strategy to protect coastlines and reduce flood costs, as well as a principal method for habitat provision and compensation for historic losses of intertidal habitat, in line with the EU Habitats Directive (Esteves 2013; Brady and Boda 2017; ABPmer 2021). Saltmarsh carbon storage and climate benefits have neither been a reason for restoration, nor a criterion to measure the success, or failure, of MR schemes in the UK (Burden et al. 2013). However, the role of blue carbon as a means for emission reductions in line with national targets has gained popularity in climate policy and relies on both the successful management and restoration of coastal wetlands (Kelleway et al. 2020).

Interest in nature-based solutions to mitigate climate change are on the rise (Seddon et al. 2020; Austin et al. 2021), and the declaration of the United Nations Decade on Ecosystem Restoration (2021–30), which aims to accelerate global restoration of degraded ecosystems, in part to tackle the climate crisis (Waltham et al. 2020), complements this global movement. Currently, the implementation of blue carbon policy and management actions, including saltmarsh restoration via MR, remains underdeveloped (Macreadie et al. 2019). The growing political dialogue around blue carbon and nature-based solutions to climate change, which played a significant role in the COP26 discussions hosted by the UK, presents an opportunity for future managed realignment schemes to be designed for the maximisation of climate change mitigation benefits, or at least for the inclusion of blue carbon as a measurable co-benefit of future restoration schemes. Current knowledge of drivers that best account for variation in saltmarsh blue carbon stocks and accumulation, for example local hydrology (Macreadie et al. 2017) and particle size (Kelleway et al. 2016), can be used for predictive mapping of potential hotspots, or optimal locations to maximise blue carbon storage, as focal points for future MR schemes. Careful monitoring at MR sites is required to avoid CO_2 (and other greenhouse gas) fluxes to the atmosphere following future disturbance (Macreadie et al. 2019). Furthermore, the success of such schemes, alongside long-term monitoring, would be enhanced by a clear identification of measures of success and evaluation techniques, interdisciplinary research and collaboration, and policy alignment (Waltham et al. 2020).

Emerging policy and investment opportunities

In 2019, an administrative agreement between Defra and the Inter-American Development Bank was reached, where Defra agreed, through UK official development assistance, to contribute £12.75m to the UK Blue Carbon Fund. This cemented the UK governments' commitment to the role of blue carbon in mitigating climate change through International Climate Finance.

Today, that blue carbon commitment is reflected across the devolved nations of the UK in the Environment Strategy for Scotland, the Wales Environment Bill and the 25-year plan for England. Northern Ireland launched a consultation for their own Environment Strategy in late 2019, which reflects the policies and commitment of the other UK nations in sustainably managing natural resources and the benefits they provide, positively embracing a nature-based solutions approach.

Greenhouse gas (GHG) emissions and removals resulting from changes in saltmarsh management can be included in national emission accounting under the Land Use, Land Use Change and Forestry sector. However, they are not included in the UK GHG inventory (GHGI) at this time. The Intergovernmental Panel on Climate Change (IPCC) Wetland Supplement (IPCC 2014) includes guidelines for the quantification and accounting of GHG emissions and removals associated with the management of different wetland types, including drainage and rewetting of tidal marsh. Inclusion of saltmarshes in the UK GHGI is considered by many as a key objective that will enhance current efforts to account for, protect and restore these long-term carbon stores, realising their potential for climate change mitigation. The UK has set a precedent for their inclusion by electing to report emissions from peatlands in its national inventory for the second commitment period of the Kyoto Protocol, and by 2022 at the latest, under the obligations of the UN Framework Convention on Climate Change. By implementing restoration of saltmarsh (mostly via managed realignment) as a nature-based solution to capture carbon and therefore remove GHG from the atmosphere, saltmarsh can contribute to the UK governments' commitment (by law under the UK's Climate Change Act) to reduce GHG emissions and achieve net zero by 2050.

Conclusions

Our chapter highlights best practice in achieving national-scale carbon stock assessments for UK saltmarsh habitats, including advances in field, laboratory and data methods. National-scale carbon stock assessments can be readily achieved when (a) improved spatial estimates of present-day (and past for land-use change assessments) habitat extent is known; (b) information is available to constrain plant community (NVC) type and distributions across saltmarsh surfaces; (c) sufficient data exist to relate plant community (NVC) to underlying soil OC and soil density; (d) knowledge of the stratigraphy and soil OC content are available; and (e) appropriate upscaling statistical methods are used to incorporate uncertainty estimates. In this chapter, we also introduce the first UK-specific saltmarsh conversion for LOI estimates of soil organic matter to soil OC, which will greatly enhance the use of historical LOI data to bolster UK saltmarsh OC stock assessments. We have assessed the underlying drivers that determine the spatial distribution, magnitude and future vulnerability of UK saltmarsh habitats – these drivers are numerous and include RSLR, the availability of adjacent land for coastline migration (or, when unavailable, for generating coastal squeeze), sediment supply, coastal exposure to wave energy and so on. Long-term sea-level drivers are likely to play an important role in shaping UK coastal environments and associated carbon stocks, with the risks of carbon stock losses arising from coastal squeeze likely to be acute in some regions. However, there is great potential for management interventions that safeguard these long-term carbon stores through the protection, restoration and creation of saltmarsh habitats around the UK; some of these opportunities will require changes in coastal management policy and new investment. The realisation of the climate mitigation potential of UK saltmarsh habitats, and other blue carbon habitats around the globe, highlight an important opportunity to account for, protect and restore these long-term carbon stores in our coastal environments. The carbon stocks in UK saltmarsh habitats provide an emerging policy opportunity for their inclusion in national greenhouse gas inventory accounting; and we conclude that progress towards the implementation of this habitat into the UK greenhouse gas inventory is now overdue.

Acknowledgements

This work was supported by the Natural Environment Research Council (grant NE/R010846/1) Carbon Storage in Intertidal Environments (C-SIDE) project. We thank Dr Heather Austin for help preparing the manuscript and Dr Sally Little for constructive review comments that greatly improved the final version of this chapter.

References

ABPmer (2021) Online managed realignment guide. Available at: www.abpmer.net/omreg/ (Accessed: 10 April 2021).

Adam, P. (1990) *Saltmarsh Ecology*. Cambridge Studies in Ecology. Cambridge: Cambridge University Press. https://doi.org/10.1017/CBO9780511565328

Allen, J.R.L. (2000) Morphodynamics of Holocene saltmarshes: a review sketch from the Atlantic and southern North Sea coasts of Europe. *Quaternary Science Reviews* 19 (12): 1155–1231. https://doi.org/10.1016/S0277-3791(99)00034-7

Austin, W.E.N., Smeaton, C., Riegel, S., Ruranska, P. and Miller, L. (2021) Blue carbon stock in Scottish saltmarsh soils. *Scottish Marine and Freshwater Science* 12 (13). http://doi.org/10.7489/12372-1

Austin, W.E.N., Smeaton, C., Ladd, C.J.T. and Havelock, G.M. (2020) Carbon storage in intertidal environments (C-SIDE). IQUA Autumn Symposium 2020: Carbon Sequestration. 27 November.

Baily, B. and Pearson, A.W. (2001) Change detection mapping of saltmarsh areas of South England from Hurst Castle to Pagham Harbour. Department of Geography, University of Portsmouth report to Posford Haskoning consultants, English Nature and Environment Agency.

Balke, T., Herman, P.M.J. and Bouma, T.J. (2014) Critical transitions in disturbance-driven ecosystems: identifying windows of opportunity for recovery. *Journal of Ecology* 102: 700–8. https://doi.org/10.1111/1365-2745.12241

Barlow, N., Long, A., Saher, M., Gehrels, W., Garnett, M. and Scaife, R. (2014) Saltmarsh reconstructions of relative sea-level change in the North Atlantic during the last 2000 years. *Quaternary Science Reviews* 99: 1–16. https://doi.org/10.1016/j.quascirev.2014.06.008

Beaumont, N.J., Jones, L., Garbutt, A., Hansom, J.D. and Toberman, M. (2014) The value of carbon sequestration and storage in coastal habitats. *Estuarine, Coastal and Shelf Science* 137: 32–40. https://doi.org/10.1016/j.ecss.2013.11.022

Bianchi, T.S. and Canuel, E.A. (2011) *Chemical Biomarkers in Aquatic Ecosystems*. Princeton, NJ: Princeton University Press. https://doi.org/10.23943/princeton/9780691134147.001.0001

Boorman, L.A. (2003) Saltmarsh review. An overview of coastal saltmarshes, their dynamic and sensitivity characteristics for conservation and management. Joint Nature Conservation Committee Report No. 334, JNCC, Peterborough.

Boorman, L.A. and Hazelden, J. (2017) Managed re-alignment; a salt marsh dilemma? *Wetlands Ecology and Management* 25: 387–403. https://doi.org/10.1007/s11273-017-9556-9

Bradley, S.L., Milne, G.A., Shennan, I. and Edwards, R. (2011) An improved glacial isostatic adjustment model for the British Isles. *Journal of Quaternary Science* 26: 541–52. https://doi.org/10.1002/jqs.1481

Brady, A.F. and Boda, C.S. (2017) How do we know if managed realignment for coastal habitat compensation is successful? Insights from the implementation of the EU Birds and Habitats Directive in England. *Ocean & Coastal Management* 143: 164–74. https://doi.org/10.1016/j.ocecoaman.2016.11.013

Burden, A., Garbutt, R.A., Evans, C.D., Jones, D.L. and Cooper, D.M. (2013) Carbon sequestration and biogeochemical cycling in a saltmarsh subject to coastal managed realignment. *Estuarine, Coastal and Shelf Science* 120: 12–20. https://doi.org/10.1016/j.ecss.2013.01.014

Burden, A., Garbutt, A. and Evans, C.D. (2019) Effect of restoration on saltmarsh carbon accumulation in Eastern England. *Biology Letters* 15 (1): 20180773. https://doi.org/10.1098/rsbl.2018.0773

Burden, A., Smeaton, C., Angus, S., Garbutt, A., Jones, L., Lewis, H.D. and Rees, S.M. (2020) Impacts of climate change on coastal habitats relevant to the coastal and marine environment around the UK. *MCCIP Science Review* 2020: 228–55.

Carnell, P.E., Windecker, S.M., Brenker, M., Baldock, J., Masque, P., Brunt, K. and Macreadie, P.I. (2018) Carbon stocks, sequestration, and emissions of wetlands in south eastern Australia. *Global Change Biology* 24: 4173–84. https://doi.org/10.1111/gcb.14319

Church, J.A., Clark, P.U., Cazenave, A., Gregory, J.M., Jevrejeva, S., Levermann, A., Merrifield, M.A., Milne, G.A., Nerem, R.S., Nunn, P.D., Payne, A.J., Pfeffer, W.T., Stammer D. and Unnikrishnan, A.S. (2013) Sea level change. In T.F. Stocker, D. Qin, G.-K. Plattner, M. Tignor, S.K. Allen, J. Boschung, A. Nauels, Y. Xia, V. Bex and P.M. Midgley (eds.) *Climate Change 2013: The Physical Science Basis.* Contribution of Working Group I to the Fifth Assessment Report of the Intergovernmental Panel on Climate Change. Cambridge and New York: Cambridge University Press.

Cloern, J.E., Canuel, E.A. and Harris, D. (2002) Stable carbon and nitrogen isotope composition of aquatic and terrestrial plants of the San Francisco Bay estuarine system. *Limnology and Oceanography* 47 (3): 713–29. https://doi.org/10.4319/lo.2002.47.3.0713

Craft, C.D., Seneca, E.D. and Broome, S.W. (1991) Loss on ignition and Kjeldahl digestion for estimating organic carbon and total nitrogen in estuarine marsh soils: calibration with dry combustion. *Estuaries* 14 (2): 175–9. https://doi.org/10.2307/1351691

Dai, W., Li, H., Chen, X., Xu, F., Zhou, Z. and Zhang, C. (2020) Saltmarsh expansion in response to morphodynamic evolution: field observations in the Jiangsu coast using UAV. *Journal of Coastal Research* 95 (SI): 433–7. https://doi.org/10.2112/SI95-084.1

Doody, J.P. (2013) Coastal squeeze and managed realignment in southeast England, does it tell us anything about the future? *Ocean & Coastal Management* 79: 34–41. https://doi.org/10.1016/j.ocecoaman.2012.05.008

Doughty, C.L. and Cavanaugh, K.C. (2019) Mapping coastal wetland biomass from high resolution unmanned aerial vehicle (UAV) imagery. *Remote Sensing* 11 (5): 540. https://doi.org/10.3390/rs11050540

Doughty, C.L., Ambrose, R.F., Okin, G.S. and Cavanaugh, K.C. (2021) Characterizing spatial variability in coastal wetland biomass across multiple scales using UAV and satellite imagery. *Remote Sensing in Ecology and Conservation*. https://doi.org/10.1002/rse2.198

Duarte, C.M., Losada, I.J., Hendriks, I.E., Mazarrasa, I. and Marba, N. (2013) The role of coastal plant communities for climate change mitigation and adaptation. *Nature Climate Change* 3: 961–8. https://doi.org/10.1038/nclimate1970

Environment Agency (2003) Regulated tidal exchange: an intertidal habitat creation technique. EA contract report, Kingfisher House, Peterborough.

Environment Agency (2007) Saltmarsh management manual. EA R&D report: SC030220. Kingfisher House, Peterborough.

Esteves, L.S. (2013) Is managed realignment a sustainable long-term coastal management approach? *Journal of Coastal Research* (65): 933–8. https://doi.org/10.2112/SI65-158.1

European Union (1992) Council Directive 92/43/EEC.

Fitton, J.M., Rennie, A.F., Hansom, J.D., Muir, F.M.E. (2021) Remotely sensed mapping of the intertidal zone: A Sentinel-2 and Google Earth Engine methodology. *Remote Sensing Applications: Society and Environment* 22: 100499. https://doi.org/10.1016/j.rsase.2021.100499

FitzGerald, D.M. and Hughes, Z. (2019) Marsh processes and their response to climate change and sea-level rise. *Annual Review of Earth and Planetary Sciences* 47: 481–517. https://doi.org/10.1146/annurev-earth-082517-010255

French, J. (2006) Tidal marsh sedimentation and resilience to environmental change: exploratory modelling of tidal, sea-level and sediment supply forcing in predominantly allochthonous systems. *Marine Geology* 235 (1–4): 119–36. https://doi.org/10.1016/j.margeo.2006.10.009

Frew, C. (2012) Section 4.1.1: Coring Methods. In S.J. Cook, L.E. Clarke and J.M. Nield (eds) *Geomorphological Techniques* (online edition). London: British Society for Geomorphology. https://www.geomorphology.org.uk/geomorph_techniques

Geraldi, N.R., Ortega, A., Serrano, O., Macreadie, P.I., Lovelock, C.E., Krause-Jensen, D., Kennedy, H., Lavery, P.S., Pace, M.L., Kaal, J. and Duarte, C.M. (2019) Fingerprinting blue carbon: rationale and tools to determine the source of organic carbon in marine depositional environments. *Frontiers in Marine Science* 6: 263. https://doi.org/10.3389/fmars.2019.00263

Glew, J.R., Smol, J.P. and Last, W.M. (2002) Sediment core collection and extrusion tracking environmental change using lake sediments. In W.M. Last and J.P. Smol (eds) *Tracking Environmental Change Using Lake Sediments*, pp. 73–105. Dordrecht: Springer. https://doi.org/10.1007/0-306-47669-X_5

Gonçalves, J.A. and Henriques, R. (2015) UAV photogrammetry for topographic monitoring of coastal areas. *ISPRS Journal of Photogrammetry and Remote Sensing* 104: 101–11.

Gonneea, M.E., Maio, C.V., Kroeger, K.D., Hawkes, A.D., Mora, J., Sullivan, R., Madsen, S., Buzzard, R.M., Cahill, N. and Donnelly, J.P. (2019) Salt marsh ecosystem restructuring enhances elevation resilience and carbon storage during accelerating relative sea-level rise. *Estuarine, Coastal and Shelf Science* 217: 56–68. https://doi.org/10.1016/j.ecss.2018.11.003

Green, D.R., Hagon, J.J., Gómez, C. and Gregory, B.J. (2019) Using low-cost UAVs for environmental monitoring, mapping, and modelling: examples from the coastal zone. In R.R. Krishnamurthy, M.P. Jonathan, S. Srinivasalu and B. Glaeser (eds), *Coastal Management: Global Challenges and Innovations*, pp. 465–501. Academic Press. https://doi.org/10.1016/B978-0-12-810473-6.00022-4

Hansom, J.D., Lees, R.G., Maslen, J., Tilbrook, C. and McManus, J. (2001) Coastal dynamics and sustainable management: the potential for managed realignment in the Forth estuary. In J.E. Gordon and K.F. Lees (eds) *Earth Science and the Natural Heritage*, pp. 148–60. Edinburgh: TSC.

Haywood, B.J., Hayes, M.P., White, J.R. and Cook, R.L. (2020) Potential fate of wetland soil carbon in a deltaic coastal wetland subjected to high relative sea level rise. *Science of the Total Environment* 711: 135185. https://doi.org/10.1016/j.scitotenv.2019.135185

Haynes, T. (2016) Scottish saltmarsh survey national report. Commissioned Report No. 786, Scottish Natural Heritage.

Hengl, T. and MacMillan, R.A. (2019) *Predictive Soil Mapping with R*. Wageningen: OpenGeoHub foundation. Available at: www.soilmapper.org (Accessed: 2 September 2021).

Herbert, E.R., Windham-Myers, L. and Kirwan, M.L. (2021) Sea level rise enhances carbon accumulation in United States tidal wetlands. *One Earth* 4: 425–33. https://doi.org/10.1016/j.oneear.2021.02.011

Horton, B.P., Shennan, I., Bradley, S.L., Cahill, N., Kirwan, M., Kopp, R.E. and Shaw, T.A. (2018) Predicting marsh vulnerability to sea-level rise using Holocene relative sea-level data. *Nature Communications* 9: 2687. https://doi.org/10.1038/s41467-018-05080-0

Howard, J., Hoyt, S., Isensee, K., Telszewski, M. and Pidgeon, E. (2014) Coastal blue carbon: methods for assessing carbon stocks and emissions factors in mangroves, tidal salt marshes, and seagrasses. Conservation International, Intergovernmental Oceanographic Commission of UNESCO, International Union for Conservation of Nature. Arlington, VA.

IPCC (2014) 2013 supplement to the 2006 IPCC guidelines for national greenhouse gas inventories: wetlands, ed. T. Hiraishi, T. Krug, K. Tanabe, N. Srivastava, J. Baasansuren, M. Fukuda, and T.G. Troxler. IPCC, Switzerland.

Kelleway, J.J., Saintilan, N., Macreadie, P.I. and Ralph, P.J. (2016) Sedimentary factors are key predictors of carbon storage in SE Australian saltmarshes. *Ecosystems* 19 (5): 865–80. https://doi.org/10.1007/s10021-016-9972-3

Kelleway, J.J., Saintilan, N., Macreadie, P.I., Baldock, J.A. and Ralph, P.J. (2017) Sediment and carbon deposition vary among vegetation assemblages in a coastal salt marsh. *Biogeosciences* 14 (16): 3763–79. https://doi.org/10.5194/bg-14-3763-2017

Kelleway, J.J., Serrano, O., Baldock, J.A., Burgess, R., Cannard, T., Lavery, P.S., Lovelock, C.E., Macreadie, P.I., Masqué, P., Newnham, M. and Saintilan, N. (2020) A national approach to greenhouse gas abatement through blue carbon management. *Global Environmental Change* 63: 102083. https://doi.org/10.1016/j.gloenvcha.2020.102083

Kirwan, M.L., Temmerman, S., Skeehan, E.E., Guntenspergen, G.R. and Fagherazzi, S. (2016) Overestimation of marsh vulnerability to sea level rise. *Nature Climate Change* 6: 253–60. https://doi.org/10.1038/nclimate2909

Kirwan, M.L. and Mudd, S.M. (2012) Response of salt-marsh carbon accumulation to climate change. *Nature* 489: 550–3. https://doi.org/10.1038/nature11440

Keuskamp, J.A., Dingemans, B.J., Lehtinen, T., Sarneel, J.M. and Hefting, M.M. (2013) Tea Bag Index: a novel approach to collect uniform decomposition data across ecosystems. *Methods in Ecology and Evolution* 4 (11): 1070–5. https://doi.org/10.1111/2041-210X.12097

Kuhn, M., Wing, J., Weston, S., Williams, A., Keefer, C., Engelhardt, A., Cooper, T., Mayer, Z., Kenkel, B. and Team, R.C. (2020) Package 'caret'. *The R Journal* 223.

Ladd, C.J.T., Skov, M.W., Austin, W.E.N. and Smeaton, C. (submitted) Best practice for upscaling soil organic carbon stocks in salt marshes. *Geoderma*.

Ladd, C.J.T. (2021) Review on processes and management of saltmarshes across Great Britain. *Proceedings of the Geologists' Association* 132: 269–83. https://doi.org/10.1016/j.pgeola.2021.02.005

Ladd, C.J.T., Duggan-Edwards, M.F., Bouma, T.J., Pagès, J.F. and Skov, M.W. (2019) Sediment supply explains long-term and large-scale patterns in salt marsh lateral expansion and erosion. *Geophysical Research Letters* 46: 11178–87. https://doi.org/10.1029/2019GL083315

Lambeck, K., Smither, C. and Johnston, P. (1998) Sea-level change, glacial rebound and mantle viscosity for northern Europe. *Geophysical Journal International* 134 (1): 102–44. https://doi.org/10.1046/j.1365-246x.1998.00541.x

Lang'at, J.K.S., Kairo, J.G., Mencuccini, M., Bouillon, S., Waldron, S., Skov, M.W. and Huxham, M. (2014) Rapid losses of surface elevation following tree girdling and cutting in tropical mangroves. *PloS One*. https://doi.org/10.1371/journal.pone.0107868

Long, A.J., Barlow, N.L.M., Gehrels, W.R., Saher, M.H., Woodworth, P.L., Scaife, R.G., Brain, M.J. and Cahill, M. (2014) Contrasting records of sea-level change in the eastern and western North Atlantic during the last 300 years. *Earth and Planetary Science Letters* 388: 110–22. https://doi.org/10.1016/j.epsl.2013.11.012

Macreadie, P.I., Anton, A., Raven, J.A., Beaumont, N., Connolly, R.M., Friess, D.A., Kelleway, J.J., Kennedy, H., Kuwae, T., Lavery, P.S., Lovelock, C.E., Smale, D.A., Apostolaki, E.T., Atwood, T.B., Baldock, J., Bianchi, T.S., Chmura, G.L., Eyre, B.D., Fourqurean, J.W., Hall-Spencer, J.M., Huxham, M., Hendriks, I.E., Krause-Jensen, D., Laffoley, D., Luisetti, T., Marbà, N., Masque, P., McGlathery, K.J., Megonigal, J.P., Murdiyarso, D., Russell, B.D., Santos, R., Serrano, O., Silliman, B.R., Watanabe, K. and Duarte, C.M. (2019) The future of Blue Carbon science. *Nature Communications* 10: 3998. https://doi.org/10.1038/s41467-019-11693-w

Macreadie, P.I., Nielsen, D.A., Kelleway, J.J., Atwood, T.B., Seymour, J.R., Petrou, K., Connolly, R.M., Thomson, A.C., Trevathan-Tackett, S.M. and Ralph, P.J. (2017) Can we manage coastal ecosystems to sequester more blue carbon? *Frontiers in*

Ecology and the Environment 15 (4): 206–13. https://doi.org/10.1002/fee.1484

Maris, T., Cox, T., Temmerman, S., De Vleeschauwer, P., Van Damme, S., De Mulder, T., Van den Bergh, E. and Meire, P. (2007) Tuning the tide: creating ecological conditions for tidal marsh development in a controlled inundation area. *Hydrobiologia* 588: 31–43. https://doi.org/10.1007/s10750-007-0650-5

Marley, A.R., Smeaton, C. and Austin, W.E. (2019) An assessment of the Tea Bag Index method as a proxy for organic matter decomposition in inter-tidal environments. *Journal of Geophysical Research: Biogeosciences* 124 (10): 2991–3004. https://doi.org/10.1029/2018JG004957

Martin, V.Y., Christidis, L. and Pecl, G.T. (2016) Public interest in marine citizen science: is there potential for growth?. *BioScience* 66 (8): 683–92. https://doi.org/10.1093/biosci/biw070

Maynard, C., McManus, J., Crawford, R.M.M. and Paterson, D.M. (2011) A comparison of short-term sediment deposition between natural and trans-planted saltmarsh after saltmarsh restoration in the Eden Estuary (Scotland). *Plant Ecology & Diversity* 4: 103–13. https://doi.org/10.1080/17550874.2011.560198

Mcleod, E., Chmura, G.L., Bouillon, S., Salm, R., Björk, M., Duarte, C.M., Lovelock, C.E., Schlesinger, W.H. and Silliman, B.R. (2011) A blueprint for blue carbon: toward an improved understanding of the role of vegetated coastal habitats in sequestering CO_2. *Frontiers in Ecology and the Environment* 9: 552–60. https://doi.org/10.1890/110004

McTigue, N., Davis, J., Rodriguez, A.B., McKee, B., Atencio, A. and Currin, C. (2019) Sea level rise explains changing carbon accumulation rates in a salt marsh over the past two millennia. *Journal of Geophysical Research: Biogeosciences* 124 (10): 2945–57. https://doi.org/10.1029/2019JG005207

Moore, G.E., Burdick, D.M., Routhier, M.R., Novak, A.B. and Payne, A.R. (2021) Effects of a large-scale, natural sediment deposition event on plant cover in a Massachusetts salt marsh. *PloS One* 16 (1): p.e0245564. https://doi.org/10.1371/journal.pone.0245564

Morris, R.K.A., Reach, I.S., Duffy, M.J., Collins, T.S. and Leafe, R.N. (2004) On the loss of saltmarshes in south-east England and the relationship with *Nereis diversicolor*. *Journal of Applied Ecology* 41: 787–91. https://doi.org/10.1111/j.0021-8901.2004.00932.x

Mueller, P., Schile-Beers, L.M., Mozdzer, T.J., Chmura, G.L., Dinter, T., Kuzyakov, Y., Groot, A.V.D., Esselink, P., Smit, C., D'Alpaos, A. and Ibáñez, C. (2018) Global-change effects on early-stage decomposition processes in tidal wetlands – implications from a global survey using standardized litter. *Biogeosciences* 15 (10): 3189–202. https://doi.org/10.5194/bg-15-3189-2018

Nieuwenhuize, J., Maas, Y.E. and Middelburg, J.J. (1994) Rapid analysis of organic carbon and nitrogen in particulate materials. *Marine Chemistry* 45 (3): 217–24. https://doi.org/10.1016/0304-4203(94)90005-1

Oppenheimer, M., Glavovic, B.C., Hinkel, J., van de Wal, R., Magnan, A.K., Abd-Elgawad, A., Cai, R., Cifuentes-Jara, M., DeConto, R.M., Ghosh, T., Hay, J., Isla, F., Marzeion, B., Meyssignac, B. and Sebesvari, Z. (2019) Sea level rise and implications for low-lying islands, coasts and communities. In H.-O. Pörtner, D.C. Roberts, V. Masson-Delmotte, P. Zhai, M. Tignor, E. Poloczanska, K. Minten-beck, A. Alegría, M. Nicolai, A. Okem, J. Petzold, B. Rama and N.M. Weyer (eds) *IPCC Special Report on the Ocean and Cryosphere in a Changing Climate*.

Ouyang, X. and Lee, S.Y. (2020) Improved estimates on global carbon stock and carbon pools in wetlands. *Nature Communications* 11: 317. https://doi.org/10.1038/s41467-019-14120-2

Ouyang, X. and Lee, S.Y. (2014) Updated estimates of carbon accumulation rates in coastal marsh sediments. *Biogeosciences* 11: 5057–71. https://doi.org/10.5194/bg-11-5057-2014

Palmer, M., Howard, T., Tinker, J., Lowe, J., Bricheno, L., Calvert, D., Edwards, T., Gregory, J., Harris, G., Krijnen, J., Pickering, M., Roberts, C. and Wolf, J. (2018) UKCP18 marine report. Met Office, UK. Available at: https://ukclimateprojections.metoffice.gov.uk (Accessed: 2 September 2021).

Pedregosa, F., Varoquaux, G., Gramfort, A., Michel, V., Thirion, B., Grisel, O., Blondel, M., Prettenhofer, P., Weiss, R. and Dubourg, V. (2011) Scikit-learn: machine learning in Python. *The Journal of Machine Learning Research* 12: 2825–30.

Peltier, W.R. (2002) Global glacial isostatic adjustment: palaeogeodetic and space geodetic test of the ICE-4G (VM2) model. *Journal of Quaternary Science* 17 (5–6): 491–510. https://doi.org/10.1002/jqs.713

Pham, T.D., Xia, J., Ha, N.T., Bui, D.T., Le, N.N. and Tekeuchi, W. (2019) A review of remote sensing approaches for monitoring blue carbon ecosystems: mangroves, seagrasses and salt marshes during 2010–2018. *Sensors* 19: 1933. https://doi.org/10.3390/s19081933

Phelan, N., Shaw, A. and Baylis, A. (2011) *The Extent of Saltmarsh in England and Wales: 2006–2009*. Bristol: Environment Agency.

Pinton, D., Canestrelli, A., Wilkinson, B., Ifju, P. and Ortega, A. (2020) A new algorithm for estimating ground elevation and vegetation characteristics in coastal salt marshes from high-resolution UAV-based LiDAR point clouds. *Earth Surface Processes and Landforms* 45 (14): 3687–701. https://doi.org/10.1002/esp.4992

Pribyl, D.W. (2010) A critical review of the conventional SOC to SOM conversion factor. *Geoderma* 156: 75–83. https://doi.org/10.1016/j.geoderma.2010.02.003

Pringle, A.W. (1995) Erosion of a cyclic saltmarsh in Morecambe Bay, north-west England. *Earth Surface Processes and Landforms* 20: 387–405. https://doi.org/10.1002/esp.3290200502

Reed, D.J. (1995) The response of coastal marshes to sea-level rise: survival or submergence? *Earth Surface Processes and Landforms* 20: 39–48. https://doi.org/10.1002/esp.3290200105

Rees, S., Angus, S., Creer, J., Lewis, H. and Mills, R. (2019) Guidelines for the selection of biological SSSIs. Part 2 Detailed guidelines for Habitats and Species. Chapter 1a Coastlands (coastal saltmarsh, sand dune, machair, shingle, and maritime cliff and slopes, habitats). JNCC.

Rogers, K., Kelleway, J.J., Saintilan, N., Megonigal, J.P., Adams, J.B., Holmquist, J.R., Lu, M., Schile-Beers, L., Zawadzki, A., Mazumder, D., Woodroffe, C.D. (2019) Wetland carbon storage controlled by millennial-scale variation in relative sea-level rise. *Nature* 467: 91–5. https://doi.org/10.1038/s41586-019-0951-7

Rushby, G.T., Richards, G.T., Gehrels, W.R., Anderson, W.P., Bateman, M.D. and Blake, W.H. (2019) Testing the mid-Holocene relative sea-level highstand hypothesis in North Wales, UK. *The Holocene* 29 (9): 1491–1502. https://doi.org/10.1177/0959683619854513

Sakamaki, T. and Nishimura, O. (2006) Dynamic equilibrium of sediment carbon content in an estuarine tidal flat: characterization and mechanisms. *Marine Ecology Progress Series* 328: 29–40. https://doi.org/10.3354/meps328029

Seddon, N., Chausson, A., Berry, P., Girardin, C.A., Smith, A. and Turner, B. (2020) Understanding the value and limits of nature-based solutions to climate change and other global challenges. *Philosophical Transactions of the Royal Society B* 375 (1794): 20190120. https://doi.org/10.1098/rstb.2019.0120

Shennan, I., Bradley, S.L. and Edwards, R. (2018) Relative sea-level changes and crustal movements in Britain and Ireland since the Last Glacial Maximum. *Quaternary Science Reviews* 188: 143–59. https://doi.org/10.1016/j.quascirev.2018.03.031

Smeaton, C., Barlow, N.L.M. and Austin, W.E.N. (2020) Coring and compaction: best practice in blue carbon stock and burial estimations. *Geoderma* 364: 114180. https://doi.org/10.1016/j.geoderma.2020.114180

The Saltmarsh App (2021). Online. Available at: https://apps.apple.com/gb/app/the-saltmarsh-app/id1111609923 (Accessed: 5 June 2021).

Torio, D.D. and Chmura, G.L. (2013) Assessing coastal squeeze of tidal wetlands. *Journal of Coastal Research* 29: 1049–61. https://doi.org/10.2112/JCOASTRES-D-12-00162.1

Trevathan-Tackett, S.M., Kelleway, J., Macreadie, P.I., Beardall, J., Ralph, P. and Bellgrove, A. (2015) Comparison of marine macrophytes for their contributions to blue carbon sequestration. *Ecology* 96 (11): 3043–57. https://doi.org/10.1890/15-0149.1

van der Wal, D. and Pye, K. (2004) Patterns, rates and possible causes of saltmarsh erosion in the Greater Thames area (UK). *Geomorphology* 61: 373–91. https://doi.org/10.1016/j.geomorph.2004.02.005

Vaughn, D.R., Bianchi, T.S., Shields, M.R., Kenney, W.F. and Osborne, T.Z. (2021) Blue carbon soil stock development and estimates within Northern Florida wetlands. *Frontiers in Earth Science* 9: 6. https://doi.org/10.3389/feart.2021.552721

Verardo, D.J., Froelich, P.N. and McIntyre, A. (1990) Determination of organic carbon and nitrogen in marine sediments using the Carlo Erba NA-1500 Analyzer. *Deep Sea Research Part A. Oceanographic Research Papers* 37 (1): 157–65. https://doi.org/10.1016/0198-0149(90)90034-S

Waltham, N.J., Elliott, M., Lee, S.Y., Lovelock, C., Duarte, C.M., Buelow, C., Simenstad, C., Nagelkerken, I., Claassens, L., Wen, C.K. and Barletta, M. (2020) UN Decade on Ecosystem Restoration 2021–2030 – what chance for success in restoring coastal ecosystems? *Frontiers in Marine Science* 7: 71. https://doi.org/10.3389/fmars.2020.00071

Created Coastal Wetlands as Carbon Stores: Potential Challenges and Opportunities

HANNAH L. MOSSMAN, MARTIN J.P. SULLIVAN, RACHEL M. DUNK, STUART RAE, ROBERT SPARKES, JAMES TEMPEST and NIGEL PONTEE

Abstract

Intertidal habitats are widely created with the aim of benefiting biodiversity. The pressing need to mitigate greenhouse gas emissions and the increasing recognition that coastal wetlands can provide secondary benefits by storing carbon may provide additional incentives for their creation. However, there are a number of uncertainties in the carbon budget of these wetlands, including the magnitude of carbon accumulation and the relative scale of this compared with the carbon costs of site construction. Here, we explore the carbon accumulation potential of hypothetical intertidal wetland sites of different sizes and shapes created by managed realignment and made with construction material sourced from different locations. We combine different combinations of values of sedimentation and carbon content reported from created intertidal wetlands in the literature. We find that there is large variability in potential carbon accumulation rates, with sedimentation rates being the dominant control on carbon accumulation. When carbon accumulation rates are high, all hypothetical site designs paid off the carbon cost of embankment construction within a year, but when carbon accumulation rates were low and material for embankments was transported from off site, debts took about ten years to pay. Our analysis provides a broad indication of the balance between carbon accumulation and construction carbon costs in created intertidal wetlands, but further work is needed to develop a more complete carbon budget. We highlight five key research challenges that need to be addressed to better understand the potential for created intertidal wetlands to accumulate carbon.

Keywords: blue carbon, carbon budget, intertidal habitat, managed realignment, saltmarsh restoration

Correspondence: H.Mossman@mmu.ac.uk

Background

Large areas of intertidal wetlands have been lost owing to past land claim (Davidson 1991), with losses potentially continuing owing to coastal erosion, especially where landward migration of coastal habitats in response to sea-level rise is prevented by topography or

Hannah L. Mossman, Martin J.P. Sullivan, Rachel M. Dunk, Stuart Rae, Robert Sparkes, James Tempest and Nigel Pontee, 'Created Coastal Wetlands as Carbon Stores: Potential Challenges and Opportunities' in: *Challenges in Estuarine and Coastal Science*. Pelagic Publishing (2022). © Hannah L. Mossman, Martin J.P. Sullivan, Rachel M. Dunk, Stuart Rae, Robert Sparkes, James Tempest and Nigel Pontee. DOI: 10.53061/NQGO7715 coastal defences (coastal squeeze; see Pontee et al., Chapter 8 in this book). This risk of habitat loss has motivated efforts to create new habitat, with no net loss policy commitments in multiple countries requiring compensatory habitat creation (Zedler 2004; European Commission 2007; Xu et al. 2019). New intertidal habitat can be created by reinstating tidal flow to low-lying land behind sea defences. This can happen by managed realignment, where the seawall is breached to allow tidal inundation, or by regulated tidal exchange, where sluices control the entry and exit of water. A total of 2,841 ha of intertidal habitat have been created in the UK to date through managed realignment or regulated tidal exchange (Fig. 10.1), with around 15,000 ha created in Europe up to 2015 (Esteves and Williams 2017). There is considerable potential to increase the amount of intertidal habitat created in the future; it has been estimated that a further 3,350 ha could be created in the UK if shoreline management plans for managed realignment were adopted (Committee on Climate Change, 2013), while Dickie et al. (2015) estimate that 22,000 ha of saltmarsh could be created in the UK.

To date, the primary motivation for intertidal wetland creation has been habitat creation to benefit biodiversity, but intertidal wetlands also provide secondary benefit, such as

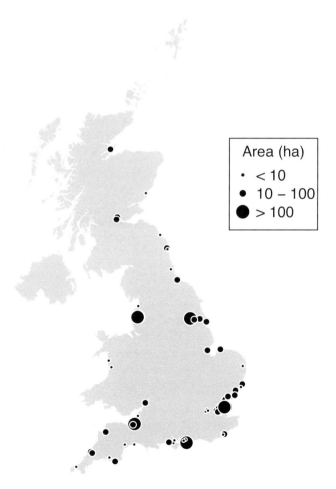

Figure 10.1 Locations of intertidal wetlands created by managed realignment or regulated tidal exchange in the United Kingdom (ABPMer 2021). In addition to the 2,841 ha covered by these sites, 45 ha have been created through beneficial use of dredge material and 317 ha through unmanaged realignments.

the provision of flood defence functions (Vuik et al. 2016; Zhu et al. 2020) and by storing and sequestering carbon (Mcleod et al. 2011). Barriers to intertidal habitat creation exist, including public opposition (Esteves and Thomas 2014) and financial costs (Stewart-Sinclair et al. 2020), and better understanding of the secondary benefits, such as carbon storage, could help overcome these barriers.

In the future, benefits such as carbon storage and sequestration may become increasingly important as motivating factors of intertidal wetland creation, both because of the need for actions to mitigate climate change and because of uncertainties in the policy environment around compensatory habitat creation (e.g. following Brexit in the UK, there is fear among conservation stakeholders that the EU Habitats and Birds Directives will face amendment or removal (Burns et al. 2019)). Funding mechanisms through verified carbon codes that translate carbon stored by habitats into purchasable credits exist for other habitats (Cevallos, Grimault and Bellassen 2019), and have motivated habitat creation. Potentially applicable standards exist elsewhere (e.g. Emmer et al. 2015) but the data to implement them in Europe are lacking, and so there is currently no verified blue (marine and coastal) carbon code in Europe.

At present, we know relatively little about the blue carbon storage potential of created coastal wetlands (but see Moritsch et al. (2021) for Australia), and these uncertainties hamper national-scale assessments of the potential for climate mitigation (Beechener et al. 2021). Based on carbon accumulation rates quantified by Burden, Garbutt and Evans ((2019), 1.04 t C ha^{-1} yr^{-1} for the first 20 years, then 0.65 t C ha^{-1} yr^{-1} after that) and Wollenberg, Ollerhead and Chmura ((2018), 13.29 t C ha^{-1} yr^{-1}), and using creation dates and areas from the OMReg database (ABPMer 2021), intertidal wetlands created in the UK could have accumulated between ~35,000 t C and 450,000 t C up to the year 2020. The huge uncertainty in this estimated accumulation rate hinders our ability to quantify the role created coastal wetlands play in climate change mitigation. Furthermore, the creation of coastal wetlands also requires substantial construction activities (e.g. Pontee 2007), with associated carbon emissions. These emissions need to be considered when quantifying the potential for carbon storage, since if they are high relative to carbon accumulation, then sites may take a long time to pay back their construction carbon debt.

In this chapter, we quantify the potential carbon accumulation rates of hypothetical managed realignment sites of different sizes and shapes based on data from the literature, and balance this with the estimated carbon costs of different hypothetical designs of embankment. We derive carbon accumulation rates by considering the carbon content of sediments and the volume of sediment that accumulates within the managed realignment sites (i.e. vertical accretion × site area). We consider the carbon costs of constructing new embankments, likely the largest constructed element, from both site-derived material and off-site material. Finally, we examine how long the carbon debt for embankments takes to repay under different site designs. While there are many other factors that influence the full carbon budget of created intertidal habitat, these calculations give a broad indication of the potential balance between carbon accumulation and construction for managed realignments, and thus an indication of where the full carbon budget may lie.

A simple assessment of carbon accumulation in created intertidal wetlands

Carbon fixed by vegetation during photosynthesis is usually rapidly returned to the atmosphere when plants die and decompose. However, through being buried by sediment this carbon can instead be transferred to a longer-lived carbon pool (Fig. 10.2; Macreadie et al. 2019). While there are many processes in the full carbon budget of created intertidal wetlands (Fig. 10.2), a major (potentially dominant) component is the rate of accumulation of

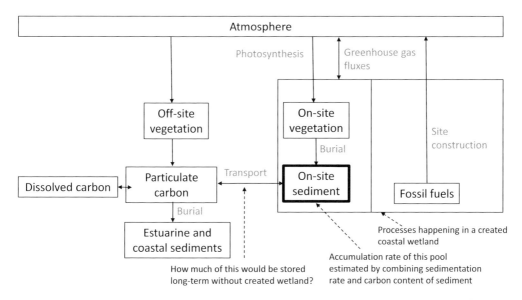

Figure 10.2 Simplified carbon budget of created intertidal wetlands. Boxes show carbon pools and solid lines show processes governing flows between pools. See Alongi (2020) for a more comprehensive carbon budget.

the sediment carbon pool. Sediment carbon accumulation is a function of net sedimentation (sediment deposited minus sediment eroded), the carbon content of deposited sediment and the carbon content of eroded sediment. If the latter are assumed to be equivalent, and sediment carbon content is assumed to be relatively homogeneous across deposited sediment, then carbon accumulation is simply the product of net sedimentation and sediment carbon content.

Only a few studies have quantified the rate of carbon accumulation in created intertidal wetlands, but a number of studies have either quantified net sedimentation or sediment carbon content. We have briefly reviewed the literature to obtain indicative values of these, and by combining these have calculated a potential realistic carbon accumulation space of created intertidal habitat (Table 10.1). Sedimentation rates vary within sites, but tend to be highest in areas of lower elevations that are inundated more regularly and are not subject to strong currents (Pontee 2014). At any one location, sedimentation rates tend to decrease over time as bed level rises and the area is inundated less frequently. Some reported sedimentation values (Table 10.1) are averages across sites while others are ranges from different elevations. We note that while obtained sedimentation rates are linear, over sufficiently long timescales sedimentation rates would be expected to decline over time as bed levels across sites fill (Pethick 1981). We did not include data where loss on ignition was measured to denote the carbon fraction, as this can be an unreliable measure of organic carbon in marine sediments (Leong and Tanner 1999).

Carbon accumulation rates (C, t C.ha^{-1}.yr^{-1}) were thus calculated as:

$$C = S \times F \times B \times 10{,}000$$

where S is net sedimentation (m.yr^{-1}), F is the fraction of sediment that is carbon on a dry weight basis and B is sediment bulk density (g.cm^{-3}). Multiplying by 10,000 converts values to be per hectare. Carbon accumulation rates were calculated for each combination of sedimentation rate and carbon content across the range of values found in the literature (Fig. 10.3). Bulk density was taken as 1 g.cm^{-3} for all calculations as this was the overall mean across values reported in the literature, calculated using the inverse variance method

Table 10.1 Literature sources used to indicate net sedimentation and sediment carbon content. Square brackets indicate codes used for sites in Figure 10.3.

Source	Sites	Net sedimentation		% Carbon	Bulk density
		Method	Time since restoration		
Liu, Fagherazzi and Cui (2021)	Meta-analysis [MA] intertidal wetlands in North America, Europe and South-East Asia from 52 papers.	Various – mean accretion across studies		N	N
Wollenberg et al. (2018)	Bay of Fundy, Canada [BF]	LiDAR (site-wide mean)	First 6 y	Y – organic	N
Clap (2009)	Paull Holme Strays, UK [PH]	Relative height from fixed marker	Years 3–5	N	Y
Manson & Pinnington (2012)	Chowder Ness, UK [CN]	LiDAR (site-wide mean)	First 5 y	N	N
Burgess, Kilkie and Callaway (2016)	Medmerry, UK [ME]	Relative height from fixed marker	First year	N	N
Brown et al. (2007)	Freiston Shore, UK [FR]	Relative height from fixed marker	First 4 y	N	N
Lawrence (2018)	Steart, UK [ST]	Relative height from fixed marker	First 3 y	N	N
Garbutt (2018)	Tollesbury, UK [TO]	Relative height from fixed marker	First 12 y	N (but available from Burden et al. 2019)	N
Burden et al. (2019)	Tollesbury [TO], Orplands [OP] and Northey Island [NH] managed realignments and nine reference natural saltmarshes [Nat] in eastern UK	NA		Y	Y
Spencer et al. (2017)	Tollesbury [TO], Orplands [OP] and Northey Island [NH]	NA		N	Y
Blackwell, Yamulki and Bol (2010)	Torridge Estuary [No Fig. 3 site code as only used for bulk density calculation]	NA		N	Y

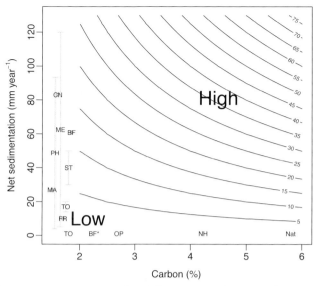

Figure 10.3 Carbon accumulation rate for different combinations of net sedimentation and carbon content. Contours show carbon accumulation rate (tC.ha^{-1}.yr^{-1}) for each combination of net sedimentation and carbon content, assuming a dry bulk density of 1.0 g.cm^{-3}. Values of net sedimentation and carbon content obtained from the literature are shown on the y and x axes respectively, which are either shown as single point estimates for a site or the range of values. See Table 10.1 for sources and site name abbreviations. Low and high refer to example sediment accumulation scenarios used in later analyses. *All % carbon measurements are of total carbon with the exception of BF, which is % organic carbon.

for pooling using the metamean function of the meta R package (Balduzzi, Rücker and Schwarzer 2019; 95% CI = 0.9 – 1.1).

Based on these data, we find that a threefold variation in sediment carbon content leads to >35-fold variation in the potential carbon accumulation rates of restored intertidal habitats (Fig. 10.3). Net sedimentation rate is a major constraint on the accumulation rates, with overall carbon accumulation potentially being very low (<5 t C.ha^{-1}.yr^{-1}), even when sediment carbon content is relatively high at 6%. In contrast, high carbon accumulation rates are possible at low carbon fractions as long as sedimentation rates are high (Fig. 10.3). While carbon accumulation rates of 70 t C.ha^{-1}.yr^{-1} are theoretically possible, we consider them unlikely, as this would require exceptionally high sedimentation rates in combination with high carbon contents equivalent to those reported for natural marshes.

While variation in net sedimentation dominates the rate at which carbon can accumulate on created intertidal wetlands, sediment carbon content will be a major factor in determining the total carbon stocks that can be achieved at a site. This is because sites are expected to gradually fill to a planar surface (which for saltmarshes is typically around mean high water spring tides – MHWS), with subsequent accretion tracking sea-level rise where sufficient sediment is available (Pontee 2014). Net sedimentation rates determine how long it takes to reach this stage, but the resultant carbon stocks will be a function of the total elevation change needed to reach this point and the carbon content of sediment. The total sediment that can be accumulated thus depends on sediment supply to site (maximum in high estuaries with high suspended sediment concentration) but also on 'accommodation space' for sediment within the site. Sites lower in the tidal frame require a greater height of sediment to reach MHWS level.

Carbon costs of embankment construction for a new managed realignment

Creating intertidal wetlands by managed realignment involves construction activities that emit carbon. These emissions include the construction of embankments around the site, movement of construction workers to and from the site, construction of additional structures such as culverts and haul roads, and loss of carbon from soil and vegetation carbon pools owing to disturbance. We focus on the former as we expect embankment construction to incur the greatest carbon cost, but note that all need to be considered for a full carbon budget of intertidal wetland creation.

While local site conditions such as topography and land ownership will affect the shape of realigned areas, we can calculate the construction carbon costs of embankments under a set of simplifying assumptions. First, we assume that a new landward sea defence that completely encloses the site on three sides is needed, as retreat to higher ground (where only the two side walls perpendicular to the coast would be required) is rare. We then consider two shapes, square or in rectangular form, the latter with the long axis parallel to the coast (length/width ratio taken as 2.4), each in three sizes of site (10 ha, 60 ha and 200 ha) to give six hypothetical managed realignment scenarios.

The carbon cost in the construction of these embankments is broadly dependent on the volume of the material that needs to be moved and the source of that material. For each of our scenarios we calculated the length of new embankment required and converted this to a volume, based on the average cross-sectional area obtained from typical embankment dimensions at Hesketh Out Marsh East (lat, lon: 53.73, −2.87) and Steart (51.20, −3.05) managed realignments (Hesketh: height above surrounding ground levels 3.4 m and cross sectional area of 63 m²; Steart: 3.0 m and 51 m²).

In order to reduce construction costs and environmental impacts associated with transporting material, embankments within managed realignment sites are commonly constructed from sediments obtained within the site. Best practice typically involves creating a neutral cut/fill balance with the materials for embankments being constructed from excavated areas. These excavated areas can be stand-alone borrow pits (which become pools or lagoons within the new site), linear channels (which become creek networks) and/or more widespread site lowering (which may create broader low-lying areas for the establishment of mudflats). If there are no or insufficient suitable sources of sediment within the site, then materials must be sourced off site and transported.

In our calculations we consider two scenarios for sourcing of embankment construction materials, with all materials either being sourced on or off site. Site-derived material is assumed to have no embodied carbon (Highways England 2015) except for fuel use from excavation, on-site transport and construction. We use a value of $0.0025\,t\,CO_2e.m^{-3}$ material based on fuel use calculations from the construction of the embankment at Steart managed realignment (Jacobs unpublished data). For material derived off site, the embodied carbon will also include transport to the site, which will vary between sites. To give a general indication of this, we used the cradle-to-site embodied carbon value of general rammed soil of $0.024\,t\,CO_2e.$ per tonne of material assuming $1.7\,t.m^{-3}$ (ICE, 2019). While embankment construction costs for material won on site implicitly include construction of creeks and lagoons, off-site material construction costs do not include creation of these features, which would incur additional carbon emissions not considered here.

Our calculations show that rectangular shapes with the longest axis parallel to the coast require a lower length of new embankment compared with square sites of the same area, thus incurring lower carbon costs. However, these differences are small compared with differences owing to the size of the site (Table 10.2). The biggest factor in the carbon cost of

Table 10.2 Carbon costs of construction of new sea defence embankments of different site design scenarios.

Site size (ha)	Site shape	Length of new embankment (m)	Volume of new embankment (m³)	On-site derived material (tCO₂e)	Off-site derived material (tCO₂e)
200	Square	4,200	239,229	598	9,761
	Rectangle	4,000	227,838	570	9,296
60	Square	2,325	132,430	331	5,403
	Rectangle	2,200	125,310	313	5,113
10	Square	300	17,088	43	697
	Rectangle	285	16,233	41	662

construction is the source of the material used for the embankment. When on-site derived material is used, the largest sites in our model have lower carbon costs compared with the smallest sites (20 times smaller area) using off-site material (Table 10.2).

Carbon costs for off-site material will be highly dependent on the distance travelled from the site source. While no transport distance is stated in the Inventory of Carbon & Energy database embodied carbon value, dividing the difference between off-site and on-site emissions per tonne by heavy goods vehicle emissions per km (Department for Transport 2012), assuming one fully laden journey and one empty return journey) gives an assumed travel distance of 154 km from off-site material sources. There is therefore substantial potential to reduce the carbon costs of construction with off-site derived material compared with values in Table 10.2 if material is obtained from close to the site. Calculating actual transport costs for both on-site and off-site derived material will therefore substantially refine calculations of carbon costs compared with using a single reference embodied carbon value.

How long will it take to pay off the embankment construction carbon debt?

We take two scenarios from the possible annual carbon accumulation space (Fig. 10.3); 'low' carbon accumulation scenario with 2% carbon fraction and an average net sedimentation rate of 10 mm y⁻¹ (2 tC ha⁻¹ y⁻¹), and 'high' with 4.5% carbon fraction and average net sedimentation of 80 mm y⁻¹ (36 tC ha⁻¹ y⁻¹). While higher carbon accumulation rates are theoretically possible, these scenarios encompass the full range of values we find in the literature. Using the high and low scenarios, we calculate how long it would take each hypothetical site to pay off its carbon construction debt.

Carbon accumulation rates for the different site-scenario combinations vary over three orders of magnitude from 73 – 26,400 t.CO₂e.y⁻¹, but the translation of this into pay-off time is highly dependent on the origin of material used to construct the embankments (Table 10.3). Under high carbon accumulation rates, all sites offset the carbon from construction of the embankment within one year, regardless of where the material was won. Likewise, when the material was derived on site, all sites offset the carbon within one year, even under the low carbon accumulation scenario. Under the low carbon accumulation scenario with off-site derived material the sites take six to twelve years to accumulate the carbon to offset construction costs (Table 10.3). The time taken is dependent on the perimeter to area ratio of the site design, with the 60 ha site taking the longest.

Table 10.3 Carbon accumulation rates for different site design scenarios under low and high carbon accumulation rates, and the time taken for carbon accumulation to offset construction of the new embankment needed for the site.

| Site size (ha) | Site shape | Carbon accumulation ($tCO_2e\ y^{-1}$) | | Time to accumulate carbon from construction of new embankment (y) | | | |
| | | | | Low | | High | |
		Low	High	On-site	Off-site	On-site	Off-site
200	Square	1,467	26,400	0.4	6.7	<0.1	0.4
	Rectangle			0.4	6.3	<0.1	0.4
60	Square	440	7,920	0.8	12.3	<0.1	0.7
	Rectangle			0.7	11.6	<0.1	0.6
10	Square	73	1,320	0.6	9.5	<0.1	0.5
	Rectangle			0.6	9.1	<0.1	0.5

Future research challenges

Our simple modelling exercise demonstrates how wetlands created by managed realignment can rapidly accumulate enough carbon to pay off their embankment construction carbon debt under a range of design and carbon accumulation scenarios; we anticipate that this will be the largest portion of the total construction costs. However, there are a number of challenges involved in moving from this simple model to a full carbon budget of created intertidal wetlands.

Challenge 1: Predicting variation in sedimentation rates across space and time

Variation in sedimentation rates causes substantial variation in the rate at which carbon accumulates. Sedimentation rates are driven by numerous factors including sea level, tides, waves, fluvial flows and sediment supply. Sediment supply is the key factor in the long term expansion and erosion of natural saltmarsh (Ladd et al. 2019) and variation in suspended sediment concentrations (SSC) in the wider estuary/coastal area is expected to be a major driver of sedimentation, with higher sedimentation (and hence faster carbon accumulation) in estuaries with higher SSCs. There is also variation in SSCs within estuaries, with sites located nearer to the turbidity maxima within estuaries, and hence with higher SSCs, expected to have faster carbon accumulation. Site design is also expected to affect sedimentation. Rates would be expected to be higher inside managed realignment sites where walls have not been removed, since they have lower levels of wave and tidal energy. Reduced tidal exchange will likely lower sedimentation rates owing to fewer opportunities for sediment import. For example, Oosterlee et al. (2020) report net sedimentation rates approximately 30 times lower in a regulated tidal exchange than in an adjacent site with full tidal exchange. However, management of regulated tidal exchange sites may influence sedimentation rates; for example, holding sediment rich tidal waters on the site for a few tidal cycles may increase sedimentation rates. Finally, sedimentation rates will vary within sites, with faster rates at lower elevations as these areas experience more tidal inundations than higher areas (Pontee 2014) and in vegetated areas, where vegetation reduces wave and current energy, allowing sediment to settle out of suspension (Temmerman et al. 2005). The variation in sedimentation rates with elevation creates temporal variation in site-wide sedimentation rates. As sites increase in elevation, sedimentation rates will decline, until eventually the majority of the site becomes a level plain accreting in line with sea-level rise, assuming there is sufficient sediment supply (Schuerch et al. 2018). This trajectory of decreasing sedimentation can be assessed using sites of different ages as a space-for-time substitution (Burden, Garbutt and Evans 2019), but patterns are likely to vary between estuaries with different inundation regimes and SSC.

While predicting what the sedimentation (and hence carbon accumulation) rates will be before site creation is desirable to leverage carbon credit funding, these can be monitored after a site has been created. Rates of sedimentation can be measured by using fixed markers, or by taking repeated geo-referenced measurements of surface sediment elevation using differential GPS on the ground or aerial LiDAR flights. Understanding the comparability of these different methods is important in order to standardise the quantification of carbon accumulation. High sedimentation rates can mean that plates or stakes are rapidly buried; for example, Oosterlee et al. (2020) lost seven out of eleven of their surface elevation tables owing to extremely high sedimentation rates. This creates a bias to measurements of sedimentation in locations where rates are lower (i.e. data are missing not at random), so site-wide rates are likely to be underestimated. Airborne LiDAR measurements overcome the almost unavoidable biases associated with field sampling by being able to measure locations that would be too dangerous to reach in the field or where plates and stakes would be lost. However, completely filtering out low-growing vegetation from LiDAR-derived digital terrain models is challenging (Pontee and Serato 2019), which could lead to overestimation of sedimentation rates. This would introduce a more significant error in estuaries where the signal is lower owing to low sedimentation rates, and is most important when comparing areas that were unvegetated and where vegetation became established between the LiDAR surveys. Field sampling and airborne LiDAR are potentially most powerful when used in combination. LiDAR can be used to assess whether sedimentation rates measured with sediment pins are representative of the wider site, while field measurements can be used to validate LiDAR measurements.

Challenge 2: Filling the gaps in the carbon budget of created intertidal wetlands

Monitoring the growth rate of the sediment carbon pool provides a broad approximation for the carbon budget of created intertidal wetlands, but only if the assumption that other elements of the carbon budget (Fig. 10.2) are small is met.

One uncertainty is the source of the carbon that accumulates on created intertidal wetlands, and whether this represents an additional store of carbon or whether it would have been stored in the absence of managed realignment schemes (Macreadie et al. 2019). Allochthonous (off-site) carbon is not eligible for offset credits under the Verified Carbon Standard Methodology VM0033 to avoid the risk of double accounting carbon sequestered by other habitats (Emmer et al. 2015). However, created intertidal wetlands may still contribute to the long-term storage of carbon that, while fixed by vegetation in other habitats, would have been returned to the atmosphere in the absence of burial in sediment. While the carbon residence time in vegetation is typically up to 50 years (Yizhao et al. 2015), carbon remains in saltmarsh sediments for hundreds to thousands of years (Luk et al. 2021).

A second uncertainty is the extent to which greenhouse gas fluxes from created intertidal wetlands offset the carbon accumulation in sediment. While the potential for intertidal wetlands to be sources of carbon dioxide, methane and nitrous oxide is well known, there is an insufficient evidence base globally to quantify their impacts (Macreadie et al. 2019). A further complication is that fluxes vary in space and time as they are influenced by vegetation and salinity (Martin and Moseman-Valtierra 2015; Sheng, Wang and Wu 2015). In the UK, there is contrasting evidence for the importance of greenhouse gas fluxes from created intertidal wetlands. While Burden et al. (2013) found almost no methane or nitrous oxide release from natural or managed realignments (neither at high or low elevations),

Adams, Andrews and Jickells (2012) found that both natural saltmarsh and managed realignments were small methane and nitrous oxide sources, and that this was sufficient to offset carbon sequestration rates by around 25%. Further research is needed to understand whether greenhouse gas emissions are often large enough to need routine consideration in site-scale carbon budgets, or whether their effects can safely be assumed to be small.

Challenge 3: Accounting for all aspects of construction costs

Our calculations are based on the simplifying assumption that embankment construction is the major carbon cost during site construction. However, a full carbon budget requires considering all aspects of construction, such as worker transport, concrete for sluices, fencing and haul roads, as well as other factors such as loss of carbon from vegetation and soils due to disturbance during construction. The relative importance of the carbon emitted from these other activities will be greater where retreating higher ground means that embankment construction is substantially reduced. It is likely that retreating to higher ground would have lower carbon costs, but the emissions from other activities need to be better quantified to be sure of this. Even when considering embankment construction costs, refinements are desirable. Our analysis indicates the importance of where site material is obtained, with sites where material is obtained on site having a >15-fold lower time to paying off the construction carbon debt than when carbon is obtained off site, but this difference is likely to be substantially reduced if off-site material is obtained close to the site. Importantly, the large amount of material needed for embankment construction means that small differences in the embodied carbon per unit volume scale to large differences at site level. We therefore strongly recommended tracking actual fuel consumption or transport differences to refine estimates of carbon costs whenever possible.

Challenge 4: Assessing trade-offs and synergies between design for carbon accumulation and design for biodiversity

Because sedimentation rates are highest at lower elevations in the tidal frame, creating sites at low elevations in the tidal frame would be expected to maximise carbon accumulation potential. Similarly, pools act to trap sediment and accrete rapidly, so would be beneficial for carbon accumulation. Creation of features such as pools also introduce trade-offs and synergies with biodiversity. Frequently inundated pools rapidly accumulate carbon, but this rapid sedimentation means they fill rapidly and become vegetated, losing their value for wading birds. Pools created in areas that are less frequently flooded are likely to persist for longer, but will contribute less to carbon accumulation. Locally raised features such as mounds are beneficial for nesting birds (Ausden et al. 2019) and for some plant species (Mossman, Grant and Davy 2020), but as they are less frequently inundated they are poor for carbon accumulation. Finally, from a carbon budget perspective, creation of pools and creeks is best carried out prior to breaching because dug material can be used for embankment construction and the resultant low features rapidly accumulate carbon. Importantly, construction of such features 'in the dry', as opposed to in a tidally inundated area, also has fewer health and safety risks associated, and is usually less costly financially. However, this rapid sedimentation means the features (and their biodiversity benefits) are rapidly lost, so would require maintenance activities post-breach to reinstate these features, which would need consideration of the fate of the carbon in the sediment removed. Understanding the relative costs and benefits of different features for different outcomes is important in order to better advise managers.

Challenge 5: Valuing carbon accumulation versus eventual carbon stocks

While variation in net sedimentation rates dominates how quickly carbon accumulates, the eventual asymptote of sedimentation as sites increase in elevation means that eventual carbon stocks are not dependent on sedimentation rates, except in terms of whether sedimentation is sufficient to keep pace with sea-level rise (Schuerch et al. 2018). Instead, they will be influenced by site elevation in the tidal frame and overall size (and hence to volume of sediment that can accumulate before the marsh is a level plain), and sediment carbon content. The different properties governing carbon accumulation rates and stocks means that the most valuable sites for carbon accumulation will not necessarily be the most valuable for long-term carbon stocks. Deciding how to value accumulation versus stocks will have important implications for where investment for site creation should be invested. Should investment in new sites avoid areas with slow rates of carbon accumulation (e.g. Essex, where sites take 100 years to attain the carbon stocks of natural sites (Burden, Garbutt and Evans 2019)), and focus instead on macrotidal, sediment-rich systems (e.g. the Bay of Fundy or Severn Estuary), where carbon accumulates at an order of magnitude faster rate? The answer to this question is important for determining how a carbon credit system would work.

Conclusion

While there is consistent evidence that created intertidal wetlands rapidly accumulate carbon, the rate at which this happens is highly variable, depending mainly on sedimentation rates, creating national-scale uncertainties of an order of magnitude. Our analysis indicates that under a wide range of scenarios carbon accumulation rapidly pays off construction costs, providing support for intertidal wetland creation for climate mitigation. However, refining these values to produce a full carbon budget presents several challenges, including deriving better, scheme-specific estimates of carbon emissions from construction, greenhouse gas fluxes of the new wetlands, and the additionality of accumulated carbon (i.e. what would happen to it without the scheme). Further research can help understand how important these remaining uncertainties are, and thus inform the extent they need to be evaluated when monitoring carbon offsetting by created intertidal wetlands. Finally, the potential contribution of intertidal wetland creation to national-scale carbon budgets also depends on the extent of habitat that can be created (Beechener et al. 2021). Actual habitat creation in the UK remains small compared with estimates of the potential area, and better valuation of the carbon benefits could provide incentives to create more schemes.

References

ABPMer. (2021) *Online Marine Registry (OMReg): Adapting shorelines*. ABPMer, Southampton. Available at: http://www.omreg.net (Accessed: 24 April 2021).

Adams, C.A., Andrews, J.E. and Jickells, T. (2012) Nitrous oxide and methane fluxes vs. carbon, nitrogen and phosphorous burial in new intertidal and saltmarsh sediments. *Science of the Total Environment* 434: 240–51. https://doi.org/10.1016/j.scitotenv.2011.11.058

Alongi, D.M. (2020) Carbon balance in salt marsh and mangrove ecosystems: a global synthesis. *Journal of Marine Science and Engineering* 8 (10): 767. https://doi.org/10.3390/jmse8100767

Ausden, M., Hirons, G., White, G. and Lock, L. (2019) Wetland restoration by the RSPB – what has it achieved for birds? *British Birds* 112: 315–36.

Balduzzi, S., Rücker, G. and Schwarzer, G. (2019) How to perform a meta-analysis with R: a practical tutorial. *Evidence-based Mental Health* 22 (4): 153–60. https://doi.org/10.1136/ebmental-2019-300117

Beechener, G., Curtis, T., Fulford, J., MacMillan, T., Mason, R., Massie, A., McCormack, C., Shanks, W., Sheane, R., Smith, L., Warner, D. and Vennin, S. (2021) *Achieving Net Zero: A review of the evidence behind potential carbon offsetting approaches*. Bristol: Environment Agency.

Blackwell, M.S.A., Yamulki, S. and Bol, R. (2010) Nitrous oxide production and denitrification rates in estuarine intertidal saltmarsh and managed realignment zones. *Estuarine, Coastal and Shelf Science* 87 (4): 591–600. https://doi.org/10.1016/j.ecss.2010.02.017

Brown, S.L., Pinder, A., Scott, L., Bass, J., Rispin, E., Brown, S., Garbutt, A., Thomson, A., et al. (2007) Wash Banks Flood Defence Scheme Freiston Environmental Monitoring 2002–2006. Report to Environment Agency, Peterborough. Dorchester: Centre for Ecology and Hydrology.

Burden, A., Garbutt, A. and Evans, C.D. (2019) Effect of restoration on saltmarsh carbon accumulation in Eastern England. *Biology Letters* 15 (1): 20180773. https://doi.org/10.1098/rsbl.2018.0773

Burden, A., Garbutt, R.A., Evans, C.D., Jones, D.L. and Cooper, D.M. (2013) Carbon sequestration and biogeochemical cycling in a saltmarsh subject to coastal managed realignment. *Estuarine, Coastal and Shelf Science* 120: 12–20. https://doi.org/10.1016/j.ecss.2013.01.014

Burgess, H., Kilkie, P. and Callaway, T. (2016) Understanding the physical processes occurring within a new coastal managed realignment site, Medmerry, Sussex, UK. In *Coastal Management*, pp. 263–72. London: ICE Publishing. https://doi.org/10.1680/cm.61149.263

Burns, C., Gravey, V., Jordan, A. and Zito, A. (2019) De-Europeanising or disengaging? EU environmental policy and Brexit. *Environmental Politics* 28 (2): 271–92. https://doi.org/10.1080/09644016.2019.1549774

Cevallos, G., Grimault, J. and Bellassen, V. (2019) Domestic carbon standards in Europe: overview and perspectives. Available at: https://www.i4ce.org/wp-core/wp-content/uploads/2020/02/0218-i4ce3153-DomecticCarbonStandards.pdf (Accessed: 27 May 2021).

Clapp, J. (2009) Managed realignment in the Humber estuary: factors influencing sedimentation. PhD thesis. The University of Hull.

Committee on Climate Change (2013) Managing the land in a changing climate – Adaptation Sub-Committee progress report. Chapter 5 Supporting Data and Research. Available at: https://www.theccc.org.uk/wp-content/uploads/2013/07/Chapter-5_exhibits_webcopy.xls (Accessed: 21 April 2021).

Davidson, N.C., Laffoley, D.d'A., Doody, J.P., Way, L.S., Gordon, J., Key, R., Drake, C.M., Pienkowski, M.W., Mitchell, R. and Duff, K.L. (1991) *Nature Conservation and Estuaries in Great Britain*. Peterborough: Nature Conservancy Council. http://dx.doi.org/10.13140/2.1.3522.9448

Department for Transport (2012) Domestic road freight activity (RFS01). Data about the road freight domestic activity, produced by Department for Transport. Available at: https://www.gov.uk/government/statistical-data-sets/rfs01-goods-lifted-and-distance-hauled (Accessed: 26 May 2021).

Dickie, I., Cryle, P., Anderson, S., Provins, A., Krisht, S., Koshy, A., Doku, A., Maskell, L., Norton, L., Walmsley, S., Scott, C., Fanning, T. and Nicol, S. (2015) *The Economic Case for Investment in Natural Capital in England: Final report to the Natural Capital Committee*. London: Economics for the Environment Consultancy Ltd.

Emmer, I., Needelman, B., Emmett-Mattox, S., Crooks, S., Megonigal, P., Myers, D., Oreska, M. and McGlathery, K. (2015) *VM0033 Methodology for tidal wetland and seagrass restoration. Version 1.0.* Washington, DC: Verra. Verified Carbon Standard.

Esteves, L.S. and Thomas, K. (2014) Managed realignment in practice in the UK: results from two independent surveys. *Journal of Coastal Research* 70 (sp1): 407–13. https://doi.org/10.2112/SI70-069.1

Esteves, L.S. and Williams, J.J. (2017) Managed realignment in Europe: a synthesis of methods, achievements and challenges. In Bilkovic, D.M., Mitchell, M.M., Toft, J.D. and La Peyre, M.K. (eds) *Living Shorelines: The Science and Management of Nature-based Coastal Protection*, pp. 157–80. New York: CRC Press/Taylor & Francis Group.

European Commission (2007) Guidance document on Article 6(4) of the 'Habitats Directive' 92/43/EEC. EC, Luxembourg.

Garbutt, A. (2018) Bed level change within the Tollesbury managed realignment site, Blackwater estuary, Essex, UK between 1995 and 2007. NERC Environmental Information Data Centre.

Highways England (2015) Carbon emissions calculation tool: Highways England. Available at: https://www.gov.uk/government/publications/carbon-tool (Accessed: 21 April 2021).

ICE (2019) Inventory of Carbon & Energy (ICE) database V3.0. https://circularecology.com/embodied-carbon-footprint-database.html (Accessed 21 April 2021).

Ladd, C.J.T., Duggan-Edwards, M.F., Bouma, T.J., Pagès, J.F. and Skov, M.W. (2019) Sediment supply explains long-term and large-scale patterns in salt marsh lateral expansion and erosion. *Geophysical Research Letters* 46 (20): 11178–87. https://doi.org/10.1029/2019GL083315

Lawrence, P.J. (2018) How to create a saltmarsh: understanding the roles of topography, redox and nutrient dynamics. PhD thesis. Manchester Metropolitan University.

Leong, L.S. and Tanner, P.A. (1999) Comparison of methods for determination of organic carbon in marine sediment. *Marine Pollution Bulletin* 38 (10): 875–9. https://doi.org/10.1016/S0025-326X(99)00013-2

Liu, Z., Fagherazzi, S. and Cui, B. (2021) Success of coastal wetlands restoration is driven by sediment availability. *Communications Earth & Environment* 2 (1): 1–9. https://doi.org/10.1038/s43247-021-00117-7

Luk, S.Y., Todd-Brown, K., Eagle, M., McNichol, A.P., Sanderman, J., Gosselin, K. and Spivak, A.C. (2021) Soil organic carbon development and turnover in natural and disturbed salt marsh environments. *Geophysical Research Letters* 48 (2): e2020GL090287. https://doi.org/10.1029/2020GL090287

Macreadie, P.I., Anton, A., Raven, J.A., Beaumont, N., Connolly, R.M., Friess, D.A., Kelleway, J.J.,

Kennedy, H., Kuwae, T., Lavery, P.S., Lovelock, C.E., Smale, D.A., Apostolaki, E.T., Atwood, T.B., Baldock, J., Bianchi, T.S., Chmura, G.L., Eyre, B.D., Fourqurean, J.W., Hall-Spencer, J.M., Huxham, M., Hendricks, I.E., Krause-Jensen, D., Laffoley, D., Luisetti, T., Marba, N., Masque, P., McGlathery, K.J., Megonigal, J.P., Murdiyarso, D., Russell, B.D., Santos, R., Serrano, O., Silliman, B.R., Watanbe, K. and Duarte, C.M. (2019) The future of Blue Carbon science. *Nature Communications* 10 (1): 1–13. https://doi.org/10.1038/s41467-019-11693-w

Manson, S. and Pinnington, N. (2012) *Chowder Ness (Humber Estuary). Measure Analysis in the Framework of the Interreg IVB Project TIDE.* Hull.

Martin, R.M. and Moseman-Valtierra, S. (2015) Greenhouse gas fluxes vary between *Phragmites australis* and native vegetation zones in coastal wetlands along a salinity gradient. *Wetlands* 35 (6): 1021–31. https://doi.org/10.1007/s13157-015-0690-y

Mcleod, E., Chmura, G.L., Bouillon, S., Salm, R., Björk, M., Duarte, C.M., Lovelock, C.E., Schlesinger, W.H. and Silliman, B.R. (2011) A blueprint for blue carbon: toward an improved understanding of the role of vegetated coastal habitats in sequestering CO2. *Frontiers in Ecology and the Environment* 9 (10): 552–60. https://doi.org/10.1890/110004

Moritsch, M.M., Young, M., Carnell, P., Macreadie, P.I., Lovelock, C., Nicholson, E., Raimondi, P.T., Wedding, L.M. and Ierodiaconou, D. (2021) Estimating blue carbon sequestration under coastal management scenarios. *Science of The Total Environment* 777: 145962. https://doi.org/10.1016/j.scitotenv.2021.145962

Mossman, H.L., Grant, A. and Davy, A.J. (2020) Manipulating saltmarsh microtopography modulates the effects of elevation on sediment redox potential and halophyte distribution. *Journal of Ecology* 108 (1): 94–106. https://doi.org/10.1111/1365-2745.13229

Oosterlee, L., Cox, T.J.S., Temmerman, S. and Meire, P. (2020) Effects of tidal re-introduction design on sedimentation rates in previously embanked tidal marshes. *Estuarine, Coastal and Shelf Science* 244: 106428. https://doi.org/10.1016/j.ecss.2019.106428

Pethick, J.S. (1981) Long-term accretion rates on tidal salt marshes. *Journal of Sedimentary Research* 51 (2): 571–7. https://doi.org/10.1306/212F7CDE-2B24-11D7-8648000102C1865D

Pontee, N. (2014) Accounting for siltation in the design of intertidal creation schemes. *Ocean & Coastal Management* 88: 8–12. https://doi.org/10.1016/j.ocecoaman.2013.10.014

Pontee, N.I. (2007) Realignment in low-lying coastal areas: UK experiences. *Proceedings of the Institution of Civil Engineers – Maritime Engineering* 160 (4): 155–66. https://doi.org/10.1680/maen.2007.160.4.155

Pontee, N.I. and Serato, B. (2019) Nearfield erosion at the Steart marshes (UK) managed realignment scheme following opening. *Ocean & Coastal Management* 172: 64–81. https://doi.org/10.1016/j.ocecoaman.2019.01.017

Schuerch, M., Spencer, T., Temmerman, S., Kirwan, M.L., Wolff, C., Lincke, D., McOwen, C.J., Pickering, M.D., Reef, R., Vafeidis, A.T., Hinkel, J., Nicholls, R.J. and Brown, S. (2018) Future response of global coastal wetlands to sea-level rise. *Nature* 561 (7722): 231–4. https://doi.org/10.1038/s41586-018-0476-5

Sheng, Q., Wang, L. and Wu, J. (2015) Vegetation alters the effects of salinity on greenhouse gas emissions and carbon sequestration in a newly created wetland. *Ecological Engineering* 84: 542–50. https://doi.org/10.1016/j.ecoleng.2015.09.047

Spencer, K.L., Carr, S.J., Diggens, L.M., Tempest, J.A., Morris, M.A. and Harvey, G.L. (2017) The impact of pre-restoration land-use and disturbance on sediment structure, hydrology and the sediment geochemical environment in restored saltmarshes. *Science of the Total Environment* 587: 47–58. https://doi.org/10.1016/j.scitotenv.2016.11.032

Stewart-Sinclair, P.J., Purandare, J., Bayraktarov, E., Waltham, N., Reeves, S., Statton, J., Sinclair, E.A., Brown, B.M., Shribman, Z.I. and Lovelock, C.E. (2020) Blue restoration – building confidence and overcoming barriers. *Frontiers in Marine Science* 7: 748. https://doi.org/10.3389/fmars.2020.541700

Temmerman, S., Bouma, T.J., Govers, G., Wang, Z.B., De Vries, M.B. and Herman, P.M.J. (2005) Impact of vegetation on flow routing and sedimentation patterns: three-dimensional modeling for a tidal marsh. *Journal of Geophysical Research: Earth Surface*, 110, F04019. https://doi.org/10.1029/2005JF000301

Vuik, V., Jonkman, S.N., Borsje, B.W. and Suzuki, T. (2016) Nature-based flood protection: the efficiency of vegetated foreshores for reducing wave loads on coastal dikes. *Coastal Engineering* 116: 42–56. https://doi.org/10.1016/j.coastaleng.2016.06.001

Wollenberg, J.T., Ollerhead, J. and Chmura, G.L. (2018) Rapid carbon accumulation following managed realignment on the Bay of Fundy. *PLoS One* 13 (3): e0193930. https://doi.org/10.1371/journal.pone.0193930

Xu, W., Fan, X., Ma, J., Pimm, S.L., Kong, L., Zeng, Y., Li, X., Xiao, Y., Zheng, H., Liu, J., Wu, B., An, L., Zhang, L., Wang, X. and Ouyang, Z. (2019) Hidden loss of wetlands in China. *Current Biology* 29 (18): 3065–71. https://doi.org/10.1016/j.cub.2019.07.053

Yizhao, C., Jianyang, X., Zhengguo, S., Jianlong, L., Yiqi, L., Chengcheng, G. and Zhaoqi, W. (2015) The role of residence time in diagnostic models of global carbon storage capacity: model decomposition based on a traceable scheme. *Scientific Reports* 5 (1): 16155. https://doi.org/10.1038/srep16155

Zedler, J.B. (2004) Compensating for wetland losses in the United States. *Ibis* 146: 92–100. https://doi.org/10.1111/j.1474-919X.2004.00333.x

Zhu, Z., Vuik, V., Visser, P.J., Soens, T., van Wesenbeeck, B., van de Koppel, J., Jonkman, S.N., Temmerman, S. and Bouma, T.J. (2020) Historic storms and the hidden value of coastal wetlands for nature-based flood defence. *Nature Sustainability* 3 (10): 853–62. https://doi.org/10.1038/s41893-020-0556-z

CHAPTER 11

Coastal Habitat Restoration, Invasive Species and Remote Monitoring Solutions

SOPHIE WALKER, NATHAN WALTHAM, CHRISTINA BUELOW and JORDAN ILES

Abstract

The success of ecosystem restoration depends on a multitude of factors related to both the properties of the ecosystem and resources available for the restoration. The ecological and management characteristics of the site, including the presence of invasive species or existing and continued management practices, may influence the success of restoration. Key to the ability to assess success of restoration is monitoring, which is increasingly being undertaken using remote sensing technology. In this chapter, we focus on invasive species management to restore intertidal wetland vegetation, and present a case study from North Queensland, Australia, where exclusion fences have been constructed to control feral pigs and restore intertidal vegetation. Physical vegetation sampling using quadrats and unmanned aerial vehicle (UAV) imagery were used to determine vegetation change at five sites using image classification and ground truthing. Wetlands were fenced in an attempt to prevent feral pigs accessing the sites, but this was ultimately unsuccessful owing to existing management practices allowing cattle to graze in fenced wetlands, which presumably afforded feral pig access. Learnings here suggest that restoration initiatives cannot be implemented without first considering existing land management practices and how the two interact or compete with each other. While the use of UAV technology was useful for monitoring, restoration projects still need to be carefully designed with the ecological and management practices in mind in order to be successful.

Keywords: intertidal, monitoring, remote sensing, wetlands, invasive species

Correspondence: sophie.walker1@my.jcu.edu.au

Introduction

Ecosystem restoration may take place for a variety of reasons: following industrial activity such as mining (Cross et al. 2019), as part of an environmental plan to improve an area (Lv, Zhou and Zhao 2018), to enhance ecosystem and human health (Canning et al. 2021; Waltham et al. 2021) or as part of an offset scheme (Sapkota and White 2020). Regardless of the reason for restoration, the premise is to return the state of an ecosystem to a historical benchmark (Zhao et al. 2016; Cross et al. 2019). However, ecosystems are not static, and

Sophie Walker, Nathan Waltham, Christina Buelow and Jordan Iles, 'Coastal Habitat Restoration, Invasive Species and Remote Monitoring Solutions' in: *Challenges in Estuarine and Coastal Science*. Pelagic Publishing (2022). © Sophie Walker, Nathan Waltham, Christina Buelow and Jordan Iles. DOI: 10.53061/BDDZ5865 even once restored to a benchmark state there will still be pressure from outside phenomena such as urbanisation (DeVore, Shine and Ducatez 2020), climate change (Boorman and Garbutt 2011) and invasive species (Adame et al. 2019). Invasive species can hinder the success of restoration action, and there is little literature about the influence of invasive species on restoration success, though it is acknowledged as a clear problem (Tanner et al. 2002). This is especially the case in intertidal areas given their location at the intersection of the terrestrial and marine environment (Zedler and Kercher 2004). Invasive species may impact sites in a variety of ways, from changing habitat structure to population loss through predation to loss as a cultural resource (Fordham, Georges and Corey 2006; McGregor et al. 2010; Weilhoefer et al. 2017).

It is widely held that in order to distinguish whether restoration has been successful, some form of project monitoring and evaluation is needed, as without monitoring there is no capacity to measure success (Williams and Grosholz 2008; Boorman and Garbutt 2011). This is to ensure a project continues to meet its goals during restoration and after the project has been finalised (Bayraktarov et al. 2020; Diefenderfer et al. 2021). Typically, monitoring has been undertaken through physical sampling techniques such as flora and fauna surveys, soil assessment, and/or water quality monitoring (Shuman and Ambrose 2003). However, with rapid advances in technological development, researchers and managers alike have leveraged these innovations to monitor and investigate habitats (Kimball et al. 2021). Radar, thermal cameras and motion trigger camera traps are examples of technologies that have revolutionised data collection, particularly in the realm of wildlife monitoring (Chabot and Bird 2015). Satellite-based remote sensing enables a finer scale understanding of the way ecosystems operate than was previously possible with physical sampling (Duffy et al. 2018), and its use in on the rise (Lyu et al. 2017; Castellanos-Galindo et al. 2019; Harris et al. 2019). Even once a project is 'finished', sporadic monitoring in the long term is still required to ensure the habitat both maintains its state and to pick up on any potential changes or negative impacts such as invasive species (Boorman and Garbutt 2011).

Monitoring is crucial for the objective of managing invasive species. Where an invasion cannot be prevented, the quick detection of its presence means that control actions can be implemented (Adams et al. 2015). The earliest stage of an invasion in an area carries the highest success rate for eradication (Williams and Grosholz 2008). The dynamic nature of intertidal areas paradoxically makes control action more difficult but detection easier. For example, the use of chemical control for invasive vegetation can either be diluted or carried to other parts of an ecosystem by tidal flushing where it may impact native vegetation (Guerra-García et al. 2018; Weidlich et al. 2020). However, invasive species may germinate or flower at different times to native vegetation, making them more easily detectable at certain times of year (Hill et al. 2017).

This chapter will identify the ways in which invasive species impact the success of restoration and evaluate how technology can be leveraged in restoration monitoring and evaluation, including the use of remote sensing and how satellite or unmanned aerial vehicle (UAV) imagery can be integrated into physical sampling programmes. A brief overview of three key monitoring techniques is given, including how remote sensing methodologies can be used to evaluate restoration progress and success. Intertidal habitats monitored include saltmarsh, mangroves, mudflats and wetlands, termed here 'intertidal wetland environments'. Invasive species are here taken to mean non-native species to the environment they are detected in with a focus on those organisms that have some form of negative impact (Jeschke et al. 2014). This chapter ends with a case study from a coastal wetland system in tropical northern Australia, where a combination of physical sampling (vegetation surveys) and imagery analysis derived from UAV surveys were used to assess vegetation changes in five wetland sites.

The cost of invasive flora and fauna and their impact on intertidal wetland environments

Intertidal wetland environments are uniquely vulnerable to invasive species, for several reasons. These habitats accumulate material from multiple sources, both marine and terrestrial, through the transmission of plant matter, sediment and nutrients from upstream and from nearshore coastal waters (Zedler and Kercher 2004). Anthropogenic impacts also make wetlands more vulnerable to invasion, as these ecosystems are located at the intersection of many potential transmission pathways such as urbanised areas, port facilities or recreation areas (Zedler and Kercher 2004; Williams and Grosholz 2008). Though periodically high salinity levels in intertidal areas can also exclude invasive weeds (Bick et al. 2020), this results in a higher chance of invasion if hydrology of an area is impacted, for example through reduced or changed flow regimes as a result of urban expansion. Disturbed or impacted environments are believed to favour invasive species (Strayer 2010), a characteristic that is likely to continue to favour them as the climate changes, especially as invasive species have a tolerance for a wider range of climatic conditions (Hellmann et al. 2008; Havel et al. 2015). Invasive species modify habitats by replacing food sources, through predation or changing habitat structure resulting in changes to the abiotic environment (Leonard, Wren and Beavers 2002; McGregor et al. 2010).

Invasive plants may replace stands of native vegetation, as is the case in northern Australia, where the invasive weed Para Grass *Urochloa mutica* creates dense monotypic stands replacing native Wild Rice *Oryza* sp. and Water Chestnut *Eleocharis dulcis* on monsoonal floodplains (Boyden et al. 2019). Both species are a key part of the aquatic food web (Boyden et al. 2019), and their replacement with Para Grass results in a reduced food source for ecologically important species such as Magpie Geese *Anseranas semipalmata* (McGregor et al. 2010). Formation of monotypic stands of plants may result in changes to canopy cover, which has subsequent impacts on water and soil temperature. For instance, stands of Reed Canary Grass *Phalaris arundinacea* in a Portland wetland created taller canopies, which led to increases in soil and water temperatures reducing vegetation and arthropod diversity (Weilhoefer et al. 2017).

Invasive fauna may be equally as destructive; feral pigs (Wild Boar *Sus scrofa*), for example, impact the environment predominantly through rooting behaviour while feeding, and this turnover in soil can increase turbidity in the water column (Mitchell 2010). Pig rooting behaviour is concentrated around water sources, and this activity increases the area of bare soil and open water (Doupe et al. 2010; Mitchell 2010). These changes in biophysical water properties and plant communities create stressful conditions for aquatic life. For example, feral pigs have been implicated in the declines of at least two freshwater turtle species in Cape York: the Jardine River Painted Turtle *Emydura subglobosa* and the Northern Snake-necked Turtle *Chelodina oblonga* (Fordham, Georges and Corey 2006; Schaffer, Doupé and Lawler 2009). Feral pig damage has been also shown to damage wetland aquatic vegetation and water quality, which can have implications on the thermal and hypoxia exposure conditions for fish (Waltham and Schaffer 2018).

Management and control measures

Invasion is a common problem in restoration projects, as ecosystems are already in a degraded state and therefore are less resilient to disturbance (Tanner et al. 2002; Alvarez-Taboada, Paredes and Julián-Pelaz 2017). Being able to anticipate and plan for an invasive species scenario is thus key to the ongoing success of a project (Tanner et al. 2002). In some cases, pre-site selection will avoid the problem entirely by choosing a location without an existing invasive species; prevention of invasion in the first instance is the preferred management strategy (Williams and Grosholz 2008). As this is not always possible owing to the expanding distributions of some invasive species (Boyden et al. 2019), alternative control measures need to be put in place, supplemented by monitoring. Where invasive species are discovered, quick detection and quick management action following that detection mean that eradication at the site may be possible (Williams and Grosholz 2008). However, intertidal-specific factors, poorly planned projects and poorly designed monitoring schemes makes this type of scenario unrealistic in many cases (Zedler and Kercher 2004; Lovelock and Brown 2019). Along with continued monitoring, there is consensus that information on an invasion, including distribution, tolerance limits, ecology and dispersal methods, are key to its management (Adams et al. 2015).

In addition to prevention, post-invasion control measures can be implemented to manage invasive species. For example, containment via exclusion fencing may be used, though this only works on a small scale (Reddiex et al. 2006). Control measures may also be lethal in nature; these include chemical control (Reddiex et al. 2006; Mitchell 2010) and manual removal (Bengsen et al. 2014). Such measures may be used in combination with education in order to prevent the species spreading to other areas, for instance through vegetation that can be carried by recreational boats or 4WD tyres (Adame et al. 2019). A combination of methods is usually most effective to control invasive populations, although their application is reliant on resources available, including ongoing funding, geographic range and technical capability of the management team (Bengsen et al. 2014).

What makes control difficult?

Controlling invasions in intertidal wetland environments is a significant challenge for managers (Zedler and Kercher 2004). Intertidal habitats are prone to invasions from both marine and terrestrial flora and fauna, requiring monitoring for a diverse range of invasive species (Reddiex et al. 2006). This may mean that an organisation undertaking restoration in intertidal areas may need to spend significant time gathering knowledge and developing and deploying a diverse range of control techniques. This comes at a time, labour and financial cost (Reddiex et al. 2006). Practically, the treatment of invasive vegetation with chemical pesticide is an often-used method for supressing invasive populations (Guerra-García et al. 2018). However, the dynamic nature of intertidal areas due to regular flushing from the tide may dilute the application or carry it to other parts of the ecosystem where it may impact native vegetation (Guerra-García et al. 2018; Weidlich et al. 2020). The impacts of climate change have the potential to make control measures more labour and resource intensive in intertidal wetland environments. Where control measures are conducted seasonally (e.g. manual removal of plant matter in summer months), changed climate conditions such as increased temperatures and longer warm intervals mean that control measures need to be implemented more frequently and for longer periods (Rahel and Olden 2008).

Restoration efforts in intertidal wetland environments

The value and services provided by coastal wetlands are well known (Mitsch and Gosselink 1993). They are among the most biodiverse ecosystems on the planet (Mitsch and Gosselink 1993; Gopal 2013; Weinstein and Litvin 2016; Schuerch et al. 2018), supporting rich aquatic floral and faunal communities (Borja et al. 2010; Elliott and Whitfield 2011; Jiang et al. 2015). However, because of their position on the coast these wetlands are also often targeted for some level of modification, either for road transport, housing estates, heavy industry, agricultural production or recreation (Spalding 2007; Arkema et al. 2015; Agboola et al. 2016). This results in the complete loss of habitats in some regions, while other areas suffer from disturbance and fragmentation (McKinney, Raposa and Kutcher 2009; Waltham and

Sheaves 2015; Weber and Wolter 2017). Efforts by managers to restore intertidal wetland environments is increasing (Waltham et al. 2020), though access to data demonstrating restoration success are limited, which becomes fundamental when attempting to assess biodiversity return for the funding invested by government or private sector markets (Weinstein and Litvin 2016; Waltham et al. 2019; Canning et al. 2021).

What makes restoration successful?

A combination of factors makes restoration successful, ranging from project financing and planning to stakeholder engagement and environmental condition (Stewart-Sinclair et al. 2020). A baseline knowledge of an ecosystem is needed before projects begin in order to have a standard with which to measure success (Boorman and Garbutt 2011; Bellgrove et al. 2017; Abeysinghe et al. 2019). Clear, specific and realistic goal setting is essential to ensure that progress or success can be measured (Beck et al. 2011). This includes quantifiable goals beyond the initial life of the project and targets surrounding abiotic factors, floral coverage and composition, and faunal abundance and diversity (Lundquist and Granek 2005; Boorman and Garbutt 2011; Cross, Bateman and Cross 2020). Realistically, continued management throughout and after a project is largely dependent on project financing and, for government-based projects, stability of the government (Lundquist and Granek 2005; Lovelock and Brown 2019).

Restoration activities may impact nearby people and communities; thus, stakeholder engagement is key to the success of restoration (Lundquist and Granek 2005; Cadier et al. 2020; Stewart-Sinclair et al. 2020). Without human engagement, even the best placed project may fail if use of intertidal resources by people continues (Lovelock and Brown 2019). Anthropogenic use of an ecosystem that is being restored may also lead to a slower recovery (Beck et al. 2011), for example where the wash of recreational boats harms oyster reefs (Garvis, Sacks and Walters 2015). Habitats may also have cultural values that conflict with restoration plans, such as when projects by necessity exclude people from an area (Torpey-Saboe et al. 2015). Stakeholder consultation is important at all stages: engaging with community members can help inform site selection through the transfer of local knowledge, which can ensure long-term sustainability of sites by creating community buy-in to the project (Lundquist and Granek 2005).

Part of the success of a restoration project lies in the initial site selection combined with baseline knowledge about ecosystem condition (Lewis 2005). This may include the hydrology or abundance of fauna or the existence of pressures such as invasive species or surrounding land-uses. However, projects often lack baseline data or only have a nearby reference site (Taddeo and Dronova 2019). This is particularly pertinent in estuarine and tropical environments, where seasonal and diurnal changes in tide and hydrology can result in fluctuations in ecosystem condition. Wet seasons bring influxes of freshwater and dry seasons can lead to periodic changes to wetted extent within the wetlands as the margins dry (Blanchette, Rousseau and Poulin 2018). If the goal is to restore or rehabilitate an existing degraded ecosystem, instead of creating new habitat, the underlying reasons for that degradation need to be assessed and accounted for when starting a project (Waltham et al. 2021). If surrounding land-uses stay the same over the life of the project, it is unlikely to succeed (Lewis 2005; Lovelock and Brown 2019).

Historically, most restoration, especially in terrestrial or mangrove environments, has focused on restoration of vegetation through planting seedlings (Lewis 2005; Macdonald et al. 2015; Yip et al. 2019). Poor outcomes can occur when invasive species are used for planting, or if area targets are the primary restoration goal, placing less emphasis on long-term species persistence (Yip et al. 2019). However, practices have evolved both as regulatory requirements for restoration projects counteracting industrial development have changed and as new knowledge regarding the importance of connectivity has come to light

(Macdonald et al. 2015; Taddeo and Dronova 2019). This has led to a greater inclusion of abiotic factors and targets around floral and faunal diversity in the long term in metrics measuring restoration success (Cross et al. 2019; Cross, Bateman and Cross 2020).

Monitoring and evaluation

Why monitor?

Monitoring is necessary to determine if a project has met its goals and is required to ensure that an area will continue to do so. Beyond avoiding the influx of an invasive species, however, monitoring of restoration projects to detect invasive species depends on the spatial and temporal resolution of the programme. In intertidal environments in particular, the overall length of the monitoring programme is important, as these ecosystems often take longer to recover than their terrestrial counterparts (Beck et al. 2011).

Monitoring may take several forms, each of which operates at different spatial scales and has its own advantages and disadvantages that may make it suitable for a restoration monitoring programme (Table 11.1). This is important for intertidal wetland environments because they are both spatially and temporally diverse, which can be difficult to access and sample (Chabot and Bird 2015).

Other considerations

Fundamental gaps in knowledge about the ecology of systems, how they should be restored, and the ecology of invasive species makes management decision-making difficult (Williams and Grosholz 2008; Bellgrove et al. 2017). Therefore, another concept underpinning monitoring at different spatial scales is greater interaction between science and management (Cadier et al. 2020). There is an extensive body of knowledge on invasive species, including their impact on intertidal areas (Zedler and Kercher 2004; Williams and Grosholz 2008; Alvarez-Taboada, Paredes and Julián-Pelaz 2017; Abeysinghe et al. 2019; Adame et al. 2019). However, there is little literature that looks at applying scientific knowledge to the assessment and management of restoration projects from the side of a practitioner (Williams and Grosholz 2008; Cadier et al. 2020).

Applying research to the management of invasive species-impacted restoration projects could be helpful in several ways. In light of the advances in technology that may increase cost effectiveness, robustness or accuracy of monitoring, it is important to test those methods (Cross, Bateman and Cross 2020). This is a strength of scientific research, which is adept at evaluating problems, formulating system models and testing assumptions and paradigms (Gosselin and Gosselin 2009; Samhouri et al. 2014). When science is included in the management of habitats, for example through adaptive management approaches, it ensures that the information contributing to management policy is relevant, credible and legitimate (Arkema et al. 2015; Stephenson 2019; Gillan, Karl and Van Leeuwen 2020). Restoration projects, particularly long-term ones, provide ample datasets for scientists to deepen knowledge on habitats (Williams and Grosholz 2008).

Adding remote sensing to a restoration manager's toolbox allows for the development of robust, accurate tracking of changes in restoration project sites (Duffy et al. 2018). It is likely that a range of tools will be needed to gather the information needed at different spatial scales (Alvarez-Taboada, Paredes and Julián-Pelaz 2017): UAVs to map within habitat changes, satellites to delineate habitat boundaries and in situ measurements to ground truth remotely sensed imagery and gather even finer scale biotic and abiotic data (Alvarez-Taboada, Paredes and Julián-Pelaz 2017; Duffy et al. 2018; Judge, Choi and Helmuth 2018). Using multiple method types at different scales (satellite, UAV and in situ measurements)

Table 11.1 Brief overview of the advantages and disadvantages of different sampling methods for restoration monitoring as identified in the literature

Data collection method	Advantages	Disadvantages
Satellite	Can create highly accurate maps of large areas (Belluco et al. 2006; Müllerová et al. 2017; Boyden et al. 2019) Different spatial and temporal resolution options with different satellites (Gray et al. 2018) Advances in automated classification and machine learning are making processing imagery faster (Dronova et al. 2012)	Lower spatial scale in general than UAV imagery (Duffy et al. 2018) Time consuming to digitise or manually identify patches of vegetation (Tanner et al. 2002; Garvis, Sacks and Walters 2015) Small patches of vegetation are difficult to identify when sub-pixel size (Belluco et al. 2006; Alvarez-Taboada et al. 2017) Still requires ground-truthed data (Belluco et al. 2006) Measurement errors compound when calculating change from multiple image classifications as error compounds over time (Boyden et al. 2019) Cannot map underneath canopy cover (Arroyo et al. 2010)
Unmanned Aerial Vehicle (UAV)	Higher spatial resolution than satellite imagery (Alvarez-Taboada et al. 2017; Müllerová et al. 2017; Abeysinghe et al. 2019) Fine scale allows intra-habitat to be detected (Duffy et al. 2018; Dale et al. 2020) Flexible deployment means you can tailor sampling design to ecology of species (Müllerová et al. 2017; Duffy et al. 2018) Flexible deployment means you can take advantage of ideal weather conditions to maximise data quality, e.g. glare or tides (Duffy et al. 2018; Gray et al. 2018) More cost effective over small areas (Gray et al. 2018; Castellanos-Galindo et al. 2019; Dale et al. 2020) Captures more data faster than fieldwork (Gray et al. 2018) and has the ability to become automated (Gomez and Green 2017)	No standardised methods easily available (Alvarez-Taboada et al. 2017; Castellanos-Galindo et al. 2019) Using consumer equipment is not precise enough to turn data into management (Alvarez-Taboada et al. 2017; Hill et al. 2017; Müllerová et al. 2017) Still requires some ground-truthed data (Abeysinghe et al. 2019) Generates large volumes of data – lots of processing power required (Castellanos-Galindo et al. 2019) High start-up cost (Gray et al. 2018) Small pixel size can result in lots of error through misclassification (Alvarez-Taboada et al. 2017; Müllerová et al. 2017) Constrained by local legislation (Müllerová et al. 2017)
Floral and faunal sampling	Can use inexpensive equipment (Castellanos-Galindo et al. 2019) Easily standardised (Tanner et al. 2002; Chainho et al. 2010) Can use multiple gears to sample a broad range of fauna – and standardise for comparison (e.g. catch per unit effort) (Silver et al. 2017; Lechêne et al. 2018) If access is possible can spread sampling points to get good spatial coverage of sites (Chainho et al. 2010; Mazik et al. 2010; Grizzle et al. 2018) If many measurements are taken – can detect small-scale changes in abiotic factors (e.g. temperature, chlorophyll) (Chainho et al. 2010; Grizzle et al. 2018)	Gear selectivity, e.g. based on mesh size (Lechêne et al. 2018) Can be problems applying a small sample (e.g. a quadrat) to a whole site (Mander, Marie-Orleach and Elliott 2013; Duffy et al. 2018) Time consuming to map boundaries of vegetation by hand/using a GPS (Tanner et al. 2002) Can be difficult to access all areas for sampling, e.g. fast-flowing or deep water (Mackay, James and Arthington 2010; Silver et al. 2017) Small spatial scale required to observe changes in intertidal environments not always possible to observe with physical sampling design (Chainho et al. 2010; Mander, Marie-Orleach and Elliott 2013; Lv et al. 2018) If long-term programme and equipment or method change, it is very hard to compare results – cannot resample in the past (Chainho et al. 2010)

where possible will allow the capture of accurate information over large spatial scales (Belluco et al. 2006; Alvarez-Taboada, Paredes and Julián-Pelaz 2017).

As previously mentioned, seasonality in intertidal areas can make invasive species control difficult, but it may also make detection easier (Müllerová et al. 2017). Monitoring, particularly through remote sensing, operates best when knowledge is available on plant phenology, biology and ecology, in order to find the greatest 'exploitable difference' between an invasive and native species (Bolch et al. 2020). Studies identifying invasive Giant Hogweed *Heracleum mantegazzianum* and Japanese Knotweed *Fallopia japonica* through UAV imagery in the Czech Republic used distinct differences in flower and stem colours between the invasive and native species to map extents of the invasive vegetation (Müllerová et al. 2017). Similarly, the use of a UAV and manual image interpretation for the invasive Yellow Flag Iris *Iris pseudacorus* in British Columbia provided a more accurate map of the invasion than in situ mapping owing to its distinctive flower colour and leaf shape (Hill et al. 2017). In both examples, exploitable differences between native and invasive species could be leveraged to best design monitoring and control programmes.

Intertidal wetland environments restoration case study: Round Hill

In November 2016, the Burnett Mary Regional Group designated restoration sites in and around Round Hill Reserve, a coastal wetland in the upper reaches of Round Hill creek, located in the Baffle catchment, North Queensland, Australia (Fig. 11.1).

Fences were constructed around one site (RH1) in order to prevent feral pig access to wetlands, four additional reference sites were included in the study (RH2–RH5). The primary objective of this restoration activity was to assess exclusion fencing as a control method for feral pig damage by examining the vegetation community in the fenced wetland in Round Hill Reserve and nearby (unfenced) wetlands. Feral pigs are considered both an invasive and pest species in Australia, and have since been listed as a key threatening process to Australia's natural environment under the Environment Protection and Biodiversity Conservation Act 1999 (Australia 2017). Both UAV imagery and quadrat vegetation sampling were used to monitor the vegetation communities at all five sites and surveys were conducted at six-monthly intervals in July 2018 and January and June 2019 (Table 11.2). Round Hill creek was identified as a restoration priority because it is an important Fish Habitat Area (FHA) and a Directory of Important Wetlands in Australia (DIWA) listed wetland (Department of Environment and Science 2013). A full report of the methods and results has been published (Waltham, Doupé and Lawler 2020). Here, we present results most relevant to the impact of invasive species and use of technology in monitoring restoration sites.

The results of this study showed that land-cover change for wetland environments can be assessed using UAVs as classification of land-cover classes showed change across all three time periods (e.g. RH2: Figs 11.2, 11.3). Overall, bare soil and water were the predominant land-cover categories at most sites and high levels of variation were present in all land-cover types between sampling events. The high variability of land-cover between time periods was unsurprising since wetlands in north-eastern Australia are known to be changeable with the seasons (Doupe 2010). What was surprising was the lower prominence of vegetation as a category, especially since aquatic plants were only identified at one site, as this type of vegetation is a crucial part of the wetland biome (Grasset et al. 2017). While this could be owing to lack of skill on the part of the producer in creating training samples (i.e. aquatic plants were simply not identified correctly), it could also be a clue that the wetland vegetation was not recovering.

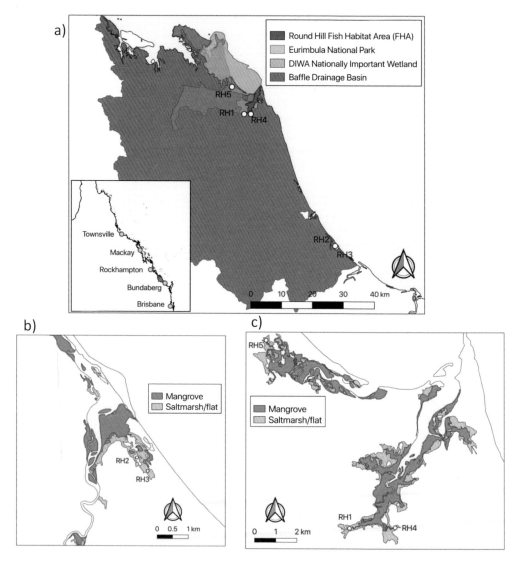

Figure 11.1 (a) Baffle Drainage Basin (in red) and five survey locations: Round Hill reserve (RH1), Littabella reference (RH2 and RH3), Round Hill reference (RH4), Eurimbula NP reference (RH5); (b) Inset map of RH1, RH4 and RH5; (c) Inset map of RH2 and RH3.

The accuracy of classification was assessed by assigning 100 randomly distributed points per image. Each point was then ground truthed by inspecting the image to confirm it had been classified correctly. A confusion matrix comparing the number of correctly classified points to the number of incorrectly classified for each site and survey was developed to quantify the amount of disagreement between classification and ground truthing. The kappa coefficient for each site across all survey dates is presented in Fig. 11.4.

Of the five wetlands surveyed, the fenced wetland in Eurimbula National Park (RH1) supported a more diverse floral community than other sites visited. This was mostly because of the presence of freshwater species that were absent elsewhere. There were some limitations in the vegetation sampling design that made it difficult to compare UAV-derived vegetation types with what was measured on the ground during quadrat surveys. Quadrats were generally placed close to or on boundaries between vegetation types; this was

Table 11.2 Restored and reference wetland location descriptions.

Code	Location	Management or Disturbance
RH1	Round Hill reserve 24°13′55.35″S 151°49′55.45″E	Declared FHA and listed in the DIWA. Adjacent to Eurimbula National Park. Feral pig exclusion fencing installed November 2016 (Queensland Parks and Wildlife, Burnett Mary Regional Group); however, seasonal cattle grazing permitted.
RH2	Littabella Down reference 24°36′35.72″S 152°7′25.90″E	State-owned land with road causing tidal restriction, cattle grazing and feral pig disturbance.
RH3	Littabella Up reference 24°36′49.40″S 152°7′38.17″E	State-owned land with road causing tidal disconnection, cattle grazing and feral pig disturbance.
RH4	Round Hill reference 24°13′54.17″S 151°51′11.47″E	Privately owned land with cattle grazing and feral pig disturbance.
RH5	Eurimbula NP reference 24°9′11.89″S 151°47′34.90″E	Located within Eurimbula National Park, with cattle grazing and feral pig disturbance.

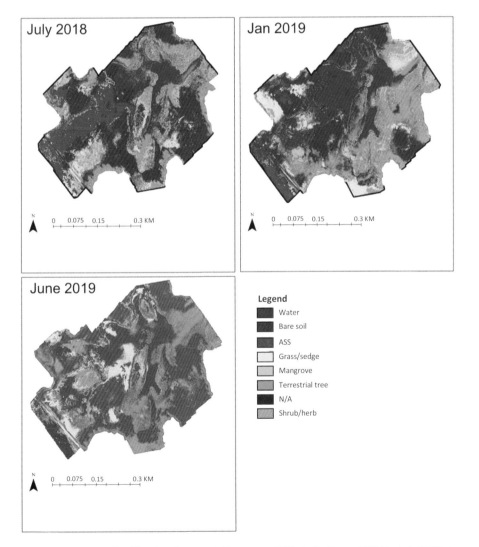

Figure 11.2 Vegetation classification from aerial imagery at Littabella Down (RH2) in July 2018, January 2019 and June 2019. Colours indicate areas classified as bare soil, grass, mangrove, shrubs, terrestrial trees, water and acid sulphate soil.

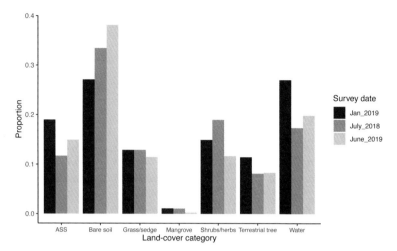

Figure 11.3 Proportion of land-cover calculated from classified imagery (maximum likelihood classification performed in ArcGIS) at site Littabella Down (RH2) taken from UAV imagery. Graph shows comparison between imagery taken in July 2018, January 2019 and June 2019.

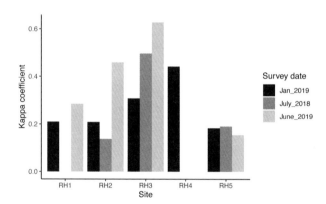

Figure 11.4 Comparison of kappa coefficients across all sites and survey dates. Kappa coefficient produced using the compute confusion matrix tool in ArcGIS.

especially the case with 'high' sites that were placed at the saltmarsh–mangrove boundary. Mangroves over quadrats were not recorded, as the vegetation survey was focused on what was within the quadrat on the ground. This may have been because the trunk itself was not within the quadrat, while branches and foliage were overhead.

UAV imagery and field observations identified higher vegetation cover in marsh areas outside the fenced enclosure in areas where cattle were unable to access, owing to the natural barrier of a bend in the creek. There was some initial indication that the vegetation community was returning following the fencing. However, in the subsequent years of this study, it was apparent that protection of wetland values was limited by the continual access to the fenced wetland by cattle as part of land tenure arrangements with adjacent landholders. The effort of installing fences for fauna movement and any other implemented control measures (e.g. baiting) needs to be supported by changes to the land tenure arrangement, to ensure that the values and services aspired to are reached in order to invest in this sensitive wetland habitat. Overall, the accuracy of the UAV imagery classification was low, except for areas that were spectrally distinct enough from other categories to contain exploitable differences (e.g. acid sulphate soil in the red band). Despite this, the identification of general trends (i.e. that land-cover has changed) remains useful information, and the failings identified as part of this study in the ability to quantify land-cover change still represent a springboard for the advancement of wetland classification. The use of UAV imagery allowed the extension of sampling site wide, instead of relying solely on vegetation identified in the quadrats. Sampling in this case could not have occurred solely with a drone; the inability to identify vegetation to a species level from imagery showed that quadrat sampling was still essential in delineating biota. This shows that while UAVs have the ability to become an important tool for environmental management (and are on their way to becoming one), there is still work to be done in ensuring that the benefits of UAVs can be implemented in a useful and accurate manner.

Conclusion

As restoration projects increase in both number and scale to account for growing impacts on intertidal communities, a diverse array of tools will be required to monitor projects. As seen in Table 11.1, all methods have advantages and disadvantages that may make them suited to different types of sites, habitats or situations. It is up to managers to assess projects and best design monitoring programmes to assess progress in a way that is robust, accurate and cost effective. Invasive species are a continuing and growing problem for intertidal areas, and the dispersed nature and increasing extent of many invasive species means that monitoring is required on a larger scale. The effects of climate change are likely to exacerbate this problem, both in allowing for the expansion of invasive species' ranges and in making control action more difficult. Filling fundamental gaps in knowledge regarding the ecology of both native and invasive species will make discovering and making use of exploitable differences in invasive species more effective. Long-term monitoring projects with realistic goals will enable detection of invasive species quicker, leading to more effective control measures. The Round Hill case study demonstrates that existing land management practices need to be considered when attempting to restore wetland sites. UAV technology enabled monitoring at a larger spatial scale than would have been possible with physical sampling alone. While more work needs to be completed to integrate the UAV work and physical sampling to provide the highest monitoring benefit, UAVs in this case still presented a clear benefit in demonstrating the ways in which land-cover changed throughout the monitoring period. There is further work to be done in integrating this technology into land practices, especially in monitoring regimes. While research using UAVs is highly innovative, there is a gap in the way they are used in land management (as the Round Hill case study demonstrated). Future research should look for ways to leverage these powerful tools so that they can be added to the environmental manager's toolbox. This process can only be completed by taking the needs of managers into account, and working with them to discover ways these technologies can be implemented to make data collection for monitoring more robust.

References

Abeysinghe, T., Milas, A.S., Arend, K., Hohman, B., Reil, P., Gregory, A. and Vázquez-Ortega, A. (2019) Mapping invasive *Phragmites australis* in the Old Woman Creek estuary using UAV remote sensing and machine learning classifiers. *Remote Sensing* 11: 1380–403 https://doi.org/10.3390/rs11111380

Adame, M.F., Arthington, A.H., Waltham, N., Hasan, S., Selles, A. and Ronan, M. (2019) Managing threats and restoring wetlands within catchments of the Great Barrier Reef, Australia. *Aquatic Conservation: Marine and Freshwater Ecosystems* 29: 829–39. https://doi.org/10.1002/aqc.3096

Adams, V.M., Petty, A.M., Douglas, M.M., Buckley, Y.M., Ferdinands, K.B., Okazaki, T., Ko, D.W. and Setterfield, S.A. (2015) Distribution, demography and dispersal model of spatial spread of invasive plant populations with limited data. *Methods in Ecology and Evolution* 6: 782–94. https://doi.org/10.1111/2041-210X.12392

Agboola, J.I., Ndimele, P.E., Odunuga, S., Akanni, A., Kosemani, B. and Ahove, M.A. (2016) Ecological health status of the Lagos wetland ecosystems: implications for coastal risk reduction. *Estuarine, Coastal and Shelf Science* 183: 73–81. https://doi.org/10.1016/j.ecss.2016.10.019

Alvarez-Taboada, F., Paredes, C. and Julián-Pelaz, J. (2017) Mapping of the invasive species Hakea sericea using Unmanned Aerial Vehicle (UAV) and worldview-2 imagery and an object-oriented approach. *Remote Sensing* 9: 913–30. https://doi.org/10.3390/rs9090913

Arkema, K.K., Verutes, G.M., Wood, S.A., Clarke-Samuels, C., Rosado, S., Canto, M., Rosenthal, A., Ruckelshaus, M., Guannel, G. and Toft, J. (2015) Embedding ecosystem services in coastal planning leads to better outcomes for people and nature. *Proceedings of the National Academy of Sciences* 112: 7390–5. https://doi.org/10.1073/pnas.1406483112

Arroyo, L.A., Johansen, K., Armston, J. and Phinn, S. (2010) Integration of LiDAR and QuickBird imagery for mapping riparian biophysical parameters and land cover types in Australian tropical savannas. *Forest Ecology and Management* 259: 598–606. https://doi.org/10.1016/j.foreco.2009.11.018

Australia (2017) Threat abatement plan for predation, habitat degradation, competition and disease transmission by feral pigs (*Sus scrofa*) (2017): background document. Available at: https://www.environment.gov.au/system/files/resources/b022ba00-ceb9-4d0b-9b9a-54f9700e7ec9/files/tap-feral-pigs-2017-background-document.pdf (Accessed: 3 September 2021).

Bayraktarov, E., Brisbane, S., Hagger, V., Smith, C.S., Wilson, K.A., Lovelock, C.E., Gillies, C., Steven, A.D. and Saunders, M.I. (2020) Priorities and motivations of marine coastal restoration research. *Frontiers in Marine Science* 7: 484–98 https://doi.org/10.3389/fmars.2020.00484

Beck, M.W., Brumbaugh, R.D., Airoldi, L., Carranza, A., Coen, L.D., Crawford, C., Defeo, O., Edgar, G.J., Hancock, B., Kay, M.C., Lenihan, H.S., Luckenbach, M.W., Toropova, C.L., Zhang, G. and Guo, X. (2011) Oyster reefs at risk and recommendations for conservation, restoration, and management. *BioScience* 61: 107–16. https://doi.org/10.1525/bio.2011.61.2.5

Bellgrove, A., Mckenzie, P.F., Cameron, H. and Pocklington, J.B. (2017) Restoring rocky intertidal communities: Lessons from a benthic macroalgal ecosystem engineer. *Marine Pollution Bulletin* 117: 17–27. https://doi.org/10.1016/j.marpolbul.2017.02.012

Belluco, E., Camuffo, M., Ferrari, S., Modenese, L., Silvestri, S., Marani, A. and Marani, M. (2006) Mapping salt-marsh vegetation by multispectral and hyperspectral remote sensing. *Remote Sensing of Environment* 105: 54–67. https://doi.org/10.1016/j.rse.2006.06.006

Bengsen, A.J., Gentle, M.N., Mitchell, J.L., Pearson, H.E. and Saunders, G.R. (2014) Impacts and management of wild pigs *Sus scrofa* in Australia. *Mammal Review* 44: 135–47. https://doi.org/10.1111/mam.12011

Bick, E., de Lange, E.S., Kron, C.R., da Silva Soler, L., Liu, J. and Nguyen, H.D. (2020) Effects of salinity and nutrients on water hyacinth and its biological control agent, *Neochetina bruchi*. *Hydrobiologia* 847: 3213–24. https://doi.org/10.1007/s10750-020-04314-x

Blanchette, M., Rousseau, A.N. and Poulin, M. (2018) Mapping wetlands and land cover change with Landsat archives: the added value of geomorphologic data: cartographie de la dynamique spatio-temporelle des milieux humides à partir d'archives Landsat: la valeur ajoutée de données géomorphologiques. *Canadian Journal of Remote Sensing* 44: 337–56. https://doi.org/10.1080/07038992.2018.1525531

Bolch, E.A., Santos, M.J., Ade, C., Khanna, S., Basinger, N.T., Reader, M.O. and Hestir, E.L. (2020) Remote detection of invasive alien species. In: J. Cavender-Bares, J.A. Gamon and P.A. Townsend (eds) *Remote Sensing of Plant Biodiversity*. Springer International. https://doi.org/10.1007/978-3-030-33157-3_12

Boorman, L.A. and Garbutt, A. (2011) 10.09 – Restoration Strategies for Intertidal Salt Marshes. In: Baird, D., & Mehta, A. (eds) *Treatise on Estuarine and Coastal Science*, pp. 189–215. Academic Press, an imprint of Elsevier. https://doi.org/10.1016/B978-0-12-374711-2.01009-3

Borja, Á., Dauer, D.M., Elliott, M. and Simenstad, C.A. (2010) Medium-and long-term recovery of estuarine and coastal ecosystems: patterns, rates and restoration effectiveness. *Estuaries and Coasts* 33: 1249–60. https://doi.org/10.1007/s12237-010-9347-5

Boyden, J., Wurm, P., Joyce, K.E. and Boggs, G. (2019) Spatial dynamics of invasive para grass on a monsoonal floodplain, Kakadu National Park, Northern Australia. *Remote Sensing* 11 (18): 2090–12. https://doi.org/10.3390/rs11182090

Cadier, C., Bayraktarov, E., Piccolo, R. and Adame, M.F. (2020) Indicators of coastal wetlands restoration success: a systematic review. *Frontiers in Marine Science* 7 (7): 1017–28. https://doi.org/10.3389/fmars.2020.600220

Canning, A., Javis, D., Costanza, R., Finisdore, J., Smart, J.C.R., Hasan, S., Lovelock, C., Marr, H., Beck, M., Stephenson, K., Gillies, C., Wilson, P. and Waltham, N.J. (2021) Financial incentives for large scale wetland restoration: beyond markets to common asset trusts. *One Earth* 4 (7): 937–50. https://doi.org/10.1016/j.oneear.2021.06.006

Castellanos-Galindo, G.A., Casella, E., Mejía-Rentería, J.C. and Rovere, A. (2019) Habitat mapping of remote coasts: Evaluating the usefulness of lightweight unmanned aerial vehicles for conservation and monitoring. *Biological Conservation* 239: 73–83. https://doi.org/10.1016/j.biocon.2019.108282

Chabot, D. and Bird, D.M. (2015) Wildlife research and management methods in the 21st century: where do unmanned aircraft fit in? *Journal of Unmanned Vehicle Systems* 1 (1): 137–5. https://doi.org/10.1139/juvs-2015-0021

Chainho, P., Silva, G., Lane, M.F., Costa, J.L., Pereira, T., Azeda, C., Almeida, P.R., Metelo, I. and Costa, M.J. (2010) Long-term trends in intertidal and subtidal benthic communities in response to water quality improvement measures. *Estuaries and Coasts* 33: 1314–26. https://doi.org/10.1007/s12237-010-9321-2

Cross, S.L., Bateman, P.W. and Cross, A.T. (2020) Restoration goals: Why are fauna still overlooked in the process of recovering functioning ecosystems and what can be done about it? *Ecological Management and Restoration* 21: 4–8. https://doi.org/10.1111/emr.12393

Cross, S.L., Tomlinson, S., Craig, M.D., Dixon, K.W. and Bateman, P.W. (2019) Overlooked and undervalued: the neglected role of fauna and a global bias in ecological restoration assessments. *Pacific Conservation Biology* 25: 331–41. https://doi.org/10.1071/PC18079

Dale, J., Burnside, N.G., Strong, C.J. and Burgess, H.M. (2020) The use of small-unmanned aerial systems for high resolution analysis for intertidal wetland restoration schemes. *Ecological Engineering* 143: 260–8. https://doi.org/10.1016/j.ecoleng.2019.105695

DeVore, J.L., Shine, R. and Ducatez, S. (2020) Urbanization and translocation disrupt the relationship between host density and parasite abundance. *Journal of Animal Ecology* 89: 1122–33. https://doi.org/10.1111/1365-2656.13175

Diefenderfer, H.L., Steyer, G.D., Harwell, M.C., Loschiavo, A.J., Neckles, H.A., Burdick, D.M., Johnson, G.E., Buenau, K.E., Trujillo, E. and Callaway, J.C. (2021) Applying cumulative effects to strategically advance large-scale ecosystem restoration. *Frontiers in Ecology and the Environment* 19: 108–17. https://doi.org/10.1002/fee.2274

Doupe, R.G., Mitchell, J., Knott, M.J., Davis, A.M. and Lymbery, A.J. (2010) Efficacy of exclusion fencing to protect ephemeral floodplain lagoon habitats from feral pigs (*Sus scrofa*). *Wetlands Ecology Management* 18: 69–78. https://doi.org/10.1007/s11273-009-9149-3

Dronova, I., Gong, P., Clinton, N.E., Wang, L., Fu, W., Qi, S. and Liu, Y. (2012) Landscape analysis of wetland plant functional types: The effects of image segmentation scale, vegetation classes and classification methods. *Remote Sensing of Environment* 127: 357–69. https://doi.org/10.1016/j.rse.2012.09.018

Duffy, J.P., Pratt, L., Anderson, K., Land, P.E. and Shutler, J.D. (2018) Spatial assessment of intertidal seagrass meadows using optical imaging systems and a lightweight drone. *Estuarine, Coastal and Shelf Science* 200: 169–80. https://doi.org/10.1016/j.ecss.2017.11.001

Elliott, M. and Whitfield, A.K. (2011) Challenging paradigms in estuarine ecology and management. *Estuarine, Coastal and Shelf Science* 94: 306–14. https://doi.org/10.1016/j.ecss.2011.06.016

Fordham, D., Georges, A. and Corey, B. (2006) Compensation for inundation-induced embryonic diapause in a freshwater turtle: achieving predictability in the face of environmental stochasticity. *Functional Ecology* 20: 670–7. https://doi.org/10.1111/j.1365-2435.2006.01149.x

Garvis, S.K., Sacks, P.E. and Walters, L.J. (2015) Formation, movement, and restoration of dead intertidal oyster reefs in Canaveral National Seashore and Mosquito Lagoon, Florida. *Journal of Shellfish Research* 34: 251–58. https://doi.org/10.2983/035.034.0206

Gillan, J.K., Karl, J.W. and Van Leeuwen, W.J.D. (2020) Integrating drone imagery with existing rangeland monitoring programs. *Environmental Monitoring and Assessment* 192 (5): 1–20. https://doi.org/10.1007/s10661-020-8216-3

Gomez, C. and Green, D.R. (2017) Small unmanned airborne systems to support oil and gas pipeline monitoring and mapping. *Arabian Journal of Geosciences* 10 (9): 202–19. https://doi.org/10.1007/s12517-017-2989-x

Gopal, B. (2013) Future of wetlands in tropical and subtropical Asia, especially in the face of climate change. *Aquatic Sciences* 75: 39–61. https://doi.org/10.1007/s00027-011-0247-y

Gosselin, F. and Gosselin, F. (2009) Management on the basis of the best scientific data or integration of ecological research within management? Lessons learned from the northern spotted owl saga on the connection between research and management in conservation biology. *Biodiversity and Conservation* 18: 777–93. https://doi.org/10.1007/s10531-008-9449-6

Grasset, C., Levrey, L.H., Delolme, C. et al. (2017) The interaction between wetland nutrient content and plant quality controls aquatic plant decomposition. *Wetlands Ecology and Management* 25: 211–19. https://doi.org/10.1007/s11273-016-9510-2

Gray, P., Ridge, J., Poulin, S., Seymour, A., Schwantes, A., Swenson, J. and Johnston, D. (2018) Integrating drone imagery into high resolution satellite remote sensing assessments of estuarine environments. *Remote Sensing (Basel, Switzerland)* 10: 1257–81. https://doi.org/10.3390/rs10081257

Grizzle, R.E., Rasmussen, A., Martignette, A.J., Ward, K. and Coen, L.D. (2018) Mapping seston depletion over an intertidal eastern oyster (*Crassostrea virginica*) reef: Implications for restoration of multiple habitats. *Estuarine, Coastal and Shelf Science* 212: 265–72. https://doi.org/10.1016/j.ecss.2018.07.013

Guerra-García, A., Barrales-Alcalá, D., Argueta-Guzmán, M., Cruz, A., Mandujano, M.C., Arévalo-Ramírez, J.A., Milligan, B.G. and Golubov, J. (2018) Biomass allocation, plantlet survival, and chemical control of the invasive chandelier plant (*Kalanchoe delagoensis*) (Crassulaceae). *Invasive Plant Science and Management* 11: 33–9. https://doi.org/10.1017/inp.2018.6

Harris, J.M., Nelson, J.A., Rieucau, G. and Broussard III, W.P. (2019) Use of drones in fishery science. *Transactions of the American Fisheries Society* 148: 687–97. https://doi.org/10.1002/tafs.10168

Havel, J.E., Kovalenko, K.E., Thomaz, S.M., Amalfitano, S. and Kats, L.B. (2015) Aquatic invasive species: challenges for the future. *Hydrobiologia* 750: 147–70. https://doi.org/10.1007/s10750-014-2166-0

Hellmann, J.J., Byers, J.E., Bierwagen, B.G. and Dukes, J.S. (2008) Five potential consequences of climate change for invasive species. *Conservation Biology* 22: 534–43. https://doi.org/10.1111/j.1523-1739.2008.00951.x

Hill, D.J., Tarasoff, C., Whitworth, G.E., Baron, J., Bradshaw, J.L. and Church, J.S. (2017) Utility of unmanned aerial vehicles for mapping invasive plant species: a case study on yellow flag iris (*Iris pseudacorus* L.). *International Journal of Remote Sensing* 38: 2083–105. https://doi.org/10.1080/01431161.2016.1264030

Jeschke, J.M., Bacher, S., Blackburn, T.M., Dick, J.T.A., Essl, F., Evans, T., Gaertner, M., Hulme, P.E., Kühn, I., Mrugała, A., Pergl, J.A.N., Pyšek, P., Rabitsch, W., Ricciardi, A., Richardson, D.M., Sendek, A., Vilà, M., Winter, M. and Kumschick, S. (2014) Defining the impact of non-native species. *Conservation Biology* 28: 1188–94. https://doi.org/10.1111/cobi.12299

Jiang, T.-T., Pan, J.-F., Pu, X.-M., Wang, B. and Pan, J.-J. (2015) Current status of coastal wetlands in China: degradation, restoration, and future management. *Estuarine, Coastal and Shelf Science* 164: 265–75. https://doi.org/10.1016/j.ecss.2015.07.046

Judge, R., Choi, F. and Helmuth, B. (2018) Recent advances in data logging for intertidal ecology. *Frontiers in Ecology and Evolution* 6: 213–31. https://doi.org/10.3389/fevo.2018.00213

Kimball, M.E., Connolly, R.M., Alford, S.B., Colombano, D.D., James, W.T., Kenworthy, M.D., Norris, G.S., Ollerhead, J., Ramsden, S., Rehage, J.S., Sparks, E.L., Waltham, N.J., Worthington, T.A. and Taylor, M.D. (2021) Novel applications of technology for advancing tidal marsh ecology. *Estuaries and Coasts* 44: 1568–78. https://doi.org/10.1007/s12237-021-00939-w

Lechêne, A., Boët, P., Laffaille, P. and Lobry, J. (2018) Nekton communities of tidally restored marshes: a whole-estuary approach. *Estuarine, Coastal and Shelf Science* 207: 368–82. https://doi.org/10.1016/j.ecss.2017.08.038

Leonard, L.A., Wren, P.A. and Beavers, R.L. (2002) Flow dynamics and sedimentation in *Spartina alterniflora* and *Phragmites australis* marshes of the Chesapeake Bay. *Wetlands* 22: 415–24. https://doi.org/10.1672/0277-5212(2002)022[0415:FDASIS]2.0.CO;2

Lewis, R.R. (2005) Ecological engineering for successful management and restoration of mangrove forests. *Ecological Engineering* 24: 403–18. https://doi.org/10.1016/j.ecoleng.2004.10.003

Lovelock, C.E. and Brown, B.M. (2019) Land tenure considerations are key to successful mangrove restoration. *Nature Ecology & Evolution* 3: 1135–35. https://doi.org/10.1038/s41559-019-0942-y

Lundquist, C.J. and Granek, E.F. (2005) Strategies for successful marine conservation: integrating socioeconomic, political, and scientific factors. *Conservation Biology* 19: 1771–8. https://doi.org/10.1111/j.1523-1739.2005.00279.x

Lv, W., Zhou, W. and Zhao, Y. (2018) Macrobenthos functional groups as indicators of ecological restoration in reclaimed intertidal wetlands of China's Yangtze Estuary. *Regional Studies in Marine Science* 22: 93–100. https://doi.org/10.1016/j.rsma.2018.06.003

Lyu, P., Malang, Y., Liu, H.H.T., Lai, J., Liu, J., Jiang, B., Qu, M., Anderson, S., Lefebvre, D.D. and Wang, Y. (2017) Autonomous cyanobacterial harmful algal blooms monitoring using multirotor UAS. *International Journal of Remote Sensing* 38: 2818–43. https://doi.org/10.1080/01431161.2016.1275058

Macdonald, S.E., Landhäusser, S.M., Skousen, J., Franklin, J., Frouz, J., Hall, S., Jacobs, D.F. and Quideau, S. (2015) Forest restoration following surface mining disturbance: challenges and solutions. *New Forests* 46: 703–32. https://doi.org/10.1007/s11056-015-9506-4

Mackay, S.J., James, C.S. and Arthington, A.H. (2010) Macrophytes as indicators of stream condition in the wet tropics region, Northern Queensland, Australia. *Ecological Indicators* 10: 330–40. https://doi.org/10.1016/j.ecolind.2009.06.017

Mander, L., Marie-Orleach, L. and Elliott, M. (2013) The value of wader foraging behaviour study to assess the success of restored intertidal areas. *Estuarine, Coastal and Shelf Science* 131: 1–5. https://doi.org/10.1016/j.ecss.2013.07.010

Mazik, K., Musk, W., Dawes, O., Solyanko, K., Brown, S., Mander, L. and Elliott, M. (2010) Managed realignment as compensation for the loss of intertidal mudflat: a short term solution to a long term problem? *Estuarine, Coastal and Shelf Science* 90: 11–20. https://doi.org/10.1016/j.ecss.2010.07.009

Mcgregor, S., Lawson, V., Christophersen, P., Kennett, R., Boyden, J., Bayliss, P., Liedloff, A., Mckaige, B. and Andersen, A.N. (2010) Indigenous wetland burning: conserving natural and cultural resources in Australia's World Heritage-listed Kakadu National Park. *Human Ecology* 38: 721–9. https://doi.org/10.1007/s10745-010-9362-y

McKinney, R.A., Raposa, K.B. and Kutcher, T.E. (2009) Use of urban marine habitats by foraging wading birds. *Urban Ecosystems* 13: 191–208. https://doi.org/10.1007/s11252-009-0111-1

Mitchell, J. (2010) Experimental research to quantify the environmental impact of feral pigs within tropical freshwater ecosystems. Final report to the

Department of the Environment, Water, Heritage and the Arts. In Australian Government, Department of Environment and Energy (2017) Threat abatement plan for predation, habitat degradation, competition and disease transmission by feral pigs (*Sus scrofa*): background document. Available at: https://www.environment.gov.au/system/files/resources/b022ba00-ceb9-4d0b-9b9a-54f9700e7ec9/files/tap-feral-pigs-2017-background-document.pdf (Accessed: 3 September 2021).

Mitsch, W.J. and Gosselink, J.G. (1993) *Wetlands*. New York: Van Nostrand Reinhold.

Müllerová, J., Brůna, J., Bartaloš, T., Dvořák, P., Vítková, M. and Pyšek, P. (2017) Timing is important: unmanned aircraft vs. satellite imagery in plant invasion monitoring. *Frontiers in Plant Science* 8: 887–900. https://doi.org/10.3389/fpls.2017.00887

Rahel, F.J. and Olden, J.D. (2008) Assessing the effects of climate change on aquatic invasive species. *Conservation Biology* 22: 521–33. https://doi.org/10.1111/j.1523-1739.2008.00950.x

Reddiex, B., Forsyth, D.M., Mcdonald-Madden, E., Einoder, L.D., Griffioen, P.A., Chick, R.R. and Robley, A.J. (2006) Control of pest mammals for biodiversity protection in Australia. I. Patterns of control and monitoring. *Wildlife Research (East Melbourne)* 33: 691–709. https://doi.org/10.1071/WR05102

Samhouri, J.F., Haupt, A.J., Levin, P.S., Link, J.S. and Shuford, R. (2014) Lessons learned from developing integrated ecosystem assessments to inform marine ecosystem-based management in the USA. *ICES Journal of Marine Science* 71: 1205–15. https://doi.org/10.1093/icesjms/fst141

Sapkota, Y. and White, J.R. (2020) Carbon offset market methodologies applicable for coastal wetland restoration and conservation in the United States: a review. *Science of the Total Environment* 701: 121–30. https://doi.org/10.1016/j.scitotenv.2019.134497

Schaffer, J.G., Doupé, R. and Lawler, R.I. (2009) What for the future of the Jardine River Painted Turtle? *Pacific Conservation Biology* 15: 92–95. https://doi.org/10.1071/PC090092

Schuerch, M., Spencer, T., Temmerman, S., Kirwan, M.L., Wolff, C., Lincke, D., McOwen, C.J., Pickering, M.D., Reef, R. and Vafeidis, A.T. (2018) Future response of global coastal wetlands to sea-level rise. *Nature* 561: 231–234. https://doi.org/10.1038/s41586-018-0476-5

Science, DOEA (2013) Seventeen Seventy-Round Hill fish habitat area – facts and maps, WetlandInfo. Available at: https://wetlandinfo.des.qld.gov.au/wetlands/facts-maps/fish-habitat-area-seventeen-seventy-round-hill/ (Accessed: 1 February 2021).

Shuman, C.S. and Ambrose, R.F. (2003) A comparison of remote sensing and ground-based methods for monitoring wetland restoration success. *Restoration Ecology* 11: 325–33. https://doi.org/10.1046/j.1526-100X.2003.00182.x

Silver, B.P., Hudson, J.M., Lohr, S.C. and Whitesel, T.A. (2017) Short-term response of a coastal wetland fish assemblage to tidal regime restoration in Oregon. *Journal of Fish and Wildlife Management* 8: 193–208. https://doi.org/10.3996/112016-JFWM-083

Spalding, M.D. (2007) Marine ecoregions of the world: a bioregionalization of coastal and shelf areas. *Bioscience* 57: 573–83. https://doi.org/10.1641/B570707

Stephenson, P.J. (2019) Integrating Remote Sensing into Wildlife Monitoring for Conservation. *Environmental Conservation* 46: 181–3. https://doi.org/10.1017/S0376892919000092

Stewart-Sinclair, P.J., Purandare, J., Bayraktarov, E., Waltham, N., Reeves, S., Statton, J., Sinclair, E.A., Brown, B.M., Shribman, Z.I. and Lovelock, C.E. (2020) Blue restoration – building confidence and overcoming barriers. *Frontiers in Marine Science* 7: 748–61. https://doi.org/10.3389/fmars.2020.541700

Strayer, D.L. (2010) Alien species in fresh waters: ecological effects, interactions with other stressors, and prospects for the future. *Freshwater Biology* 55: 152–74. https://doi.org/10.1111/j.1365-2427.2009.02380.x

Taddeo, S. and Dronova, I. (2019) Geospatial tools for the large-scale monitoring of wetlands in the San Francisco estuary: opportunities and challenges. *San Francisco Estuary and Watershed Science* 17 (2): 9–33. https://doi.org/10.15447/sfews.2019v17iss2art2

Tanner, C.D., Cordell, J.R., Rubey, J. and Tear, L.M. (2002) Restoration of freshwater intertidal habitat functions at Spencer Island, Everett, Washington. *Restoration Ecology* 10: 564–76. https://doi.org/10.1046/j.1526-100X.2002.t01-1-02034.x

Torpey-Saboe, N., Andersson, K., Mwangi, E., Persha, L., Salk, C. and Wright, G. (2015) Benefit sharing among local resource users: the role of property rights. *World Development* 72: 408–18. https://doi.org/10.1016/j.worlddev.2015.03.005

Waltham, N.J. and Schaffer, J.R. (2018) Thermal and asphyxia exposure risk to freshwater fish in feral-pig-damaged tropical wetlands. *Journal of Fish Biology* 93: 723–8. https://doi.org/10.1111/jfb.13742

Waltham, N.J. and Sheaves, M. (2015) Expanding coastal urban and industrial seascape in the Great Barrier Reef World Heritage Area: critical need for coordinated planning and policy. *Marine Policy* 57: 78–84. https://doi.org/10.1016/j.marpol.2015.03.030

Waltham, N.J., Buelow, C.A. and Iles, J.A. (2020) Evaluating wetland restoration success: feral pig fencing for conservation of Round Hill Reserve. Report to the National Environmental Science Program. Reef and Rainforest Research Centre Limited, Cairns.

Waltham, N.J., Burrows, D., Wegscheidl, C., Buelow, C., Ronan, M., Connolly, N., Groves, P., Audas, D., Creighton, C. and Sheaves, M. (2019) Lost floodplain wetland environments and efforts to restore connectivity, habitat and water quality settings on the Great Barrier Reef. *Frontiers in Marine Science* 6: 71–85. https://doi.org/10.3389/fmars.2019.00071

Waltham, N.J., Elliott, M., Lee, S.Y., Lovelock, C., Duarte, C.M., Buelow, C., Simenstad, C.,

Nagelkerken, I., Claassens, L., Wen, C.K.-C., Barletta, M., Connolly, R.M., Gillies, C., Mitsch, W.J., Ogburn, M.B., Purandare, J., Possingham, H. and Sheaves, M. (2020) UN decade on ecosystem restoration 2021–2030 – what chance for success in restoring coastal ecosystems? *Frontiers in Marine Science* 7: 71–76 https://doi.org/10.3389/fmars.2020.00071

Waltham, N.J., Alcott, C., Barbeau, M.A., Cebrián, J., Connolly, R.M., Deegan, L., Dodds, K., Gaines, L.A.G., Gilby, B., Henderson, C.J., McLuckie, C., Minello, T.J., Norris, G., Ollerhead, J., Pahl, J., Reinhardt, J., Rezek, R., Simenstad, C., Smith, J., Sparks, E.L., Staver, L., Ziegler, S. and Weinstein, M.P. (2021) Tidal wetland restoration optimism in rapidly changing climate and seascape. *Estuaries and Coasts* 44: 1681–90 https://doi.org/10.1007/s12237-020-00875-1

Weber, A. and Wolter, C. (2017) Habitat rehabilitation for juvenile fish in urban waterways: a case study from Berlin, Germany. *Journal of Applied Ichthyology* 33: 136–43. https://doi.org/10.1111/jai.13212

Weidlich, E.W.A., Flórido, F.G., Sorrini, T.B. and Brancalion, P.H.S. (2020) Controlling invasive plant species in ecological restoration: a global review. *Journal of Applied Ecology* 57: 1806–17. https://doi.org/10.1111/1365-2664.13656

Weilhoefer, C.L., Williams, D., Nguyen, I., Jakstis, K. and Fischer, C. (2017) The effects of reed canary grass (*Phalaris arundinacea* L.) on wetland habitat and arthropod community composition in an urban freshwater wetland. *Wetlands Ecology and Management* 25: 159–75. https://doi.org/10.1007/s11273-016-9507-x

Weinstein, M. and Litvin, S. (2016) Macro-restoration of tidal wetlands: a whole estuary approach. *Ecological Restoration* 34 (1): 27–38. https://doi.org/10.3368/er.34.1.27

Williams, S.L. and Grosholz, E.D. (2008) The invasive species challenge in estuarine and coastal environments: marrying management and science. *Estuaries and Coasts* 31: 3–20. https://doi.org/10.1007/s12237-007-9031-6

Yip, L.S., Hamilton, S., Barbier, E.B., Primavera, J. and Lewis III, R.R. (2019) Better restoration policies are needed to conserve mangrove ecosystems. *Nature Ecology & Evolution* 3: 870–2. https://doi.org/10.1038/s41559-019-0861-y

Zedler, J.B. and Kercher, S. (2004) Causes and consequences of invasive plants in wetlands: opportunities, opportunists, and outcomes. *Critical Reviews in Plant Sciences* 23: 431–52. https://doi.org/10.1080/07352680490514673

Zhao, Q., Bai, J., Huang, L., Gu, B., Lu, Q. and Gao, Z. (2016) A review of methodologies and success indicators for coastal wetland restoration. *Ecological Indicators* 60: 442–52. https://doi.org/10.1016/j.ecolind.2015.07.003

CHAPTER 12

Multi-decadal Responses of Coastal Ecosystems to Climate Change, Pollution and Non-indigenous Species in the Western and Mid-English Channel

ROGER J.H. HERBERT, GUILLAUME CORBEAU, LAURENT GODET,
NICOLAS DESROY, NOVA MIESZKOWSKA, LOUISE B. FIRTH, ALICE E. HALL
and STEPHEN J. HAWKINS

Abstract

For over a century, the English Channel (La Manche) has been a strategic location for observation and measurement of changes in marine biodiversity and ecosystems. Straddling cold and warm temperate biogeographic provinces, the Channel is particularly sensitive to variable weather and climate patterns caused by Atlantic Multidecadal Oscillation cycles, which have a major influence on sea and air temperatures and wave energy influencing the whole ecosystem. Superimposed on these natural cycles is human-induced global warming which, combined with pollution and growth in international trade and waterborne traffic, has resulted in significant changes in invertebrate assemblages and the arrival of many non-indigenous species. In the context of past, present and future environmental impacts, we review broad-scale patterns in distribution of important coastal functional indicators over this period, and undertake a horizon-scanning approach to forecast how future species assemblages might influence the functioning of coastal ecosystems in the Channel and north-east Atlantic.

Keywords: climate change, non-native species, pollution, long-term time series, English Channel, Normanno-Breton Gulf, north-east Atlantic

Correspondence: rherbert@bournemouth.ac.uk

Introduction

Recognition of the importance of climate-related impacts on marine ecosystem structure and functioning and the provision of services has increased considerably over recent decades (Southward, Hawkins and Burrows 1995; Perry et al. 2005; Burrows et al. 2011; Philippart et al. 2011; Doney et al. 2012; Laffoley and Baxter 2016; Cheung et al. 2016; Hawkins et al. 2016). At the forefront of this research globally, and specifically the north-east Atlantic, has

Roger J.H. Herbert, Guillaume Corbeau, Laurent Godet, Nicolas Desroy, Nova Mieszkowska, Louise B. Firth, Alice E. Hall and Stephen J. Hawkins, 'Multi-decadal Responses of Coastal Ecosystems to Climate Change, Pollution and Non-indigenous Species in the Western and Mid-English Channel' in: *Challenges in Estuarine and Coastal Science*. Pelagic Publishing (2022). © Roger J.H. Herbert, Guillaume Corbeau, Laurent Godet, Nicolas Desroy, Nova Mieszkowska, Louise B. Firth, Alice E. Hall and Stephen J. Hawkins. DOI: 10.53061/XPLN8836 been the intense monitoring of biota and oceanographic processes of the English Channel, or La Manche. The region has a long history of study; it was pioneered by France with the establishment of marine stations (e.g. Concarneau in 1859, Roscoff in 1872, Tatihou Island, 1882, which moved to Dinard in 1935) with strong links to universities and the French Museum of Natural History. The British contribution lagged a little behind, but with the opening of the Marine Biological Association's (MBA) Laboratory at Plymouth in 1888, a strong presence developed in the Western Channel with both resident and visiting scientists from a number of UK universities as well as many gifted amateurs. The cluster in Plymouth expanded further with the establishment of the Institute of Marine Environmental Research in the 1970s and its subsequent merger with some of the MBA's activities to form the Plymouth Marine Laboratory (Hawkins et al. 2013). Universities have also made a major contribution on both sides of the Channel (Caen, Brest in France; Southampton, Plymouth, Exeter and latterly Bournemouth in England).

There has been a long-standing realisation that oceanographic processes in the English Channel are essentially driven by those in the Atlantic to the south and west, in the Bay of Biscay, Western approaches and Southern Celtic Sea (Southward et al. 2005 for review). Since the early international investigations of the first decade of the twentieth century, coordinated by the International Council for the Exploration of the Sea (ICES), it has been known that the residual water flow is from west to east, but with some complex gyres and counter-currents (Pingree and Maddock 1977; Pingree and Griffiths 1978; Salomon and Breton 1993; Southward et al. 2005). There is a transition from oceanic conditions from west to more neritic conditions in the east, with tidally mixed water occurring as the channel shallows and the tidal range increases, leading to frontal regions explored in the 1980s where mixed water abuts that which is stratified in the summer (Pingree, Mardell and Maddock 1983; Sherwin and Jonas 1994; Southward et al. 2005). There are areas with very large tidal amplitudes such as the Bay of St Malo, and areas with smaller amplitudes such as from Portsmouth to Portland Bill. Fluctuations in both offshore and inshore ecosystems have occurred over the last 150 years, driven by hydro-meteorological cycles upon which more recent rapid climate change has been superimposed (Russell et al. 1971; Southward, Hawkins and Burrows 1995; Mieszkowska et al. 2014).

Fluctuations have been recorded in hydrography, particularly sea temperature, with a warm period at the end of the nineteenth century, followed by a colder period each side of the First World War, a period of warming up until the late 1950s, an extremely cold winter of 1962/3 that heralded in a cooler period and then rapid warming from the late 1980s onwards (Fig. 12.1, HadISST1 model UK Meteorological Office; Raynor et al. 2003). The various components of the ecosystem have responded with changes in plankton (Russell 1973; Maddock, Boalch and Harbour 1981; Boalch 1987; Southward, Hawkins and Burrows 1995; Beaugrand, Ibañez and Reid 2000, 2002; Southward et al. 2005), pelagic and bottom fish (Genner et al. 2004; Southward et al. 2005; Genner et al. 2010; Heath et al. 2012) and offshore benthos (Hinz et al. 2011; Gaudin et al. 2018) as well as changes in abundance and shifts in distribution of intertidal and shallow water biota (Southward, Hawkins and Burrows 1995; Mieszkowska et al. 2005; Hawkins et al. 2009). These fluctuations became known as the Russell Cycle (Russell et al. 1971; Cushing and Dickson 1976), in what Southward (1980) described as an inconstant ecosystem ultimately driven by climate fluctuations. The alternation of warm water Pilchards *Sardina pilchardus* (also called Sardines) and cold-water Herring *Clupea harengus* have been known since the Middle Ages. The collapse of Plymouth Herring fishing in the 1930s and its replacement by Pilchards was driven by ecosystem-level changes (Southward, Boalch and Maddock 1988a, b; Southward, Hawkins and Burrows 1995). Overfishing has, however, played its part with larger bodied bottom-living fish being particularly vulnerable, but with smaller bodied fish following climate (Genner et al. 2010). Overfishing has no doubt influenced the benthos by the disturbance caused by bottom trawling (Holme 1960, 1966; Hinz, Prieto and Kaiser 2009), but there is evidence of recovery when fishing gears are excluded from marine protected areas (Sheehan et al. 2013). In recent years, there has been a realisation that the Russell Cycle is really an expression of the Atlantic Multidecadal Oscillation (AMO) (Edwards et al. 2013), a larger scale North Atlantic phenomenon. However the zooplankton time-series may not be truly cyclical (McManus, Priscilla and Coombs 2016).

It has long been realised that fluctuations in the intertidal zone track changes offshore (Southward 1963, 1980; Southward, Hawkins and Burrows 1995), and more recently that they are broadly related to the AMO (Mieszkowska et al. 2014). Thus, to keep this review manageable, we have concentrated on the intertidal and shallow subtidal zones on both sides of the western and central Channel, including the Normanno-Breton Gulf. We consider the region's past present and future vulnerability and its importance as a strategic location for monitoring environmental change. We outline future challenges for the maintenance of ecosystem functioning and delivery of ecosystem services.

A strategic monitoring location

The identification of distinct regions where biota was similar or different to other parts of the coast was first recognised in the mid-nineteenth century (Forbes 1858). Extending southwards, four main provinces were identified: Arctic, Boreal, Celtic and Lusitanian, with the Lusitanian province extending northwards from the Mediterranean to the western English Channel. Subsequent work on intertidal fauna on both sides of the Channel (Fischer-Piette 1936; Moore and Kitching 1939; Crisp and Southward 1958) confirmed the existence of a west–east gradient with largely cold-temperate (Boreal species) being prevalent to the east of the Isle of Wight on the English coast and the Cotentin Peninsula in Normandy, France, although some southern fauna occurred further east along the French coast. Forbes (1858) produced a map of the broad distributional limits of marine species of the British Isles and northern France, with a line depicting the 'general limit of southern types' (reproduced in Hiscock et al. 2004). This line, recently referred to as the 'Forbes Line' (*sensu* Firth et al. 2021) passes just to the east of the Isle of Wight and the Cotentin Peninsula. This Boreal-Lusitanian transition zone (Hiscock 1998) has been globally influential in identifying responses of marine fauna to variation in sea and air temperature, most notably in relation to the fauna of rocky shores (Keith et al. 2011).

Intertidal indicators

Climate-sensitive intertidal invertebrates are particularly useful sentinels for monitoring ecosystem impacts as they respond to both air and sea temperatures and are not exploited (Helmuth et al. 2006). Yet, owing to inherent background variability in invertebrate recruitment and mortality, to establish patterns and links with climate and other environmental parameters, it is necessary to maintain long-term time series of abundance at multiple sites (Hawkins et al. 2003; Hawkins et al. 2013b; Mieszkowska et al. 2014). Building on early data collected by Fischer-Piette (1936) and Crisp and Southward (1958), and subsequent monitoring by Southward, Hawkins and other individuals in the 1980s and 1990s, the Marine Biodiversity and Climate Change (MarClim) project established a nationwide network of monitoring sites on UK rocky shores from the late 1990s/2000 (Mieszkowska et al. 2005). Climate-related impacts on intertidal habitats and ecosystems is updated through periodic publication of a report card by the Marine Climate Change Impacts Partnership (mccip.org.uk; Mieszkowska, Burrows and Sugden 2020).

Fluctuations in sea surface temperature (Fig. 12.1) have resulted in increased abundances of southern fauna during warmer phases. For both sides of the Channel, we have updated the distribution of selected indicator species presented in Crisp and Southward (1958)

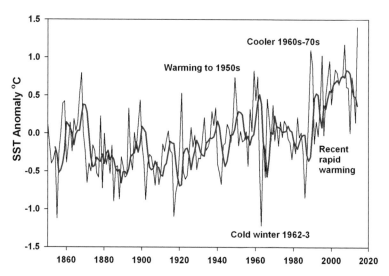

Figure 12.1 Annual Sea Surface Temperature (SST) anomalies (1960–91 baseline) and five-year smoothed mean) for English Channel Grid Cell 50°N, 1.3°W. Data extracted from HadISST for (Raynor et al. 2003; Met Office Hadley Centre observations datasets)

from the 1950s (Table 12.1). Of the 15 southern (warm-adapted) species, nine have shown range expansions on the English coast and seven on the French side. Only the Purple Sea Urchin *Paracentrotus lividus*, which had been established on the north Brittany coast in the 1970s, appears to have undergone regression, perhaps owing to over-exploitation slowing recovery when warming occurred from late 1980s (Hawkins, pers. comm. from A.J. Southward). For others, the distribution is stable or unknown. The past decade has been cooler (Fig. 12.1; see also Burrows et al. 2020), which may have arrested population growth of some of these species. Of the northern, cold-adapted species, the barnacles *Semibalanus balanoides* and *Balanus crenatus* have shown signs of decline in south-west England. The Dog Whelk *Nucella lapillus* has declined from around some English ports and harbours; however, this may be because of the legacy of tributyltin (Bryan et al. 1986; Gibbs et al. 1987). Records of the Boreo-Arctic Barnacle *Balanus balanus* warrant further investigation, as the species can be confused with other large *balanidae*.

In south-west England, the ratio of Lusitanian (warm-adapted) barnacles *Chthamalus stellatus* and *C. montagui* and the boreal (cold-adapted) counterpart *Semibalanus balanoides* were shown to be linked to temperature (Southward and Crisp, 1954, 1956; Southward 1967, 1991) with a time lag of one to two years (Southward 1967, 1991; Southward, Hawkins and Burrows 1995; Southward et al. 2005; Hawkins et al. 2009; Phillipart et al 2013). In warmer summers, *Chthamalus* spp. may have multiple broods that can result in higher recruitment (Burrows, Hawkins and Southward 1992; O'Riordan, Myers and Cross 1992; O'Riordan et al. 2004). Yet the size of the chthamalid population is mediated by release from competition with the dominant *S. balanoides* (Connell 1961), which recruits less well in warmer years (Poloczanska et al. 2008). Time-series analysis of *S. balanoides* and experiments have shown that warmer winters result in lower embryo survival and larval supply (Rognstad and Hilbish 2014; Rognstad, Wethey and Hilbish 2014; Rognstad et al. 2018) and higher water temperatures may inhibit penis development, which can result in fertilisation failure and population decline (Herrera et al. 2019) owing to low recruitment (Svensson et al. 2005). Over the past century, the southern geographic limit of *S. balanoides* on the French coast has contracted northwards by 300 km (Wethey and Woodin 2008). Modelling indicates that continued climate warming will result in the regional extinction

Table 12.1 Changes in distribution of selected southern (warm-adapted), intermediate, northern (cold-adapted) and non-native invertebrate indicators on rocky shores in the Channel from 1950s–2020. Species list based on Table 1 and Table 7 in Crisp and Southward (1958)

	England south coast		France north coast		
Species	Expansion/ Regression	Comments	Expansion/ Regression	Comments	Source
SOUTHERN					
Actinia fragacea	Stable	Locally Common in SW England east to Isle of Wight. Rare to Occasional on along Sussex coast. Eastern limit Eastbourne.	Expansion	Abundant on Jersey, Alderney, Chausey, and most of Brittany shores. Eastern limit Fecamp, 2001.	Hawkins and Herbert unpubl. data; Le Mao et al. 2019
Anemonia viridis	Expansion east from Brighton	Common to Abundant in SW England. Rare to Occasional on Sussex coast. Eastern limit Hastings, 2018.	Expansion east from Grandecamp-les-Bains	Abundant on Gulf coasts. In 2000, Common at Cap de Carteret, Frequent at Cap Levy, Port-en Bessin-Huppain; Occasional at Etretat. Eastern limit Fecamp, Abundant.	Herbert and Hall unpubl. data; Le Mao et al. 2019
Chthamalus montagui	Minor expansion east since 1970s from Bonchurch, Isle of Wight; Crisp et al. (1981)	Common to Abundant on SW rocky shores east to Swanage. Eastern limits Southsea and Bembridge (Isle of Wight) 2019.	Expansion east since 1970s from Cap de la Hague; Crisp et al. (1981)	In 2000s, Common at Cap de Carteret, Occasional along north Normandy coast. Eastern limit. Barfleur, Rare	Crisp et al. 1981; Herbert et al. 2007; Wethey et al. 2011; Hawkins et al. 2009; Hawkins and Herbert unpubl. data; Le Mao et al. 2019
Chthamalus stellatus	Minor expansion east since 1970s from Brook, Isle of Wight: Crisp et al. (1981)	Frequent to Common on SW coast Eastern limits Southsea and Bembridge.	Expansion east since 1970s from Cap de la Hague; Crisp et al. (1981)	Present in 2003 in Alderney, 2009 at Flamanville, 2013 in the Minquiers, Rare. In 2000, Frequent at Cap de Carteret and along north Normandy coast. Eastern limit Cap Levy.	Crisp et al. 1981; Herbert et al. 2007; Wethey et al. 2011; Hawkins et al. 2009; Hawkins and Herbert unpubl. data. Le Mao et al. 2019; Mieszkowska et al. 2020
Gibbula magus	Unknown	SW coast. Few records in Eastern Channel. Data lacking.	Minor Expansion	On most rocky shores of the Gulf, but also on gravel in the subtidal zone.	MarLin (2020); Le Mao et al. 2019
Haliotis tuberculata	Not present		Stable	Decline of population owing to *Vibrio* parasitosis in the 1990s, but recent recovery.	Pichon et al. 2013

England south coast			France north coast		
Species	Expansion/ Regression	Comments	Expansion/ Regression	Comments	Source
Melarhaphe neritoides	Expansion east from Bembridge, Isle of Wight	Locally Abundant on south coast. Eastern limit Suffolk and East Anglia on groynes and sea defences.	Stable	Abundant in the Gulf.	Hawkins et al. 2009; Hawkins and Herbert unpubl. data; Le Mao et al. 2019
Perforatus perforatus	Expansion east from Bembridge, Isle of Wight	Locally Frequent to Abundant on SW coast and on pier piles and structures. Eastern limit Hastings.	Expansion east from St Vaast-la-Hougue	Present on most rocky shores of the Gulf, and in circalittoral zone. In 2001, Frequent to Abundant on rocky shores between Cap Carteret along north Cotentin coast to Port-en-Bessin-Huppain. Occasional Fecamp, Dieppe. Eastern limit Le Treport.	Herbert et al. 2003; Hawkins and Herbert unpubl. data; Le Mao et al. 2019
Paracentrotus lividus	Not present	No change	Regression (St Cast) Common in 1970s from Cap Fréhel to Bréhat and CI.	Recently, present only in Alderney.	Le Mao et al. 2019
Patella depressa	Expansion east from Ventnor, Isle of Wight	Locally Abundant SW England. Frequent to Common on southern shores of Isle of Wight. Rare on Sussex coast of Channel from 2012 onwards. Eastern limit Eastbourne.	Stable	Common in 2000 east to Roch du Castel-Venden. Eastern limit Pointe de Barfleur, Rare.	Hawkins et al. 2009; Hawkins and Herbert unpubl. data; Le Mao et al. 2019
Patella ulyssiponensis	Expansion east from Ventnor, Isle of Wight	Common to Abundant on SW coast. Frequent to Common on southern shores of the Isle of Wight. Occasional to Rare on Sussex coast. Eastern limit near Dover, 2018.	Stable	Abundant in the Gulf. In 2000, Frequent Vauville eastwards to limit at Pointe de Barfleur.	Hawkins et al. 2009; Hawkins and Herbert unpubl. data; Le Mao et al. 2019
Phorcus lineatus	Expansion east from Lyme Regis	Common to Abundant on SW coast to Lyme Regis. Early 2000s expansion to east Dorset and Isle of Wight. Rare on Sussex coast from 2017. Eastern limit Beachy Head.	Expansion east from Pointe de Hoc	Common in Gulf in the 2000s–10s and the Hague in 2004. In 2000/1, Occasional at Cap de Carteret. Abundant at Baie Ecalgrain east to St Vaast. Common at Grandcamp, Rare at Port en Bennin-Huppain. Eastern limit Fecamp, Rare.	Hawkins and Herbert unpubl. data; Le Mao et al. 2019
Sabellaria alveolata	Stable	Common to Abundant in locations where sandy beaches abut rocky shores. Eastern limit Charmouth, Lyme Bay.	Stable	Very Abundant in whole Gulf.	Lecornu et al. 2016; Curd et al. 2020

England south coast			France north coast		
Species	Expansion/ Regression	Comments	Expansion/ Regression	Comments	Source
Steromphala pennanti	Not present	Not present	Stable	Abundant on every rocky shore of the Gulf 2000–10. In 2001 Abundant along north Normandy coast. Eastern limit Barfleur, Abundant.	Hawkins and Herbert unpubl. data. Le Mao et al. 2019
Steromphala umbilicalis	Expansion east from Bembridge, Isle of Wight	Abundant on SW coast. Abundant Isle of Wight and east Sussex coast. Eastern limit Joss Bay and Thanet coast 2013, Frequent.	Stable	Abundant in the Gulf. 2003 Flamanville, 2010 Alderney. In 2001, Abundant east along Cotentin. Eastern limit Le Treport. Recent data lacking.	Hawkins et al. 2009; Meiszkowska et al. 2012; Hawkins and Herbert unpubl. data; Le Mao et al. 2019
INTERMEDIATE					
Actinia equina	Stable	Common on most rocky shores on south coast of England.	Stable	Abundant on every rocky shore in the Gulf, especially on most exposed coasts.	Le Mao et al. 2019
Calliostoma zizyphinum	Unknown	Common on lower shore and subtidal on SW coasts. Occasional specimens found intertidally on Isle of Wight. Common at Bembridge in 2020.	Expansion (subtidal)	Extremely Abundant in many habitats, not only on rocky shores.	Hawkins and Herbert unpubl. data; Hinz et al. 2011; Le Mao et al. 2019
Littorina littorea	Stable	Occasional to Abundant on most rocky shores eastwards to Sussex coast.	Expansion	Expansion on CI coasts but declined around Saint-Brieuc Bay and North Cotentin. Data lacking.	Hawkins and Herbert unpubl. data; Le Mao et al. 2019
Littorina obtusata/ fabalis	Stable	Occasional to Common on most rocky shores.	Expansion	Found mostly in Fucoids. Recent colonisation of Bréhat area, Jersey and Alderney coasts.	Hawkins and Herbert unpubl. data; Le Mao et al. 2019
Patella vulgata	Stable	Common to Abundant on most rocky shores.	Minor expansion	Abundant on all the Gulf rocky shores.	Hawkins and Herbert unpubl. data; Le Mao et al. 2019
Steromphala cineraria	Stable	Occasional to Frequent on Isle of Wight shores and along Sussex coast.	Stable	Very Abundant in the whole Gulf, including subtidal zone.	Hawkins and Herbert unpubl. data; Le Mao et al. 2019
Verruca stroemia	Unknown	Occasional specimens found at Bembridge, Isle of Wight. Not seen along Sussex coast.	Stable	Formerly Abundant in subtidal zone where strong currents (passage de la déroute, Minquiers). Recently on a few Brittany shores, Granville, Jersey and Guernsey.	Hawkins and Herbert unpubl. data; Le Mao et al. 2019

England south coast		France north coast			
Species	Expansion/ Regression	Comments	Expansion/ Regression	Comments	Source
NORTHERN					
Balanus balanus	No change	Not present.	Minor expansion	Nearly absent from the Gulf, recent observations in Sark and Jersey.	Le Mao et al. 2019
Balanus crenatus	Minor regression from west	Frequent to Common on Isle of Wight and Sussex coast, getting rarer in intertidal zone in western basin in Devon and Cornwall.	Stable	Very Abundant, mostly from lower shore to –50m. Lack of recent data on the Normandy and CI coasts.	Le Mao et al. 2019
Littorina saxatilis	Stable	Common on rocky shores throughout.	Stable	Present on most rocky shores of the Gulf. In 2001, Present in east Channel. Abundant Cap Blanc Nez.	Hawkins and Herbert unpubl. data
Nucella lapillus	Minor regression	Common on most rocky shores and structures. Populations at sites affected by TBT in 1960s–80s have recovered, however some imposex remained in 2012–13. Not found at badly affected sites in Devon and Cornwall in 2017.	Expansion	Abundant on all rocky shores. Recent stations found on South Normandy coasts of the Gulf	Spence et al. 1990; Langstone et al. 2015; Hawkins et al. 2017; Le Mao et al. 2019
Semibalanus balanoides	Minor regression	Abundant in east and central Channel but getting rarer in Dorset, Devon and Cornwall.	Stable	Abundant on every rocky shore in the Gulf. Abundant Normandy east to Calais (2001–10).	Herbert et al. 2007; Wethey and Woodin 2008; Wethey et al. 2011; Le Mao et al. 2019
NON-NATIVE					
Austrominius modestus	Expansion	Abundant on structures in harbours and on moderately exposed rocky shores.	Expansion	Abundant. Continuous expansion on rocky shores and artificial structures. Present on every shore of the Gulf, CI. Abundant Normandy east to Calais.	Herbert e a. 2007; Wethey et al. 2011; Le Mao et al. 2019
Amphibalanus improvisus	Stable	Adults not commonly recorded Mostly estuaries in SW England eastwards to the Solent. Larvae very abundant in Southampton Water in 2003.	Unknown	Lack of data in Brittany estuaries.	Muxagata, Williams and Sheader 2004; Le Mao et al. 2019
Magallana gigas	Expansion	Common on artificial structures close to ports and aquaculture since early-mid 2000s. Reefs/dense aggregations establishing on mudflats and rocky shores in SW England and Solent and SE England.	Expansion	Very abundant Brittany shores, Normandy rocky shores, CI coasts.	Herbert et al. 2012,2016; Le Mao et al. 2019

of *S. balanoides* from south-west England by the 2050s, leading to the competitive release of *Chthamalus* (Poloczanska et al. 2008), which will become dominant at all tidal levels.

Following the winter of 1962/3, populations of topshells (trochidae) Flat Topshell *Steromphala umbilicalis* (formerly *Gibbula umbilicalis*) and Toothed Topshell *Phorcus lineatus* (formerly *Monodonta lineata* and *Osilinus lineatus*) were significantly reduced at range margins between Lyme Bay and the Isle of Wight (Crisp 1964). Sensitivities of newly settled spat to colder winter conditions in the eastern Channel and North Sea were thought to have previously limited their range. Since 1990, adult densities at the range margins on the Dorset and Isle of Wight coasts have now increased to maxima not previously recorded (Hawkins et al. 2009), with range extensions eastwards (Table 12.1). Prior to the 1990s, the geographic distribution of the warm-adapted Black-footed Limpet *Patella depressa* and China Limpet *P. ulyssiponensis* was also limited to locations on the west of the Isle of Wight (and Forbes' Line) on the English coast; however, they have now both increased in abundance at the range margin and extended eastwards, though not as far or as rapidly as trochidae (Hawkins et al. 2009). Together with the Boreal, or Common, Limpet *P. vulgata*, these represent functionally important grazers on rocky shores and have potential to alter the abundance of primary producers (Jenkins et al. 2005; Coleman et al. 2006), with cascading ecosystem effects (Hawkins et al. 2008, 2009, 2019). Importantly, the rate of range expansion is dependent on life-history traits coupled with habitat availability and suitability acting in concert with the influence of fast offshore currents associated with coastal headlands and islands such as Portland Bill and the Isle of Wight on the English coast, which can create significant larval dispersal barriers (Keith et al. 2011). Although *S. umbilicalis* and *P. lineatus* have spread further east along the French coast, the Cotentin Peninsula still represents a significant break point for the warm-adapted barnacles *Chthamalus* spp. (Herbert et al. 2007, 2009; Wethey et al. 2011) and *Perforatus perforatus* (Herbert et al. 2003), which have longer pelagic larval duration. Species with a short pelagic larval phase, such as trochid molluscs, may be expanding more rapidly as shorter steps along the coast may reduce larval losses and enable consolidation by back-filling (Hawkins et al. 2009; Pringle et al. 2017). Increasing sea temperatures may reduce length of life-history stages and pelagic larval duration, which could facilitate more rapid range expansion (O'Connor et al. 2007). There is evidence that some of the habitat gaps in the eastern Channel may have been crossed, aided by artificial habitats acting as stepping-stones, including piers (e.g. *Perforatus perforatus*, Herbert et al. 2003) and stone-groyne sea defences (topshells and limpets, Hawkins and Herbert unpublished observations; Hawkins et al. 2009; Keith et al. 2011). Bioclimatic envelope models still need to be improved by incorporating hydrography and habitat suitability to track range expansions more effectively (Keith et al. 2011). Dispersal barriers are thought to have so far limited the expansion of Green Ormer *Haliotis tuberculata*, Pennant's Topshell *Steromphala pennanti* and Purple Sea Urchin from the French side to the English coast. Interestingly, the latter has always occurred all up the west coast of Ireland and into the Scottish Hebrides (Southward and Crisp 1954b; Lewis 1964; Simkanin et al. 2005).

Subtidal benthos

Owing to cost and the logistical complexities of survey work, monitoring responses to environmental change of subtidal benthos in the Channel has been less frequently undertaken. High wave disturbance and strong tidal streams have resulted in coarse sediments predominating across the region, especially gravels. Yet broad-scale surveys of mainly molluscs and echinoderms (Holme 1961, 1966) revealed patterns that generally reflected the west–east temperature gradient, with southern, warm-adapted species mostly confined to the west of the Channel Islands (CI). Some species that were mainly found in the western Channel on the English coast were found to extend further eastwards on the French side, which was comparable with studies of intertidal species (Crisp and

Southward 1958). Cabioch et al. (1977) found evidence of species eastern range limits in the Channel, and yet concluded that there were clear interactions between temperature, sediment size and current velocity. A re-survey of these sites in 2012–14 (Gaudin et al. 2018) showed that the spatial occurrence of warm-water species mostly increased and cold-water species decreased, which was consistent with rise in sea bottom temperatures (SBT) of 0.07–0.54 °C per decade. Western species, such as the warm-water bivalve *Gouldia minima* (Veneridae) and cold-water Tiger Scallop *Palliolum tigerinum* (Pectinideae) had both moved significantly eastwards, and are now exposed to warmer winter conditions and higher summer temperatures. Yet unsurprisingly, some western stenohaline cold-water species, such as the Sea Urchin *Echinus esculentus*, had retreated further west. However, although range expansions had occurred these were generally small and there was a significant overall lag between species centroid distribution shifts and the SBT isotherm. Similarly, relatively small responses were found in subtidal benthos off the south coast of England (Hinz et al. 2011), suggesting lower sensitivities compared with intertidal species plus the possible influence of other impacts such as intensive towed bottom trawling and dredging. Further understanding of these patterns and processes is dependent on the maintenance of long-term (multi-decadal) time-series (Hawkins et al. 2013; Mieszkowska et al. 2014), which are subject to availability of funding, policy decisions and international collaboration.

Pulse and press disturbances

Superimposed on background responses to periodic and long-term changes in temperature, there have been notable pulse-disturbances from acute pollution incidents. In March 1967, the oil tanker *Torrey Canyon*, containing 110,000 tonnes of crude oil, foundered on rocks off the coast of Cornwall in south-west England. Where oil smothered accessible rocks and beaches, the emergency response was to spray and disperse the oil with organic solvents, which proved to be more ecologically damaging to shore life than the oil itself (Smith 1968). Research and monitoring led by the Marine Biological Association at Plymouth showed that the dispersants were particularly lethal to limpets (*Patella* spp.) and other grazers, which caused almost total mortality (Southward and Southward 1978; Hawkins and Southward 1992; Hawkins et al. 2017a, b). The majority of affected shores took up to ten years to return to a pre-spill state (Southward and Southward 1978), although heavily sprayed shores took between 13 and 15 years (Hawkins and Southward 1992). This was in stark contrast to unsprayed shores, where recovery appeared to occur within two or three years (Southward and Southward 1978; Hawkins and Southward 1992). Following the *Amoco Cadiz* oil spill on the north Brittany coast in 1978, the severity of impact on sublittoral fine sediment communities was variable and dependent on the number of sensitive species present (Dauvin 2000). Recovery of populations of the amphipod *Ampelisca* to pre-spill state occurred after about 15 years (Poggiale and Dauvin 2001).

For decades prior to legislation in the mid-1980s, coastal intertidal and subtidal habitats on both sides of the Channel were exposed to high concentrations of Tributyltin (TBT), an ingredient of anti-fouling paints – a press disturbance caused by chronic pollution. This endocrine-disruptor caused masculisation of female Dog Whelks *Nucella lapillus* leading to reproductive failure, population declines and local extinctions (Bryan et al. 1986; Gibbs et al. 1987; Spence et al. 1990), particularly around UK ports and marinas such as the Fal, Plymouth (Bryan et al. 1986, Gibbs et al. 1987) and in the Solent (Herbert 1989; Bray and Herbert 1998). At Pacific Oyster farms in France, reproductive failure and abnormal shell growths were widely reported (Alzieu 2000). In estuaries and harbours, populations of bivalves, notably Peppery Furrow Shell *Scrobicularia plana*, were also seriously affected (Langston, Burt and Chesman 2007) and in Poole Harbour, UK, the legacy remains evident at some contaminated sites (Langston et al. 2015).

One species that was badly impacted by the *Torrey Canyon* oil spill was the warm-water Hermit Crab *Clibanarius erythropus*. This species was first found on the English coast of the Channel in 1959 at the end of the warm-period that peaked in the 1950s (Southward and Southward 1978, 1988), where it persisted for some time along the coasts of Cornwall and into Devon, including sites that were not impacted by the *Torrey Canyon*. Its disappearance was linked to the colder climate from the 1960s to the mid-1980s (Southward and Southward 1988), although they speculated that lack of *Nucella* shells (its preferred home) may have become important. This species has been very slow to come back as conditions warmed, but was reported again in 2016, probably owing to recruitment from Brittany in the preceding year or so. This is clearly a species that could have been influenced by climate fluctuations and both pulse and press disturbances (Hawkins et al. 2017a). Juveniles are found in a variety of shells and although the favoured stenoglossan whelks *Nucella*, *Hinia* spp. and *Ocenebra erinacea* were all badly affected by TBT pollution, adult *Clibanarius* can be found in a variety of other shells (*Phorcus lineatus*, *Littorina littorea*, Hawkins unpublished obs.).

Non-indigenous species – south and south-west England

A consequence of increased globalisation and trade is the arrival frequency and establishment of non-indigenous species (NIS), particularly in coastal areas. On the north Atlantic coast, establishment of NIS may become more common owing to rising sea temperatures (Hawkins et al. 2016, 2019). The North American Slipper Limpet *Crepidula fornicata* has transformed subtidal benthos in many areas of the western Channel and the Solent and is now spreading further west (Hinz et al. 2011). It is thought partially responsible for the demise of Native Oyster *Ostrea edulis* beds through competition, interference with settlement, feeding, habitat modification (Blanchard 1997a, Fitzgerald 2007; Smyth et al. 2018) and larval predation (Pechenik, Blanchard and Rotjan 2004). High densities of *Crepidula* are considered a major impediment to restoration efforts of *O. edulis* in the Solent (Helmer et al. 2019). Since first recorded in Chichester Harbour and the Solent in 1947, the Australasian Barnacle *Austrominius modestus* (formerly *Elminius modestus*) has now spread throughout the Channel (Table 12.1), around the British Isles and along European coasts (Crisp 1958; Herbert et al. 2007), yet significant negative ecological impacts on rocky shores have not yet been documented. Depending on the location of native populations, rising sea temperatures can exacerbate the impact of non-native species on ecosystems. The Pacific Oyster *Magallana gigas* (formerly *Crassostrea gigas*) is native to the north-west Pacific and Sea of Japan and has been extensively introduced and cultivated across the globe for over a century. Along the Atlantic coast of Europe there has been a significant poleward shift in European populations owing to the expansion of the species, thermal niche and an increase in coastal phytoplankton (Thomas et al. 2016). In southern England, reports of successful breeding and spat-falls from cultivated stocks were relatively uncommon and intermittent until the mid-2000s. However, the current warming trend has increased reproductive output, and wild populations have established in a variety of intertidal and shallow water habitats (Herbert et al. 2012, 2016). Small areas of mudflats in the Fal and Yealm estuaries in south-west England have been transformed into Pacific Oyster reefs, with densities upwards of 200 ind m^2. Similarly, wild settlement on rocky shores and artificial habitats is widespread in regions close to ports and harbours, and efforts are being made to control these populations through 'oyster bashing' (McKnight and Chudleigh 2015). Firth et al. (2021b) found that the dead shells of *M. gigas* appear to facilitate *A. modestus* in natural rocky shore habitats where the non-indigenous barnacles are less abundant and competition for space with native barnacles is greater. Conversely, *M. gigas* shells appeared to facilitate native barnacles in artificial habitats where competition for space with non-indigenous barnacles is greater. Pacific Oyster reefs have developed on rocky shores in south Brittany (Lejart and Hily 2011), but these have yet to occur in south-west England. However, under increased warming scenarios, the persistence of wild populations is uncertain as higher winter temperatures could result in greater mortality linked to complex interactions between temperature, salinity, eutrophication, physiological status and transmission of pathogens (Thomas and Bacher 2018). Exposure to projected increases in warming and seawater acidification may also cause metabolic stress and reduced feeding rates (Lemasson et al. 2018). It is possible that offshore aquaculture of Pacific Oysters may occur in future decades, which could result in further population spread and range expansion and interaction with native habitats.

The Manila Clam *Ruditapes philippinarum*, which originates in the Philippines, has naturalised in Poole Harbour, where it was introduced for aquaculture in 1987, as spawning was considered unlikely owing to cooler temperatures. There has been expansion into the Solent estuaries, yet introductions to the Exe estuary to the west have not yet resulted in wild populations (Humphreys et al. 2015). Although sea temperatures were originally thought limiting to the species' successful breeding in British waters, the evidence suggests that establishment of wild populations is more complex and is likely to depend on propagule pressure and levels of community disturbance, including abundance of predators. Although increasing sea temperatures may result in larger populations that could threaten native community composition and functionality, the species may be a beneficial source of food for birds should Boreal clam species retreat northwards (Caldow et al. 2007).

The establishment of high-risk species such as the Asian Shore Crab *Hemigrapsus sanguineus*, and others known to be on the horizon including the Veined Rapa Whelk *Rapana venosa* and American Lobster *Hommarus americanus* (Roy et al. 2014), could transform shallow waters and ecosystem services.

Non-indigenous species in the Normanno-Breton Gulf

The Normanno-Breton Gulf is a mega-tidal gulf located in the western part of the English Channel, covering more than 14,000 km², bordered by a 450 km coastline (France and CI) and characterised by a mosaic of benthic habitats dominated by coarse sediments (Cabral et al. 2015). Shellfish farming, mainly including Blue Mussel *Mytilus edulis*, Pacific Oyster and, to a lesser extent, Native Oyster and Manilla Clam, is particularly developed, with several production zones: Guernsey and Jersey in the CI; Chausey and Western Cotentin in Normandy; Cancale, Fresnaye, Arguenon, Saint-Brieuc bays in Brittany, as well as the Paimpol area. Marinas are also well developed, with up to 1,800 berths in Saint-Malo, Cherbourg and Saint-Peter. However, the main ports of the Gulf are not considered to be commercially important, with most of the freight transport being localised in Cherbourg and the CI.

The benthic fauna database (Le Mao et al. 2019) used here is an extensive compilation of all data (both from scientists and amateurs) related to all macrozoobenthic species collected or observed from 1768 to 2013. The data come from nearly 1,000 different sources, from the first naturalist expeditions of the nineteenth century (Charles-Alexis-Adrien Duhérissier de Gerville in 1829; Joshua Gosselin from 1810), stations sampled during surveys by Cabioch and others in 1971 and 1976, which were resampled in 2012 and 2014 (see Gaudin et al. 2018) and much data from amateurs (for more details on these data and the sampling methods, see Le Mao et al. 2019). It comprises a total of 97,651 records from 7,456 different stations and 2,152 species (including arthropods (618 species), molluscs (477), annelids (405), bryozoans (192), cnidarians (145), porifera (139), chordata (69), echinoderms (46) and other minor groups).

Among the 2,152 species, 68 species are considered to be non-indigenous species (NIS). Here, NIS are classed as any species introduced from a different biogeographic realm by human vectors, which might survive and subsequently reproduce and spread (IUCN 2000). If the expansion of an NIS has negative economic and/or ecological consequence, it is considered as an invasive species. In the Gulf, the majority of species are far from being invasive. Moreover, although the native ranges of a few remain unknown and are cryptogenic, we considered them as NIS.

Among the 68 NIS, Table 12.2 shows the 21 species with more than 10 records. These species come mainly from three geographical areas: the American coasts of the North Atlantic, Oceania and Asia.

The Slipper Limpet *Crepidula fornicata* dominates the NIS database with 1,215 records. This species has become invasive along the French coasts, and has been extensively studied (Blanchard et al. 1997; Dupont et al. 2006; Riquet et al. 2016). It is the only species in the Gulf to be really considered as an invasive, as it tends to reduce the abundance of species of economic interest, such as the Great Scallop *Pecten maximus* in the Bay of Saint-Brieuc (Thouzeau et al. 2000). Moreover, it locally controls the primary production, as in the Bay of Mont-Saint-Michel (Cugier et al. 2010; Androuin et al. 2018). The large number of records of the species is explained by its exponential population expansion, but also by species-specific surveys (Dupouy and Latrouite 1979; Hamon and Blanchard 1994; Blanchard 1997b, Blanchard et al. 2009). As this species represents 52% of the NIS data for the Gulf, it was

Table 12.2 Synthesis of the main Non-Indigenous Species (NIS) in the Normanno-Breton Gulf

Species (mentions ≥10)	Number of mentions	Biogeographical origin	Mode of introduction in the Channel	First occurrence in the Channel
Crepidula fornicata	1,215	Eastern coast of North America	Oyster transfers via the British Isles	19th century
Magallana gigas	206	Pacific Ocean	Aquaculture after collapse of *O. edulis*	1970s
Austrominius modestus	150	Oceania	Fouling and ballast water	1930s
Styela clava	138	Korea & Japan	Fouling	1960s
Monocorophium sextonae	106	New Zealand	Accidental	early 20th century
Ruditapes philippinarum	55	Indo-Pacific	Aquaculture	late 1970s
Hemigrapsus sanguineus	40	China & Korea	Ballast water	late 1990s
Dendrodoris limbata	33	Mediterranean sea	Accidental	early 1990s
Watersipora subtorquata	32	Indo-Pacific	Accidental	early 2000s
Homaxinella subdola	28	Pacific NW	Unknown	1860s
Phallusia mammillata	28	Cryptogenic	Unknown	early 1920s
Ficopomatus enigmaticus	27	Australia	Ship (fouling or ballast)	early 1920s
Balanus improvisus	22	Eastern coast of North America	Fouling	16th–17th century
Diplosoma listerianum	21	Cryptogenic	Unknown	late 19th
Perophora japonica	16	Pacific NW	Ship (fouling or ballast)	early 1980s
Mya arenaria	14	Eastern coast of North America	Fouling	16th century
Urosalpinx cinerea	14	Eastern coast of North America	Accidental	late 1920s
Potamopyrgus antipodarum	14	New Zealand	Accidental	mid-19th century
Ocinebrellus inornatus	14	Pacific NW	Oyster transfers	late 1990s
Molgula socialis	12	Eastern coast of North America	Unknown	early 1970s
Teredo navalis	10	Cryptogenic	Unknown	early 19th century

removed from the following analyses, because any result would be driven by this species alone. The Manilla Clam was introduced in 1972 under the control of the Institut Français des Pêches Maritimes. First cultures were developed in 1989 in the Chausey archipelago (Baffreau et al. 2018). This species has now constituted wild populations, which replaced the indigenous Grooved Carpet Shell *Ruditapes decussatus* in many areas. Other frequently recorded species are Mud Shrimp *Monocorophium sextonae*, a small arthropod from New Zealand, with 106 records; Stalked Sea Squirt *Styela clava*, a subtidal ascidian tunicate from Korea, with 138 records; *Austrominius modestus*, a barnacle from New Zealand, with 150 records; Pacific Oyster, introduced in the 1960s from the Pacific ocean to replace *Ostrea edulis* when its population collapsed, with 206 records (Le Mao et al. 2019).

Fig. 12.2 shows the first record of each NIS in the Gulf. The year of the very first observation is not clear as the oldest sources are not precisely dated (e.g. the mollusc species *Lyrodus bipartitus*, *Lyrodus pedicellatus*, *Spathoteredo spatha* and *Teredothyra excavata* recorded in Guernsey in the first half of the nineteenth century). However, we consider that the first NIS recorded in the Gulf is the Red-mouthed Rock Shell *Stramonita haemastoma*, which occurs further south in Europe and throughout Macaronesia and was observed in 1820 in St Peter Port (Guernsey) by Sir Edward Macculoch (Jeffreys 1867). The bivalve *Mya arenaria* and the Naval Shipworm *Teredo navalis* were recorded around Cherbourg in 1825 (Berry 1815; De Gerville 1825). Throughout the nineteenth century, seven new NIS were recorded every 30 years. NIS were also reported from the Northern CI (Guernsey, Sark, Alderney) until the beginning of the twentieth century, including arthropods *Newmanella radiata*, *Amphibalanus improvisus* and *Conchoderma auritum*. From the 1900s, the new NIS records were mainly observed in the southern part of the Gulf, especially in the Rance estuary (e.g. the annelid *Ficopomatus enigmaticus*, the arthropod *Amphibalanus amphitrite* and bryozoan *Victorella pavida*). In the second half of the twentieth century, new records of NIS came also from Bréhat, in Brittany. Thereafter, the number of records increased considerably (with more than 20 NIS discovered in 10 years) from the beginning of the twenty-first century, from all over the Gulf: Jersey, Chausey, Western Cotentin coast, Mont-Saint-Michel Bay and Roches-Douvres. Finally, the mollusc *Nassarius corniculum* is the most recent arrival, being discovered in the Trieux estuary in 2013.

Interestingly, the distribution of some NIS declined over time. For example, *Mya arenaria* experienced a period of high abundance, with a very large number of observations in the Bay of Paimpol in the 1880s, whereas recent records of individuals remain rare. This species is difficult to sample, and abundance may be under-estimated with grabs. The same applies to the gastropod *Urosalpinx cinerea* with no recent records: first record 1983 in Jersey (Chambers 2008); last record 1985 in the bay of Saint-Brieuc (Thouzeau, 1989). Another example is the small mollusc *Teredo navalis*, which was the most common shipworm species in the CI and in the Bay of Saint-Malo in the early twentieth century, but which seems to be scarce and restricted to Jersey in the last years. Some NIS did not really manage to settle and spread. It is particularly true for the first NIS of the Gulf, the Red-mouthed Rock Shell *Stramonita haemastoma*, which was probably introduced by a French ship coming from the Bay of Biscay, but without any subsequent settlement.

An important question is how benthic communities have been dominated by these species in space and time. We used an Exogenous Index (EI), measuring the dominance of NIS among the whole community (EI varies from 0 = no NIS among the community to 1 = all the species of the community are NIS), calculated as follows

$$EI = \frac{\text{Number of 'NonIndigenous' Species observations}}{\text{Total number of observations}}$$

As explained above, *Crepidula fornicata* dominates the number of observations, so this was removed from the EI calculation.

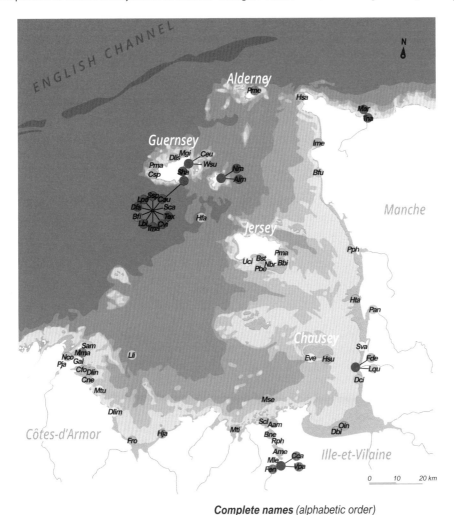

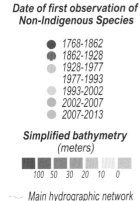

**Date of first observation of
Non-Indigenous Species**

- 1768-1862
- 1862-1928
- 1928-1977
- 1977-1993
- 1993-2002
- 2002-2007
- 2007-2013

**Simplified bathymetry
(meters)**

100 50 30 20 10 0

~ Main hydrographic network

Complete names (alphabetic order)

Aam-Amphibalanus amphitrite
Aim-Amphibalanus improvisus
Ame-Austrominius modestus
Bbi-Bankia bipennata
Bfi-Bankia fimbriatula
Bfu-Bugula fulva
Bne-Bugula neritina
Bst-Bugula stolonifera
Csp-Chelonibia sp.
Cau-Conchoderma auritum
Cvi-Conchoderma virgatum
Cca-Cordylophora caspia
Ceu-Corella eumyota
Cfo-Crepidula fornicata
Cne-Cyclope neritea
Dlim-Dendrodoris limbata
Dci-Diadumene cincta
Dlin-Diadumene lineata
Dbi-Diopatra biscayensis
Dlis-Diplosoma listerianum
Dfa-Dosima fascicularis
Eve-Eriphia verrucosa

Fde-Fenestrulina delicia
Fen-Ficopomatus enigmaticus
Fro-Fusinus rostratus
Gal-Gibbula albida
Hja-Haminoea japonica
Hsa-Hemigrapsus sanguineus
Hta-Hemigrapsus takanoi
Hfa-Hesperibalanus fallax
Hsu-Homaxinella subdola
Ime-Idotea metallica
Lli-Limnoria lignorum
Lqu-Limnoria quadripunctata
Lbi-Lyrodus bipartitus
Lpe-Lyrodus pedicellatus
Mgi-Magallana gigas
Mti-Megabalanus tintinnabulum
Mtu-Megabalanus tulipiformis
Mma-Molgula manhattensis
Mse-Monocorophium sextonae
Mar-Mya arenaria
Mle-Mytilopsis leucophaeata
Nco-Nassarius corniculum

Nbr-Neodexiospira brasiliensis
Nra-Newmanella radiata
Oin-Ocinebrellus inornatus
Pma-Paralaeospira malardi
Pja-Perophora japonica
Pph-Petricolaria pholadiformis
Pma-Phallusia mammillata
Pbe-Pileolaria berkeleyana
Pan-Potamopyrgus antipodarum
Pme-Psiloteredo megotara
Rph-Ruditapes philippinarum
Sam-Scruparia ambigua
Ssp-Spathoteredo spatha
Sha-Stramonita haemastoma
Sva-Streptosyllis varians
Sca-Styela canopus
Scl-Styela clava
Tna-Teredo navalis
Tma-Teredora malleolus
Tex-Teredothyra excavata
Uci-Urosalpinx cinerea
Vpa-Victorella pavida
Wsu-Watersipora subtorquata

Figure 12.2 Map of the first occurrence of Non-Indigenous Species in the Normanno-Breton Gulf (1700–2013). Figure © Guillaume Corbeau – Background map from Figure 1, p. 12, *Atlas of the Marine Invertebrate Fauna of the Normanno-Breton Gulf*, vol. 1 (Le Mao et al. 2019).

The EI in the Gulf was mapped within 83 polygons created using the K-means clustering method (Figs 12.3a, 12.3b): a number of clusters were defined, then the stations were grouped into clusters according to their spatial proximity. A few polygons were then merged to improve the clarity of the representation. This method improves the readability compared to regular grid representation methods, especially to avoid the classical problem of grids along the coastline, which are mostly composed of terrestrial areas. The data were divided into two periods, before and after 1981, which is the median year of the time period considered.

The dominance of NIS increased over time: prior to 1981 (Fig. 12.3a), there was only one polygon with an EI greater than 0.024, whereas after 1981 (Fig. 12.3b), almost all the

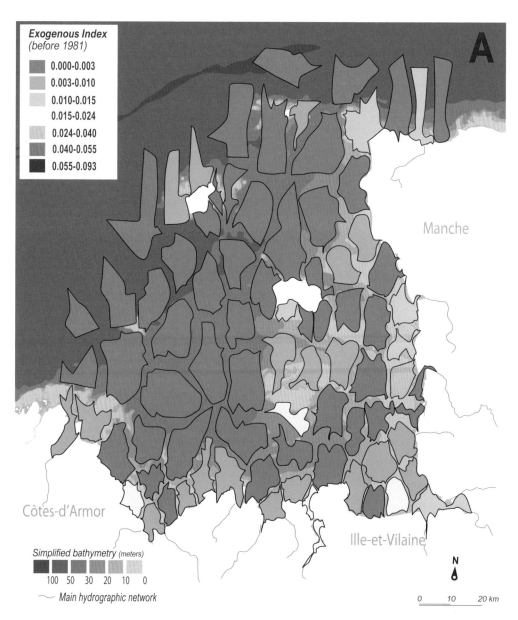

Figure 12.3 Map of the Exogenous Index in the Gulf (A) before 1981, (B) after 1981. Figure © Guillaume Corbeau – Background map from Figure 1, p. 12, *Atlas of the Marine Invertebrate Fauna of the Normanno-Breton Gulf*, vol. 1 (Le Mao et al. 2019).

polygons located along the coast had an EI greater than 0.016. Among the 83 polygons, 53 experienced an increase in the EI between the two periods and 22 did not experience any change as they did not host any NIS in any period. Note that these 22 polygons are the farthest from the shore. Lastly, only eight polygons experienced a decrease in the EI. In the first period (Fig. 12.3a), the high EI values were driven by a lack of data (e.g. around Agon-Coutainville the low number of records combined with large amount of data for the barnacle *Austrominius modestus* led to a high EI). The same was true for the EI of the polygon located at the west part of the Chausey archipelago, which had few data and where the EI was driven by a single mention of the tunicate *Molgula manhattensis*. The same applied to the

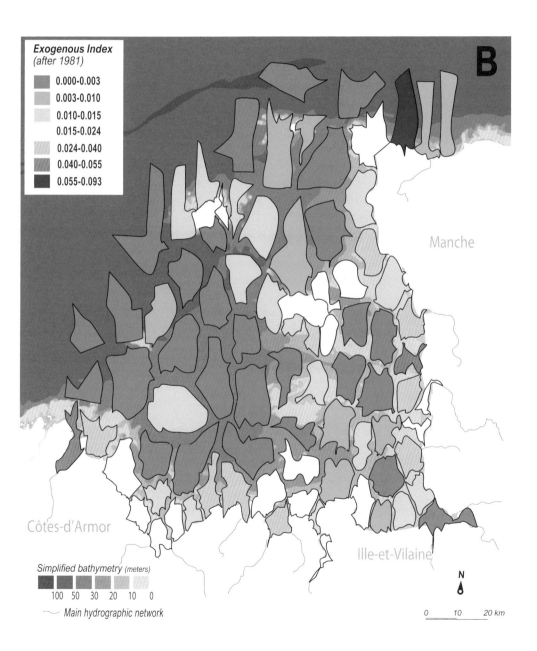

polygon located in the north-west of the Bay of Saint-Brieuc, where EI values were driven by a few records of *A. modestus*. In the second period (Fig. 12.3b), such biases are rare. It should be noted, however, that the polygons covering the bay of Mont-Saint-Michel and the one south to the Chausey archipelago had a low EI because *Crepidula fornicata*, which dominated the NIS community in these areas, was excluded from our analysis. The high value in the northern Cotentin, west of Cherbourg, was also an artefact, because records of indigenous species were rare, whereas records of the crab *Hemigrapsus sanguineus* and tunicate *Styela clava* have inflated the EI.

The analysis of such a database, built from heterogeneous data, needs to take into account biases linked to the sampling effort being unevenly distributed, the relatively poor reliability of the number of observations as a species frequency proxy, the EI values driven by a few records in some polygons and the dominance of *Crepidula fornicata* records, which makes the rest of the NIS dataset difficult to read. Nevertheless, a clear increase both in the number of NIS and in their spatial distribution is demonstrated. The Gulf is relatively favourable to the introduction and settlement of NIS. Indeed, the waters in the Gulf, and particularly the existence of gyres, favour the confinement of larvae of species with a benthopelagic cycle. Additionally, human activities potentially introducing NIS are easy to identify: shellfish farming, recreational and commercial maritime traffic and fishing activity. Indeed, the Gulf supports major shellfish production areas. For each of these regions (i.e. Bréhat region, Bay of Saint-Brieuc, Bays of Fresnaye and Arguenon, Bay of Cancale, Chausey archipelago, west Cotentin, Jersey and Guernsey), the corresponding EI increased by a factor of ten. These are not the only areas that experienced such an increase, but the role played by shellfish farming seems to be crucial.

This initial overview on the NIS of the Normanno-Breton Gulf raises two main research perspectives. The first one is to understand how the NIS may contribute to biotic homogenisation, and at what spatial and temporal trend. The second perspective is to understand how human activities, including their different practices over space and time, may control the introduction and the dispersal of the NIS. Interviews with shellfish farmers are needed to better understand the transfer of living materials, the history of their concessions as well as the changes in their practices.

Eco-engineering of coastal structures

A societal adaptation to impacts of rising and stormier seas resulting from climate change has been a proliferation of new structures and expansion of existing hard sea defences along urban coastlines and Channel ports, described as ocean sprawl (*sensu* Duarte et al. 2015). In response to predicted rise in sea levels under current emissions scenarios, new seawalls and breakwaters have been constructed to 'hold the line'. These structures have increased the area of rocky intertidal and subtidal habitat throughout the Channel. However this can result in so-called coastal squeeze and the loss of intertidal habitat (Pontee 2011). New structures may facilitate the establishment and spread of non-native species and provide stepping-stones for spread and range expansion (Airoldi and Bulleri 2011; Mineur et al. 2012; Firth et al. 2016). They can also bridge gaps between rocky habitats, driving expansions of native species responding to climate at leading edges of their range (reviewed in Firth et al. 2016), such as the trochid molluscs *Phorcus lineatus* and *Steromphala umbilicalis* and limpet *Patella depressa* (Hawkins et al. 2009).

Artificial habitats are relatively poor for marine biodiversity in comparison to natural habitats (Moschella et al. 2005; Firth et al. 2013a) and there is a growing interest in ecological enhancement to improve habitat quality and loss owing to coastal squeeze (Firth et al. 2013b; Firth et al. 2014; Dafforn et al. 2015; Evans et al. 2015; Hall et al. 2017, 2019; Morris et al. 2019; O'Shaughnessy et al. 2020). At Shaldon near Teignmouth in south-west England, holes and pits made in seawalls increased diversity and density of littorinids (Naylor et al. 2011; Firth et al. 2014b). The installation of artificial rock pools on a seawall at Bouldnor on the Isle of Wight (Hall et al. 2019) was shown to be beneficial at increasing biodiversity of the wall, and some species normally found on the lower shore colonised higher levels on the wall. Holes and grooves cut into limestone rock armour in Poole Bay increased biodiversity over two years (Hall et al. 2017), yet these features do not always survive long term in softer rocks. Similarly, over time, smaller pits that once provided refuge for myriad species (Firth et al. 2014b) can also become dominated by single individuals (Firth et al. 2020).

These interventions now come under a broad definition of 'nature-based solutions' in a rapidly growing arena of research that aims to address both the climate and biodiversity crisis. Care must be exercised when implementing such solutions, as biodiversity responses differ depending on the environmental context (Strain et al. 2021). But as Firth et al. (2020) have rightly pointed out, such approaches should only be used in already highly urbanised areas to protect lives, infrastructure and property and not be a fig leaf for coastal development in less degraded sites.

Concluding remarks

There is a climatic gradient from west to east up the Channel following the minimum winter sea temperature isotherms around the British Isles and northern France into the colder eastern Channel and southern North Sea. There is also a transition from fully oceanic to much more neritic waters up the Channel. At the end of the last Ice Age, various marine species spread northwards and eastwards from their glacial maximum refuges, with populations of some intertidal species not reaching or crossing the Channel before it opened up about 8,000 years ago (Preece 1995; Provan, Wattier and Maggs 2005; Brooks et al. 2011). Over the last 150 years of formal scientific endeavour, as climatic conditions have fluctuated, advances and retreats of species have been detected as well as changes in abundance. Since the early 1990s, observations have been closely aligned to anthropogenic ocean warming. Changes in the relative abundance of warm-water southern and cold-water northern barnacles (Southward and Crisp 1954a; Southward 1967, 1991; Southward, Hawkins and Burrows 1995) and limpets (Southward, Hawkins and Burrows 1995; Kendall et al. 2004; Hawkins et al. 2008, 2009) have been shown, with implications for biological interactions (Moore, Hawkins and Thompson 2007; Moore, Thompson and Hawkins 2007; Firth et al. 2009). Modelling studies of barnacles (Poloczanska et al. 2008) have demonstrated the interplay of direct effects of climate and its modulation by competitive interactions. In an area of much international trade and naval activity, various non-native species have successively colonised the Channel (Bishop et al. 2015; O'Shaughnessy et al. 2020). The first came mainly from North America, but more recently Australasia and Asia as trade patterns have changed. This can be viewed as a facet of global change, with the homogenisation of biota from different biogeographic realms. Such invasions are more likely in a warming and more extreme world, as many of these species are opportunist generalists with broad niches – and in many cases from eastern continental margins of North America and Asia where both extremes of summer heat and winter cold are experienced. In some cases completely new functional groups arise, such as mid-shore Pacific Oysters occurring where Native Oysters are never found; or, as in the case of *Crepidula*, adding a completely new feeding type as mucus-net feeders.

Global change interacts with regional (e.g. overfishing, eutrophication) and local scale impacts (point source chronic pollution, acute pollution incidents, habitat loss and modification owing to inappropriate coastal development or coastal defence). Some local impacts such as TBT, or proliferating sea defences in the eastern channel, can scale up to the regional impacts (Firth et al. 2013). Other impacts such as marine litter, particularly plastic, although often derived from point sources, can now be viewed as a global problem (Thompson et al. 2009; Lusher, McHugh and Thompson 2013; Steer et al. 2017). Eutrophication persists in the coastal zone, with green algal mats (*Ulva* spp.) blooming in some bays and harbours (Thornton et al. 2019; Schreyers et al. 2021); however, much pollution is under control, with improvements ultimately prompted by European Union initiatives and Directives (Water Framework Directive, Marine Strategy Framework Directive). Overfishing, although rife, is manageable. Coastal development can be somewhat compensated by eco-engineering approaches when appropriate. Thus, in the next 50–100 years, hopefully in parallel with decarbonisation of the world's economies, the priority must be on managing the interactions of climate change with those impacts that can be managed.

The English Channel, with its multiple range limits of Lusitanian warm-water and a few cold-water species, is also a natural laboratory for understanding biogeographic processes in the context of climate change. Work to date has shown the idiosyncratic nature of the responses of individual species, shaped by life-history traits, dispersal capability and habitat availability (Woodin, Wethey and Dubois 2014; Firth et al. 2021). Much can also be learnt from the spread of non-native species and the consequences for assemblage composition, community structure and ecosystem functioning (Ruesink et al. 2006; Rodriguez, 2006; Queirós et al. 2011; Xue et al. 2018). Even non-native species can provide ecosystem services such as biofiltration (Lefebvre, Barille and Clerc 2000) and habitat provision (e.g. *Magallana*, Firth et al. 2021; *Sargassum*, Kim et al. 2019). Science to both advance knowledge and inform management of the English Channel/La Manche will post-Brexit (see Hawkins 2017) need a renewed *entente cordiale* between scientists on both its north and south sides.

Acknowledgements

We are grateful to Natural Resources Wales and Natural England for funding the MarClim project.

References

Airoldi, L. and Bulleri, F. (2011) Anthropogenic disturbance can determine the magnitude of opportunistic species disturbance on marine urban infrastructures. *PLoS One* 6 (8): e22985. https://doi.org/10.1371/journal.pone.0022985

Alzieu, V. (2000) Environmental impact of TBT: the French experience. *Science of the Total Environment* 258: 99–102. https://doi.org/10.1016/S0048-9697(00)00510-6

Androuin, T., Polerecky, L., Decottignies, P., Dubois, S.F., Dupuy, C., Hubas, C., Jesus, B., Le Gall, E., Marzloff, M.P. and Carlier, A. (2018) Subtidal microphytobenthos: a secret garden stimulated by the engineer species *Crepidula fornicata*. *Frontiers in Marine Science* 5: 475. https://doi.org/10.3389/fmars.2018.00475

Baffreau, A., Pezy, J.P., Rusig, A.M., Mussio, I. and Dauvin, J.C. (2018) *Les espèces marines animals et végétales introduites en Normandie*. Bayeux: Imprimerie Moderne de Bayeux.

Beaugrand, G., Ibañez, F. and Reid, P.C. (2000) Spatial, seasonal and long-term fluctuations of plankton in relation to hydroclimatic features in the English Channel, Celtic Sea and Bay of Biscay. *Marine Ecology Progress Series* 200: 93–102. https://doi.org/10.3354/meps200093

Beaugrand, G., Reid, P.C. and Ibañez, F. (2002) Reorganization of North Atlantic marine copepod biodiversity and climate. *Science* 296: 1692–4. https://doi.org/10.1126/science.1071329

Berry, W. (1815). *The History of the Island of Guernsey: Part of the ancient duchy of Normandy, from the remotest period of antiquity to the year 1814*. London: Longman, Hurst, Rees, Orme and Brown and Hatchard.

Bishop, J.D., Wood, C.A., Lévêque, L., Yunnie, A.L. and Viard, F. (2015) Repeated rapid assessment surveys reveal contrasting trends in occupancy of marinas by non-indigenous species on opposite sides of the western English Channel. *Marine Pollution Bulletin* 95: 699–706. https://doi.org/10.1016/j.marpolbul.2014.11.043

Blanchard, M. (1997a) *Répartition et évaluation quantitative de la crépidule (Crepidula fornicata) entre le Cap Fréhel et le Mont-Saint-Michel. Rapport RST. DEL/99.05*. Brest: Ifremer.

Blanchard, M. (1997b) Spread of the slipper limpet *Crepidula fornicata* (L. 1758) in Europe. Current state and consequences. *Scientia marina* 61 (Suppl. 2): 109–18.

Blanchard, M. (2009) Recent expansion of the slipper limpet population (*Crepidula fornicata*) in the Bay of Mont-Saint-Michel (Western Channel, France) *Aquatic Living Resources* 22: 11–19. https://doi.org/10.1051/alr/2009004

Boalch, G.T. (1987) Changes in the phytoplankton of the Western English Channel in recent years. *British Phycological Journal* 22: 225–35. https://doi.org/10.1080/00071618700650291

Bray, S. and Herbert, R.J.H. (1998) A reassessment of populations of the dog-whelk *Nucella lapillus* on the Isle of Wight following legislation restricting the use of TBT antifouling paints. *Proceedings of the Isle of Wight Natural History and Archaeological Society* 14: 23–40.

Brooks, A.J., Bradley, S.L., Edwards, R.J. and Goodwyn, N. (2011) The palaeogeography of Northwest Europe during the last 20,000 years. *Journal of Maps* 7: 573–87. https://doi.org/10.4113/jom.2011.1160

Bryan, G.W., Gibbs, P.E., Hummerstone, L.G. and Burt, G.R. (1986) The decline of the gastropod *Nucella lapillus* around south-west England: evidence for the effect of tributyltin from antifouling paints. *Journal of the Marine Biological Association of the United Kingdom* 67: 525–44. https://doi.org/10.1017/S0025315400027272

Burrows, M.T., Hawkins, S.J. and Southward, A.J. (1992) A comparison of reproduction in co-occuring chthamalid barnacles, *Chthamalus stellatus* (Poli) and *Chthamalus montagui* southward. *Journal of Experimental Marine Biology and Ecology* 160: 229–49. https://doi.org/10.1016/0022-0981(92)90240-B

Burrows, M.T., Schoeman, D.S., Moore, P., Poloc-zanska, E.S., Brander, K.M., Brown, C., Bruno, J.F., Duarte, C.M., Halpern, B.S., Holding, J., Kappel, C.V., Kiessling, W., O'Connor, M.I., Pandolfi, J.M., Parmesan, C., Schwing, F.B., Sydeman, W.J. and Richardson, A.J. (2011) The pace of shifting climate in marine and terrestrial ecosystems. *Science* 334: 652–9. https://doi.org/10.1126/science.1210288

Burrows, M.T., Hawkins, S.J., Moore, J.J., Adams, L., Sugden, H., Firth, L. and Mieszkowska, N. (2020) Global-scale species distributions predict temperature-related changes in species composition of rocky shore communities in Britain. *Global Change Biology* 26: 2093–105. https://doi.org/10.1111/gcb.14968

Cabioch, L., Gentil, F., Glacon, R. and Retière, C. (1977) *Le macrobenthos des fonds meubles de la Manche: distribution générale et écologie*. In B. Keegan, P. O'Ceidigh and J. Boaden (eds) *Biology of Benthic Organisms*, pp. 115–28. Oxford: Pergamon Press. https://doi.org/10.1016/B978-0-08-021378-1.50018-X

Cabral, P., Levrel, H., Schoenn, J., Thiébaut, E., Le Mao, P., Mongruel, R., Dedieu, K., Carrier, S., Morisseau, F., Daurès, F. (2015) Marine habitats ecosystem service potential: a vulnerability approach in the Normand-Breton (Saint-Malo) Gulf, France.

Ecosystem Services 16: 306–18. https://doi.org/10.1016/j.ecoser.2014.09.007

Caldow, R.W.G., Stillman, R.A., Durrel, S.E.A. le V. dit, West, A.D., McGrorty, S., Goss-Custard, J.D., Wood, P.J. and Humphreys, J. (2007) Benefits to shorebirds from invasion of a non-native shellfish. *Proceedings of the Royal Society* (B) 274: 1449–55. https://doi.org/10.1098/rspb.2007.0072

Chambers, P. (2008) *Channel Islands Marine Molluscs: An illustrated guide to the seashells of Jersey, Guernsey, Alderney, Sark and Herm*. Royston: Charonia Media.

Cheung, W.W.L., Reygondeau, G. and Froicher, T.L. (2016) Large benefits to marine fisheries of meeting the 1.5 degrees C global warming target. *Science* 354: 1591–4. https://doi.org/10.1126/science.aag2331

Coleman, R.A., Underwood, A.J., Benedetti-Cecchi, L., Aberg, P., Arenas, F., Arrontes, J., Castro, J., Hart-noll, R.G., Jenkins, S.R., Paula, J., Della Santina, P. and Hawkins, S.J. (2006) A continental scale evaluation of the role of limpet grazing on rocky shores. *Oecologia* 147: 556–64. https://doi.org/10.1007/s00442-005-0296-9

Connell, J.H. (1961) The influence of interspecific competition and other factors on the barnacle *Chthamalus stellatus*. *Ecology* 42: 710–23. https://doi.org/10.2307/1933500

Crisp, D.J. and Southward, A.J. (1958) The distribution of intertidal organisms along the coasts of the English Channel. *Journal of the Marine Biological Association of the United Kingdom* 37: 157–208. https://doi.org/10.1017/S0025315400014909

Crisp, D.J. (1958) The spread of *Elminius modestus* Darwin in north-west Europe, *Journal of the Marine Biological Association of the United Kingdom* 37: 483–520. https://doi.org/10.1017/S0025315400023833

Crisp, D.J. (ed.) (1964) The effects of the severe winter of 1962–63 on marine life in Britain. *Journal of Animal Ecology* 33: 165–210. https://doi.org/10.2307/2355

Cugier, P., Struski, C., Blanchard, M., Mazurié, J., Pouvreau, S., Olivier, F., Trigui, J.R. and Thiébaut, E. (2010) Assessing the role of benthic filter feeders on phytoplankton production in a shellfish farming site: Mont Saint-Michel Bay, France. *Journal of Marine Systems* 82: 21–34. https://doi.org/10.1016/j.jmarsys.2010.02.013

Curd, A., Cordier, C., Firth, L.B., Bush, L., Gruet, Y., Le Mao, P., Blaze, J.A., Board, C., Bordeyne, F., Burrows, M.T., Cunningham, P.N., Davies, A.J., Desroy, N., Edwards, H., Harris, D.R., Hawkins, S.J., Kerckhof, F., Lima, F.P., McGrath, D., Meneghesso, C., Mieszkowska, N., Nunn, J.D., Nunes, F., O'Connor, N.E., O'Riordan, R.M., Power, A.M., Seabra, R., Simkanin, C. and Dubois, S. (2020). A broad-scale long-term dataset of *Sabellaria alveolata* distribution and abundance curated through the REEHAB (REEf HABitat) Project. *SEANOE*. https://doi.org/10.17882/72164

Cushing, D.H. and Dickson, R.R. (1976) The biological response in the sea to climatic changes. *Advances in Marine Biology* 14: 1–122. https://doi.org/10.1016/S0065-2881(08)60446-0

Dafforn, K.A., Glasby, T.M., Airoldi, L., Rivero, N.K., Mayer-Pinto, M. and Johnston, E.L. (2015) Marine urbanization: an ecological framework for designing multifunctional artificial structures. *Frontiers in Ecology and Environment* 13: 82–90. https://doi.org/10.1890/140050

Dauvin, J-C. (2000) The muddy fine sand *Abra alba–Melinna palmata* community of the Bay of Morlaix twenty years after the Amoco Cadiz oil spill. *Marine Pollution Bulletin* 40: 528–36. https://doi.org/10.1016/S0025-326X(99)00242-8

De Gerville, Charles-Alexis-Adrien Duhérissier (1825) Catalogue des coquilles trouvées sur lescôtes du département de la Manche, lu à la séance du 14 mars 1825. *Mémoires de la Société linnéenne du Calvados*: 169–224.

Doney, S.C., Ruckelshause, M., Duffy, J.E., Barry, J.P., Chan, F., English, C.A., Galindo, H.M., Grebmeier, J.M., Hollowed, A.B., Knowlton, N., Polovina, J., Rabalais, N.N., Sydeman, W.J. and Talley, L.D. (2012) Climate change impacts on marine ecosystems. *Annual Review of Marine Science* 4: 11–37. https://doi.org/10.1146/annurev-marine-041911-111611

Duarte, C.M., Pitt, K., Lucas, C.H., Purcell, J.E., Uye, S-i., Robinson, K., Brotz, L., Decker, M.B., Sutherland, K.R., Malej, A., Madin, L., Mianzan, H., Gilli, J.-M., Fuentes, V., Atienza, D., Pagés, F., Breitburg, D., Malek, J., Graham, W.M. and Condon, R.H. (2013) Is global ocean sprawl a cause of jellyfish blooms? *Frontiers in Ecology and the Environment* 11: 91–7. https://doi.org/10.1890/110246

Dupont, L., Richard, J., Paulet, Y.M., Thouzeau, G. and Viard, F. (2006) Gregariousness and protandry promote reproductive insurance in the invasive gastropod *Crepidula fornicata*: evidence from assignment of larval paternity. *Molecular Ecology* 15: 3009–21. https://doi.org/10.1111/j.1365-294X.2006.02988.x

Dupouy, H. and Latrouite, D. (1979) Le développement de la crépidule sur le gisement de coquilles Saint-Jacques de la baie de Saint-Brieuc. *Science et Pêche* 292: 13–19.

Edwards, M., Beaugrand, G., Helaouët, P., Alheit, J. and Coombes, S. (2013) Marine ecosystem response to the Atlantic Multidecadal Oscillation. *PLoS One* 8 (2): e57212. https://doi.org/10.1371/journal.pone.0057212

Evans, A.J., Firth, L.B., Hawkins, S.J., Morris, E.S., Goudge, H. and Moore, P.J. (2015) Drill-cored rock pools: an effective method of ecological enhancement on artificial structures. *Marine and Freshwater Research* 67: 123–30. https://doi.org/10.1071/MF14244

Firth, L.B., Crowe, T.P., Moore, P., Thompson, R.C. and Hawkins, S.J. (2009) Predicting impacts of climate-induced range expansion: an experimental framework and a test involving key grazers on temperate rocky shores. *Global Change Biology* 15: 1413–22. https://doi.org/10.1111/j.1365-2486.2009.01863.x

Firth, L.B., Mieszkowska, N., Thompson, R.C. and Hawkins, S.J. (2013a) Climate change and adaptational impacts in coastal systems: the case of sea defences. *Environmental Science: Processes & Impacts* 15: 1665–70. https://doi.org/10.1039/c3em00313b

Firth, L.B., Thompson, R.C., White, F.J., Schofield, M., Skov, M.W., Hoggart, S.P., Jackson, J., Knights, A.M. and Hawkins, S.J. (2013b) The importance of water-retaining features for biodiversity on artificial intertidal coastal defence structures. *Diversity and Distributions* 19: 1275–83. https://doi.org/10.1111/ddi.12079

Firth, L., Schofield, M., White, F.J., Skov, M.W. and Hawkins, S.J. (2014) Biodiversity in intertidal rock pools: informing engineering criteria for artificial habitat enhancement in the built environment. *Marine Environmental Research* 102: 122–30. https://doi.org/10.1016/j.marenvres.2014.03.016

Firth, L.B., Knights, A.M., Bridger, D., Evans, A.J., Mieszkowska, N., Moore, P.J., O'Connor, N.E., Sheehan, E.V., Thompson, R.C. and Hawkins, S.J. (2016) Ocean sprawl: challenges and opportunities for biodiversity management in a changing world. *Oceanography and Marine Biology: An Annual Review* 54: 193–269. https://doi.org/10.1201/9781315368597-5

Firth, L.B., Airoldi, L., Bulleri, F., Challinor, S., Chee, S.Y., Evans, A., Hanley, M.E., Knights, A.M., O'Shaughnessy, K., Thompson, R.C. and Hawkins, S.J. (2020) Greening of grey infrastructure should not be used as a Trojan horse to facilitate coastal development. *Journal of Applied Ecology* 57: 1762–8. https://doi.org/10.1111/1365-2664.13683

Firth, L.B., Harris, D., Blaze, J.A., Marzloff, M.P., Boyé, A., Miller, P.I., Curd, A., Vasquez, M., Nunn, J.D., O'Connor, N.E. and Power, A.M. (2021a) Specific niche requirements underpin multidecadal range edge stability, but may introduce barriers for climate change adaptation. *Diversity and Distributions* 27: 668–83. https://doi.org/10.1111/ddi.13224

Firth, L.B., Duff, L., Gribben, P.E. and Knights, A.M. (2021b) Do positive interactions between marine invaders increase likelihood of invasion into natural and artificial habitats? *Oikos* 130: 453–63. https://doi.org/10.1111/oik.07862

Fischer-Piette, E. 1936. Etudes sur la biogéographie intercotidale des deux rivers de la Manche. *Journal of the Linnean Society (Zool.)* 40: 181–272. https://doi.org/10.1111/j.1096-3642.1936.tb01683z.x

Fitzgerald, A. (2007) Slipper limpet utilisation and management. Final report. Port of Truro Oyster Management Group (Cornwall).

Forbes, E. (1858) The distribution of marine life, illustrated chiefly by fishes and molluscs and radiata. In A.K. Johnston (ed.) *A.K. Johnston's Physical Atlas*, pp. 99–101. Edinburgh: W. & A.K. Johnston.

Gaudin, F., Desroy, N., Dubois, S.F., Broudin, C., Cabioch, J., Fournier, J., Gentil, F., Grall, J., Houbin, C., Le Mao, P. and Thiébaut, E. (2018) Marine sublittoral benthos fails to track temperature in response to climate change in a biogeographical transition zone. *ICES Journal of Marine Science* 75: 1894–1907. https://doi.org/10.1093/icesjms/fsy095

Genner, M.J., Sims, D.W., Wearmouth, V.J., Southall, E.J., Southward, A.J., Henderson, P.A. and Hawkins, S.J. (2004) Regional climatic warming drives long-term community changes of British marine fish. *Proceedings of the Royal Society (B)* 271: 655–61. https://doi.org/10.1098/rspb.2003.2651

Genner, M.J., Sims, D.W., Southward, A.J., Budd, G.C., Masterson, P., Mchugh, M., Rendle, P., Southall, E.J., Wearmouth, V.J. and Hawkins, S.J. (2010) Body size-dependent responses of a marine fish community and fishing over a century-long scale. *Global Change Biology* 16: 517–27. https://doi.org/10.1111/j.1365-2486.2009.02027.x

Gibbs, P.E., Bryan, G.W., Pascoe, P.L. and Burt, G.R. (1987) Reproductive failure in populations of the dog-whelk *Nucella lapillus* caused by tributyltin from antifouling paints. *Journal of the Marine Biological Association of the United Kingdom* 66: 767–77. https://doi.org/10.1017/S0025315400048414

Hall, A.E., Herbert, R.J.H., Britton, J.R. and Hull, S.L. (2017) Ecological enhancement techniques to improve habitat heterogeneity on coastal defence structures. *Estuarine Coastal and Shelf Science* 210: 68–78. https://doi.org/10.1016/j.ecss.2018.05.025

Hall, A.E., Herbert, R.J.H., Britton, J.R., Boyd, I.M. and George, N.C. (2019) Shelving the coast with vertipools: retrofitting artificial rock pools on coastal structures as mitigation for coastal squeeze. *Frontiers in Marine Science* 6: 456. https://doi.org/10.3389/fmars.2019.00456

Hamon, D. and Blanchard, M. (1994) Etat de la prolifération de la crépidule (*Crepidula fornicata*) en baie de Saint-Brieuc. Institut Francais de recherche pour l'exploitation de la mer (Ifremer). Direction de l'environnement et de l'amenagement du littoral. Plouzane.

Hawkins, S.J. (2017) Ecological processes are not bound by borders: implications for marine conservation in a post-Brexit world. *Aquatic Conservation Marine and Freshwater Ecosystems* 27: 904–8. https://doi.org/10.1002/aqc.2838

Hawkins, S.J. and Southward, A.J. (1992) The *Torrey Canyon* oil spill: recovery of rocky shore communities. In *Restoring the Nation's Marine Environment. Maryland: Proceedings of the Symposium on Marine Habitat Restoration*, pp. 584–631. Sea Grant Publication, National Oceanic and Atmospheric Administration, Maryland Sea Grant College.

Hawkins, S.J., Southward, A.J. and Genner, M.J. (2003) Detection of environmental change in a marine ecosystem – evidence from the western English Channel. *Science of the Total Environment* 310: 245–6. https://doi.org/10.1016/S0048-9697(02)00645-9

Hawkins, S.J., Moore, P., Burrows, M., Poloczanska, E., Mieszkowska, N., Herbert, R.J.H., Jenkins, S.R., Thompson, S.R., Gennre, M.J. and Southward, A.J. (2008) Complex interactions in a rapidly changing world: responses of rocky shore communities to recent climate change. *Climate Research* 37: 123–33. https://doi.org/10.3354/cr00768

Hawkins, S.J., Sugden, H.E., Mieszkowska, N., Poloczanska, E., Leaper, R., Herbert, R.J.H., Genner, M.J., Moschella, P.S., Thompson, R.C., Jenkins, S.R., Southward, A.J. and Burrows, M.T. (2009) Consequences of climate-driven biodiversity changes for ecosystem functioning of North European rocky shores. *Marine Ecology Progress Series* 396: 245–59. https://doi.org/10.3354/meps08378

Hawkins, S.J., Vale, M., Firth, L.B., Burrows, M.T., Mieszkowska, N. and Frost, M. (2013a) Sustained observation of marine biodiversity and ecosystems. *Oceanography* 1 (1): 1000e101. https://doi.org/10.4172/2332-2632.1000e101

Hawkins, S.J., Firth, L.B., McHugh, M., Poloczanska, E.S., Herbert, R.J.H., Burrows, M.T., Kendall, M.A., Moore, P.J., Thompson, R.C., Jenkins, S.R. and Sims, D.W. (2013b) Data rescue and re-use: recycling old information to address new policy concerns. *Marine Policy* 42: 91–8. https://doi.org/10.1016/j.marpol.2013.02.001

Hawkins, S.J., Evans, A.J., Firth, L.B., Genner, M.J., Herbert, R.J.H., Adams, L.C., Moore, P.J., Mieszkowska, N., Thompson, R.C., Burrows, M.T. and Fenburg, P.B. (2016) Impacts and effects of ocean warming on intertidal rocky habitats. In D. Laffoley and J.M. Baxter (eds) *Explaining Ocean Warming: Causes, scale, effects and consequences*, pp. 147–76. Gland: IUCN.

Hawkins, S.J., Evans, A.J., Mieszkowska, N., Adams, L.C., Bray, S., Burrows, M.T., Firth, L.B., Genner, M.J., Leung, K.M.Y., Moore, P.J., Pack, K., Schuster, H., Sims, D.W., Whittington, M. and Southward, E.C. (2017a) Distinguishing globally-driven changes from regional- and local-scale impacts: the case for long-term and broad-scale studies of recovery from pollution. *Marine Pollution Bulletin* 124: 573–86. https://doi.org/10.1016/j.marpolbul.2017.01.068

Hawkins, S.J., Evans, A.J., Moore, J., Whittington, M., Pack, K., Firth, L.B., Adams, L.C., Moore, P.J., Masterson-Algar, P., Mieszkowska, N. and Southward, E.C. (2017b) From the *Torrey Canyon* to today: a 50 year retrospective of recovery from the oil spill and interaction with climate-driven fluctuations on Cornish rocky shores. *International Oil Spill Conference Proceedings* 1: 74–103. https://doi.org/10.7901/2169-3358-2017.1.74

Hawkins, S.J., Pack, K.E., Firth, L.B., Mieskowska, N., Evans, A.J., Herbert, R.J.H. and 39 others (2019) The intertidal zone of the north-east Atlantic region. In S.J. Hawkins, K. Bohn, L.B. Firth and G.A. Williams (eds) *Interactions in the Marine Benthos*, pp. 7–46. Cambridge: Cambridge University Press. https://doi.org/10.1017/9781108235792.003

Heath, M.R., Neat, F.C., Pinnegar, J.K., Reid, D.G., Sims, D.W. and Wright, P.J. (2012) Review of climate change impacts on marine fish and shellfish around the UK and Ireland. *Aquatic Conservation: Marine and Freshwater Ecosystems* 22: 337–67. https://doi.org/10.1002/aqc.2244

Helmer, L., Farrell, P., Hendy, I., Harding, S., Robertson, M. and Preston, J. (2019) Active management is required to turn the tide for depleted *Ostrea edulis* stocks from the effects of overfishing, disease and invasive species. *PeerJ* 7: e6431. https://doi.org/10.7717/peerj.6431

Helmuth, B., Mieszkowska, N., Moore, P. and Hawkins, S.J. (2006) Living on the edge of two changing worlds: forecasting the responses of rocky intertidal ecosystems to climate change. *Annual Review Ecology Evolution and Systematics* 37: 373–404. https://doi.org/10.1146/annurev.ecolsys.37.091305.110149

Herbert, R.J.H. (1989) A survey of the dog-whelk *Nucella lapillus* (L.) around the coast of the Isle of Wight. *Proceedings of the Isle of Wight Natural History and Archaeological Society* 8: 15–21.

Herbert, R.J.H., Hawkins, S.J., Sheader, M. and Southward, A.J. (2003) Range extension and reproduction of the barnacle *Balanus perforatus* in the eastern English Channel. *Journal of the Marine Biological Association of the United Kingdom* 83: 73–82. https://doi.org/10.1017/S0025315403006829h

Herbert, R.J.H., Southward, A.J., Sheader, M. and Hawkins, S.J. (2007) Influence of recruitment and temperature on distribution of intertidal barnacles in the English Channel. *Journal of the Marine Biological Association of the United Kingdom* 87: 487–99. https://doi.org/10.1017/S0025315407052745

Herbert, R.J.H., Southward, A.J., Clarke, R.T., Sheader, M. and Hawkins, S.J. (2009) Persistent border, an analysis of the geographic boundary of an intertidal species. *Marine Ecology Progress Series* 379: 135–50. https://doi.org/10.3354/meps07899

Herbert, R.J.H., Roberts, C., Humphreys, J. and Fletcher, S. (2012) *The Pacific oyster (Crassostrea gigas) in the UK: economic, legal and environmental issues associated with its cultivation, wild establishment and exploitation.* Report for the Shellfish Association of Great Britain.

Herbert, R.J.H., Humphreys, J., Davies, C.J., Roberts, C., Fletcher, S. and Crowe, T.P. (2016) Ecological impacts of non-native Pacific oysters (*Crassostrea gigas*) and management measures for protected areas in Europe. *Biodiversity and Conservation* 25: 2835–65. https://doi.org/10.1007/s10531-016-1209-4

Herrera, M., Wethey, D.S., Vazquez, E. and Macho, G. (2019) Climate change implications for reproductive success: temperature effect on penis development in the barnacle *Semibalanus balanoides*. *Marine Ecology Progress Series* 610: 109–23. https://doi.org/10.3354/meps12832

Hinz, H., Prieto, V. and Kaiser, M.J. (2009) Trawl disturbance on benthic communities: chronic effects and experimental predictions. *Ecological Applications* 19: 761–73. https://doi.org/10.1890/08-0351.1

Hinz, H., Capasso, E., Lilley, M., Frost, M. and Jenkins, S.R. (2011) Temporal differences across a biogeographical boundary reveal slow response of sub-littoral benthos to climate change. *Marine Ecology Progress Series* 423: 69–82. https://doi.org/10.3354/meps08963

Hiscock, K. (1998) Introduction and Atlantic-European perspective: 3–70. In K. Hiscock (ed.) *Marine Nature Conservation Review. Benthic marine ecosystems of Great Britain and the north-east Atlantic.* Peterborough: JNCC.

Hiscock, K., Southward, A., Tittley, I. and Hawkins, S.J. (2004) Effects of changing temperature on benthic marine life in Britain and Ireland. *Aquatic Conservation: Marine and Freshwater Ecosystems* 14: 333–62. https://doi.org/10.1002/aqc.628

Holme, N.A. (1961) The bottom fauna of the English Channel. *Journal of the Marine Biological Association of the United Kingdom* 41: 397–461. https://doi.org/10.1017/S0025315400023997

Holme, N.A. (1966) The bottom fauna of the English Channel. Part 2. *Journal of the Marine Biological Association of the United Kingdom* 46: 401–93. https://doi.org/10.1017/S0025315400027193

Humphreys, J., Harris, M.R., Herbert, R.J.H., Farrell, P., Jensen, A. and Cragg, S.M. (2015) Introduction, dispersal and naturalization of the Manila clam *Ruditapes philippinarum* in British estuaries, 1980–2010. *Journal of the Marine Biological Association of the United Kingdom* 95: 1163–72. https://doi.org/10.1017/S0025315415000132

IUCN SSC Invasive Species Specialist Group (2000) Guidelines for the prevention of biodiversity loss caused by alien invasive species. Fifth Meeting of the Conference of the Parties to the Convention on Biological Diversity, Nairobi, Kenya, 15–26 May 2000. https://portals.iucn.org/library/efiles/documents/Rep-2000-052.pdf.

Jeffreys, J.G. (1867) *British Conchology or an Account of the Mollusca which now Inhabit the British Isles and the Surrounding Seas: Vol. IV. In Continuation of the Gastropoda as far as the Bulla family.* London: J. Van Voorst.

Jenkins, S.R., Coleman, R.A., Della Santina, P., Hawkins, S.J., Burrow, M.T. and Hartnoll, R.G. (2005) Regional scale differences in the determinism of grazing effects in the rocky intertidal. *Marine Ecology Progress Series* 287: 77–86. https://doi.org/10.3354/meps287077

Keith, S.A., Herbert, R.J.H., Norton, P.A., Hawkins, S.J. and Newton, A.C. (2011) Individualistic species limitations of climate-induced range expansions generated by meso-scale dispersal barriers: dispersal barriers limit range expansions. *Diversity and Distributions* 17: 275–86. https://doi.org/10.1111/j.1472-4642.2010.00734.x

Kendall, M.A., Burrows, M.T., Southward, A.J. and Hawkins, S.J. (2004) Predicting the effects of marine climate change on the invertebrate prey of the birds of rocky shores. *Ibis* 146: 40–7. https://doi.org/10.1111/j.1474-919X.2004.00326.x

Kim, H.G., Hawkins, L.E., Godbold, J.A., Oh, C.W., Rho, H.S. and Hawkins, S.J. (2019) Comparison of nematode assemblages associated with *Sargassum muticum* in its native range in South Korea and as an invasive species in the English Channel. *Marine Ecology Progress Series* 611: 95–110. https://doi.org/10.3354/meps12846

Laffoley, D. and Baxter, J.M. (eds) (2016) Explaining ocean warming: causes, scale, effects and consequences. Full report, IUCN, Gland. https://doi.org/10.2305/IUCN.CH.2016.08.en

Langston, W.J., Burt, G.R. and Chesman, B.S. (2007) Feminisation of male clams *Scrobicularia plana* from estuaries in south west UK and its induction by endocrine-disrupting chemicals. *Marine Ecology Progress Series* 333: 173–84. https://doi.org/10.3354/meps333173

Langston, W.J., Pope, N.D., Davey, M., Langston, K.M., O' Hara, S.C.M., Gibbs, P.E. and Pascoe, P.L. (2015) Recovery from TBT pollution in English Channel environments: a problem solved? *Marine Pollution Bulletin* 95: 551–64. https://doi.org/10.1016/j.marpolbul.2014.12.011

Lecornu, B., Schlund, E., Basuyaux, O., Cantat, O. and Dauvin, J.C. (2016) Dynamics (from 2010–2011 to 2014) of *Sabellaria alveolata* reefs on the western coast of Cotentin (English Channel, France). *Regional Studies in Marine Science* 8: 157–69. https://doi.org/10.1016/j.rsma.2016.07.004

Lefebvre, S., Barille, L. and Clerc, M. (2000) Pacific oyster (*Crassostrea gigas*) feeding responses to a fish-farm effluent. *Aquaculture* 187: 185–98. https://doi.org/10.1016/S0044-8486(99)00390-7

Lejart, M. and Hily, C. (2011) Differential response of benthic macrofauna to the formation of novel oyster reefs (*Crassostrea gigas*, Thunberg) on soft and rocky substrate in the intertidal of the Bay of Brest, France. *Journal of Sea Research* 65: 84–93. https://doi.org/10.1016/j.seares.2010.07.004

Le Mao, P., Godet, L., Fournier, J., Desroy, N., Gentil, F., Thiébault, E. and Pourinet, L. (2019) *Atlas de la faune marine invertébrée du golfe Normano-Breton*. Roscoff: Editions de la Station biologique de Roscoff.

Lemmason, A.J., Hall-Spencer, J.M., Fletcher, S., Provstgaard-Morys, S. and Knights, A.M. (2018) Indications of future performance of native and non-native adult oysters under acidification and warming. *Marine Environmental Research* 142: 178–89. https://doi.org/10.1016/j.marenvres.2018.10.003

Lewis, J.R. (1964) *The Ecology of Rocky Shores*. London: English Universities Press.

Lusher, A.L., McHugh, M. and Thompson, R.C. (2013) Occurrence of microplastics in the gastrointestinal tract of pelagic and demersal fish from the English Channel. *Marine Pollution Bulletin* 67: 94–9. https://doi.org/10.1016/j.marpolbul.2012.11.028

Maddock, L., Boalch, G.T. and Harbour, D.S. (1981) Populations of phytoplankton in the western English Channel between 1964 and 1974. *Journal of the Marine Biological Association of the United Kingdom* 61: 565–83. https://doi.org/10.1017/S0025315400048050

MarLin (2021) The Marine Life Information Network. Available at: https://www.marlin.ac.uk (Accessed: 2 July 2021).

McKnight, W. and Chudleigh, I.J. (2015) Pacific oyster *Crassostrea gigas* control within the inter-tidal zone of the North East Kent Marine Protected Areas, UK. *Conservation Evidence* 12: 28–32.

McManus, M.C., Priscilla, L. and Coombs, S.H. (2016) Is the Russell Cycle a true cycle? Multidecadal zooplankton and climate trends in the western English Channel. *ICES Journal of Marine Science* 73: 227–38. https://doi.org/10.1093/icesjms/fsv126

Mieszkowska, N., Leaper, R., Moore, P., Kendall, M.A., Burrows, M.T., Lear, D., Poloczanska, E., Hiscock, K., Moschella, P.S., Thompson, R.C., Herbert, R.J.H., Laffoley, D., Baxter, J., Southward, A.J. and Hawkins, S.J. (2005) *Marine Biodiversity and Climate Change: Assessing and predicting the influence of climatic change using intertidal rocky shore biota*. Occasional Publication 20. Plymouth: Marine Biological Association of the United Kingdom.

Mieszkowska, N., Kendall, M.A., Hawkins, S.J., Leaper, R., Williamson, P., Hardman-Mountford, N.J. and Southward, A.J. (2006) Changes in the range of some common rocky shore species in Britain – a response to climate change? *Hydrobiologia* 555: 241–51. https://doi.org/10.1007/s10750-005-1120-6

Mieszkowska, N., Sugden, H., Firth, L.B. and Hawkins, S.J. (2014) The role of sustained observations in tracking impacts of environmental change on marine biodiversity and ecosystems. *Philosophical Transactions of the Royal Society A: Mathematical, Physical and Engineering Sciences* 372 (2025): 20130339. https://doi.org/10.1098/rsta.2013.0339

Mieszkowska, N., Burrows, M. and Sugden, H. (2020) Impacts of climate change on intertidal habitats relevant to the coastal and marine environment around the UK. *Marine Climate Change Impacts Partnership Science Review* 2020: 256–271.

Mineur, F., Cook, E.J., Minchin, D., Bohn, K., MacLeod, A. and Maggs, C.A. (2012) *Changing coasts: marine aliens and artificial structures*. In R.N. Gibson, R.J.A. Atkinson, J.D.M. Gordon and R.N. Hughes (eds) *Oceanography and Marine Biology: An Annual Review, volume 50*, pp. 189–233). Boca Raton: CRC Press Inc.

Moore, H.B. and Kitching, J.A. (1939). The biology of *Chthamalus stellatus* (Poli). *Journal of the Marine Biological Association of the United Kingdom* 23: 521–41. https://doi.org/10.1017/S0025315400014053

Moore, P., Hawkins, S.J. and Thompson, R.C. (2007) Role of biological habitat amelioration in altering the relative responses of congeneric species to climate change. *Marine Ecology Progress Series* 334: 11–19. https://doi.org/10.3354/meps334011

Moore, P., Thompson, R.C. and Hawkins, S.J. (2007) Effects of grazer identity on the probability of escapes by a canopy-forming macroalga. *Journal of Experimental Marine Biology and Ecology* 344: 170–80. https://doi.org/10.1016/j.jembe.2006.12.012

Morris, R.L., Bilkovic, D.M., Boswell, M.K., Bushek, D., Cebrian, J., Goff, J., McClenachan, G., Moody, J., Sacks, P., Shinn, J.P., Sparks, E.L., Temple, N.A., Waltres, L.J., Webb, B.M. and Swearer, S. (2019) The application of oyster reefs in shoreline protection: are we over-engineering for an ecosystem engineer? *Journal of Applied Ecology* 56: 1365–2664. https://doi.org/10.1111/1365-2664.13390

Moschella, P.S., Abbiati, M., Åberg, P., Airoldi, L., Anderson, J.M., Bacchiocchi, F., Bulleri, F., Dinesen, G.E., Frost, M., Gacia, E. and Granhag, L. (2005) Low-crested coastal defence structures as artificial habitats for marine life: using ecological criteria in design. *Coastal Engineering* 52: 1053–71. https://doi.org/10.1016/j.coastaleng.2005.09.014

Muxagata, E., Williams, J.A. and Sheader, M. (2004) Composition and temporal distribution of cirripede larvae in Southampton Water, England, with particular reference to the secondary production of *Elminius modestus*. *ICES Journal of Marine Science* 61: 585–95. https://doi.org/10.1016/j.icesjms.2004.03.015

Naylor, L.A., Venn, O., Coombes, M.A., Jackson, J. and Thompson, R.C. (2011) *Including Ecological Enhancements in the Planning, Design and Construction of Hard Coastal Structures: A process guide*. Report to the Environment Agency (PID 110461). Exeter: University of Exeter.

O'Connor, M.I., Bruno, J.F., Gaines, S.D., Halpern, B.S., Lester, S.E., Kinlan, B.P. and Weiss, J.M. (2007) Temperature control of larval dispersal and the implications for marine ecology, evolution, and conservation. *Proceedings of the National Academy of Sciences of the USA* 104: 1266–71. https://doi.org/10.1073/pnas.0603422104

O'Riordan, R.M., Myers, A.A. and Cross, T.F. (1992) Brooding in the intertidal barnacles *Chthamalus stellatus* (Poli) and *Chthamalus montagui* Southward in south-western Ireland. *Journal of Experimental Marine Biology and Ecology* 164: 135–45. https://doi.org/10.1016/0022-0981(92)90141-V

O'Riordan, R.M., Arenas, F., Arrontes, J., Castro, J.J., Cruz, T., Delany, J., Martinez, B., Fernandez, C., Hawkins, S.J., McGrath, D., Myers, A.A., Oliveros, J., Pannacciulli, F.G., Power, A-M., Relini, G., Rico, J.M. and Silva, T. (2004) Spatial variation in the recruitment of the intertidal barnacles *Chthamalus montagui* Southward and *Chthamalus stellatus* (Poli) (Crustacea: Cirripedia) over an European scale. *Journal of Experimental Marine Biology and Ecology* 304: 243–64. https://doi.org/10.1016/j.jembe.2003.12.005

Pontee, N.I. (2011) Reappraising coastal squeeze: a case study from north-west England. *Proceedings of the Institution of Civil Engineers – Maritime Engineering* 64: 127–38. https://doi.org/10.1680/maen.2011.164.3.127

Preece, R.C. (1995) *Island Britain: A quaternary perspective* (Geological Society Special Publication No. 96). London: Geological Society of London. https://doi.org/10.1144/GSL.SP.1995.096.01.01

Provan, J., Wattier, R.A. and Maggs, C.A. (2005) Phylogeographic analysis of the red seaweed *Palmaria palmata* reveals a Pleistocene marine glacial refugium in the English Channel. *Molecular Ecology* 14: 793–803. https://doi.org/10.1111/j.1365-294X.2005.02447.x

O'Shaughnessy, K.A., Hawkins, S.J., Yunnie, A.L., Hanley, M.E., Lunt, P., Thompson, R.C. and Firth, L.B. (2020) Occurrence and assemblage composition of intertidal non-native species may be influenced by shipping patterns and artificial structures. *Marine Pollution Bulletin* 154: 111082. https://doi.org/10.1016/j.marpolbul.2020.111082

Parker, I.M., Simberloff, D., Lonsdale, W.M., Goodell, K., Wonham, M., Kareiva, P.M., Williamson, M.H., Von Holle, V., Moyle, P.B., Byers, J.E. and Goldwasser, L. (1999) Impact: toward a framework for understanding the ecological effects of invaders. *Biological Invasions* 1: 3–19. https://doi.org/10.1023/A:1010034312781

Pechenik, J.A., Blanchard, M. and Rotjan, R. (2004) Susceptibility of larval *Crepidula fornicata* to predation by suspension-feeding adults. *Journal of Experimental Marine Biology and Ecology* 306: 75–94. https://doi.org/10.1016/j.jembe.2004.01.004

Perry, A., Low, P.J., Ellis, J.R. and Reynolds, J.D. (2005) Climate change and distribution shifts in marine fishes. *Science* 308: 1912–1915. https://doi.org/10.1126/science.1111322

Philippart, C.J.M., Anadón, R., Danovaro, R., Dippner, J.W., Drinkwater, K.F., Hawkins, S.J., Oguz, T., O'Sullivan, G. and Reid, P.C. (2011) Impacts of climate change on European marine ecosystems: observations, expectations and indicators. *Journal of Experimental Marine Biology and Ecology* 400: 52–69. https://doi.org/10.1016/j.jembe.2011.02.023

Pingree, R.D. and Maddock, L. (1977) Tidal eddies and coastal discharge. *Journal of the Marine Biological Association of the United Kingdom* 57: 869–75. https://doi.org/10.1017/S0025315400025224

Pingree, R.D. and Griffiths, D.K. (1978) Tidal fronts on the shelf seas around the British Isles. *Journal of Geophysical Research* 83: 4615–22. https://doi.org/10.1029/JC083iC09p04615

Pingree, R.F., Mardell, G.T. and Maddock, L. (1983) A marginal front in Lyme Bay. *Journal of the Marine Biological Association of the United Kingdom* 63: 9–15. https://doi.org/10.1017/S0025315400049754

Poggiale, J-C. and Dauvin, J-C. (2001) Long-term dynamics of three benthic *Ampelisca* (Crustacea-Amphipoda) populations from the Bay of Morlaix (western English Channel) related to their disappearance after the 'Amoco Cadiz' oil spill. *Marine Ecology Progress Series* 214: 201–9. https://doi.org/10.3354/meps214201

Poloczanska, E.S., Hawkins, S.J., Southward, A.J. and Burrows, M.T. (2008) Modelling the response of populations of competing species to climate change. *Ecology* 89: 3138–49. https://doi.org/10.1890/07-1169.1

Pontee, N.I. (2011) Reappraising coastal squeeze: a case study from north-west England. *Marine Engineering* 164: 127–38. https://doi.org/10.1680/maen.2011.164.3.127

Pringle, J.M., Byers, J.E., He, R., Pappalardo, P. and Wares, J.P. (2017) Ocean currents and competitive strength interact to cluster benthic species range boundaries in the coastal ocean. *Marine Ecology Progress Series* 567: 29–40. https://doi.org/10.3354/meps12065

Queirós, A. de M., Hiddink, J.G., Johnson, G., Cabral, H.N. and Kaiser, M.J. (2011) Context dependence of marine ecosystem engineer invasion impacts on benthic ecosystem functioning. *Biological Invasions* 13: 1059–75. https://doi.org/10.1007/s10530-011-9948-3

Rayner, N.A., Parker, D., Horton, E.B., Folland, C.K., Alexander, L.V., Rowell, D.P., Kent, E.C. and Kaplan, A. (2003) Global analyses of sea surface temperature, sea ice, and night marine air temperature since the late nineteenth century. *Journal of Geophysical Research* 108: No. D14. 4407. https://doi.org/10.1029/2002JD002670

Riquet, F., Le Cam, S., Fonteneau, E. and Viard, F. (2016) Moderate genetic drift is driven by extreme recruitment events in the invasive mollusk *Crepidula fornicata*. *Heredity* 117: 42–50. https://doi.org/10.1038/hdy.2016.24

Rognstad, R.L., Wethey, D.S. and Hilbish, T.J. (2014) Connectivity and population repatriation: limitations of climate and input into the larval pool. *Marine Ecology Progress Series* 495: 175–83. https://doi.org/10.3354/meps10590

Rognstad, R.L. and Hilbish, T.J. (2014) Temperature-induced variation in the survival of brooded embryos drives patterns of recruitment and abundance in *Semibalanus balanoides*. *Journal of Experimental Marine Biology and Ecology* 461: 357–63. https://doi.org/10.1016/j.jembe.2014.09.012

Rognstad, R.L., Wethey, D.S., Oliver, H. and Hilbish, T.J. (2018) Connectivity modelling and graph theory analysis predict recolonization in transient populations. *Journal of Marine Systems* 183: 13–22. https://doi.org/10.1016/j.jmarsys.2018.03.002

Rodriguez, L.F. (2006) Can invasive species facilitate native species? Evidence of how, when, and why these impacts occur. *Biological Invasions* 8: 927–39. https://doi.org/10.1007/s10530-005-5103-3

Roy, H., Peyton, J., Aldridge, D.C., Bantock, T., Blackburn, T.M., Britton, R., Clark, P., Cook, E. and 19 others (2014) Horizon scanning for invasive alien species with the potential to threaten biodiversity in Great Britain. *Global Change Biology* 20: 3859–71. https://doi.org/10.1111/gcb.12603

Ruesink, J.L., Feist, B.E., Harvey, C.J., Hong, J.S., Trimble, A.C. and Wisehart, L.M. (2006) Changes in productivity associated with four introduced species: ecosystem transformation of a 'pristine'

estuary. *Marine Ecology Progress Series* 311: 203–15. https://doi.org/10.3354/meps311203

Russell, F.S. (1973) A summary of the observations on the occurrence of planktonic stages of fish off Plymouth 1924–1972. *Journal of the Marine Biological Association of the United Kingdom* 53: 347–55. https://doi.org/10.1017/S0025315400022311

Russell, F.S., Southward, A.J., Boalch, G.T. and Butler, E.I. (1971) Changes in biological conditions in the English Channel off Plymouth during the last half century. *Nature* 234: 468–70. https://doi.org/10.1038/234468a0

Salomon, J.C. and Breton, M. (1993) An atlas of long-term currents in the Channel. *Oceanologica Acta* 16: 439–48.

Schreyers, L., van Emmerik, T., Biermann, L. and Le Lay, Y.F. (2021) Spotting green tides over Brittany from space: three decades of monitoring with Landsat imagery. *Remote Sensing* 13: 1408. https://doi.org/10.3390/rs13081408

Sheehan, E.V., Stevens, T.F., Gall, S.C., Cousens, S.L. and Attrill, M.J. (2013) Recovery of a temperate reef assemblage in a marine protected area following the exclusion of towed demersal fishing. *PLoS One* 8 (12): 83883. https://doi.org/10.1371/journal.pone.0083883

Sherwin, T.J. and Jonas, P.J.C. (1994) The impact of ambient stratification on marine outfall studies in British waters. *Marine Pollution Bulletin* 28: 527–33. https://doi.org/10.1016/0025-326X(94)90072-8

Simkanin, C., Power, A.M., Myers, A., McGrath, D., Southward, A.J., Mieszkowska, N., Leaper, R. and O'Riordan, R. (2005) Using historical data to detect temporal changes in the abundances of intertidal species on Irish shores. *Journal of the Marine Biological Association of the United Kingdom* 85: 1329–40. https://doi.org/10.1017/S0025315405012506

Southward, A.J. (1963) The distribution of some plankton animals in the English Channel and approaches III. Theories about long-term biological changes including fish. *Journal of the Marine Biological Association of the United Kingdom* 43: 1–29. https://doi.org/10.1017/S0025315400005208

Southward, A.J. (1967) Recent changes in abundance of intertidal barnacles in south-west England: A possible effect of climatic deterioration. *Journal of the Marine Biological Association of the United Kingdom* 47: 81–95. https://doi.org/10.1017/S0025315400033580

Southward, A.J. (1980) The western English Channel – an inconstant ecosystem? *Nature* 285: 361–6. https://doi.org/10.1038/285361a0

Southward, A.J. and Crisp, D.J. (1954a) Recent changes in the distribution of the intertidal barnacles *Chthamalus stellatus* Poli and *Balanus balanoides* L. in the British Isles. *Journal of Animal Ecology* 23: 163–77. https://doi.org/10.2307/1665

Southward, A.J. and Crisp, D.J. (1954b) The distribution of certain intertidal animals around the Irish coast. *Proceedings of the Royal Irish Academy* B 57: 1–29.

Southward, A.J. and Crisp, D.J. (1956) Fluctuations in the distribution and abundance of intertidal barnacles. *Journal of the Marine Biological Association of the United Kingdom* 35: 211–29. https://doi.org/10.1017/S0025315400009073

Southward, A.J. and Southward, E.C. (1978) Recolonization of rocky shores in Cornwall after use of toxic dispersants to clean up the *Torrey Canyon* spill. *Journal of the Fisheries Research Board of Canada* 35: 682–706. https://doi.org/10.1139/f78-120

Southward, A.J. and Southward, E.C. (1988) Disappearance of the warm-water hermit crab *Clibanarius erythropus* from south-west Britain. *Journal of the Marine Biological Association of the United Kingdom* 68: 409–12. https://doi.org/10.1017/S0025315400043307

Southward, A.J., Boalch, G.T. and Maddock, L. (1988a) Fluctuations in the herring and pilchard fisheries of Devon and Cornwall linked to change in climate since the 16th century. *Journal of the Marine Biological Association of the United Kingdom* 68: 423–45. https://doi.org/10.1017/S0025315400043320

Southward, A.J., Boalch, G.T. and Maddock, L. (1988b) Climatic change and the herring and pilchard fisheries of Devon and Cornwall. In D. Starkey (ed.) *Devon's Coastline and Coastal Waters*, pp. 33–57. Exeter: Exeter University Press.

Southward, A.J. (1991) Forty years of changes in species composition and population density of barnacles on a rocky shore near Plymouth. *Journal of the Marine Biological Association, UK* 71: 495–513. https://doi.org/10.1017/S002531540005311X

Southward, A.J., Hawkins, S.J. and Burrows, M.T. (1995) Seventy years' observations of changes in distribution and abundance of zooplankton and intertidal organisms in the western English Channel in relation to rising sea temperature. *Journal of Thermal Biology* 20: 127–55. https://doi.org/10.1016/0306-4565(94)00043-I

Southward, A.J., Langmead, O., Hardman-Mountford, N.J., Aitken, J., Boalch, G.T., Dando, P.R., Genner, M.J., Joint, I., Kendall, M.A., Halliday, N.C., Harris, R.P., Leaper, R., Mieszkowska, N., Pingree, R.D., Richardson, A.J., Sims, D.W., Smith, T., Walne, A.A. and Hawkins, S.J. (2005) Long-term oceanographic and ecological research in the western English Channel. *Advances in Marine Biology* 47: 1–105. https://doi.org/10.1016/S0065-2881(04)47001-1

Smith, J.E. (1968) Torrey Canyon Pollution and Marine Life: A Report by the Plymouth Laboratory of the Marine Biological Association of the United Kingdom. Cambridge University Press, Cambridge, UK.

Smyth, D., Mahon, A.M., Roberts, D. and Kregting, L. (2018) Settlement of *Ostrea edulis* is determined by the availability of hard substrata rather than by its nature: implications for stock recovery and restoration of the European oyster. *Aquatic Conservation: Marine and Freshwater Ecosystems* 28: 662–71. https://doi.org/10.1002/aqc.2876

Spence, S.K., Bryan, G.W., Gibbs, P.E., Masters, D., Morris, L. and Hawkins, S.J. (1990) Effects of TBT contamination on *Nucella* populations. *Functional Ecology* 4: 425–32. https://doi.org/10.2307/2389605

Steer, M., Cole, M., Thompson, R.C. and Lindeque, P.K. (2017) Microplastic ingestion in fish larvae in the western English Channel. *Environmental Pollution* 226: 250–9. https://doi.org/10.1016/j.envpol.2017.03.062

Strain, E.M., Steinberg, P.D., Vozzo, M., Johnston, E.L., Abbiati, M., Aguilera, M.A., Airoldi, L., Aguirre, J.D., Ashton, G., Bernardi, M. and Brooks, P. (2021) A global analysis of complexity–biodiversity relationships on marine artificial structures. *Global Ecology and Biogeography* 30: 140–53. https://doi.org/10.1111/geb.13202

Svensson, C-J., Jenkins, S.R., Hawkins, S.J. and Aberg, P.A. (2005) Population resistance to climate change: modelling the effects of low recruitment in open populations. *Oecologia* 142: 117–26. https://doi.org/10.1007/s00442-004-1703-3

Thomas, Y., Pouvreau, S., Alunno-Bruscia, M., Barille, L., Gohin, F., Bryere, P. and Gernez, P. (2016) Global change and climate driven invasion of the Pacific oyster (*Crassostrea gigas*) along European coasts: a bioenergetics modelling approach. *Journal of Biogeography* 43: 568–79. https://doi.org/10.1111/jbi.12665

Thomas, Y. and Bacher, C. (2018) Assessing the sensitivity of bivalve populations to global warming using an individual-based modelling approach. *Global Change Biology* 24: 4581–97. https://doi.org/10.1111/gcb.14402

Thompson, R.C., Swan, S.H., Moore, C.J., vom Saal, F.S. (2009) Our plastic age. *Philosophical Transactions of the Royal Society of London. Series B, Biological Sciences* 364: 1973–6. https://doi.org/10.1098/rstb.2009.0054

Thornton, A., Herbert, R.J.H., Stillman, R.A. and Franklin, D.J. (2019) Macroalgal mats in a eutrophic estuarine marine protected area: Implications for benthic invertebrates and wading birds. In J. Humphreys and R. Clark (eds) *Marine Protected Areas: Science, Policy and Management*, pp. 703–27. London: Elsevier. https://doi.org/10.1016/B978-0-08-102698-4.00036-8

Thouzeau, G. (1989) Déterminisme du prérecrutement de *Pecten maximus* L. en baie de Saint-Brieuc. PhD thesis, Océanographie Biologique. Université de Bretagne occidentale.

Thouzeau, G., Chauvaud, L., Grall, J. and Guérin, L. (2000) Do biotic interactions control pre-recrutement and growth of *Pecten maximus* (L.) in the Bay of Brest (France). Comptes Rendus de l'Academie des Sciences Serie 3 Sciences de la Vie.

Wethey, D.S. and Woodin, S.A. (2008) Ecological hindcasting of biogeographic responses to climate change in the European intertidal zone. *Hydrobiologia* 606: 139–51. https://doi.org/10.1007/s10750-008-9338-8

Wethey, D.S., Woodin, S.A., Hilbish, T.J., Jones, S.J., Lima, F.P. and Brannock, P.M. (2011) Response of intertidal populations to climate: effects of extreme events versus long term change. *Journal of Experimental Marine Biology and Ecology* 400: 132–44. https://doi.org/10.1016/j.jembe.2011.02.008

Woodin, S.A., Wethey, D.S. and Dubois, S.F. (2014) Population structure and spread of the polychaete *Diopatra biscayensis* along the French Atlantic coast: human-assisted transport by-passes larval dispersal. *Marine Environmental Research* 102: 110–21. https://doi.org/10.1016/j.marenvres.2014.05.006

Xue, D.X., Graves, J., Carranza, A., Sylantyev, S., Snigirov, S., Zhang, T. and Liu, J.X. (2018) Successful worldwide invasion of the veined rapa whelk, *Rapana venosa* despite a dramatic genetic bottleneck. *Biological Invasions* 20: 3297–314. https://doi.org/10.1007/s10530-018-1774-4

CHAPTER 13

Predicting the Effect of Environmental Change on Non-breeding Shorebirds with Individual-based Modelling

JOHN D. GOSS-CUSTARD and RICHARD A. STILLMAN

Abstract

For 50 years, there has been a need to investigate whether human activities in the intertidal zone – ranging from dog-walking to barrage construction and rise and fall in sea-level relative to the adjacent coast – disadvantage migratory shorebirds (mainly waders Charadrii and wildfowl Anatidae) that depend for much of their annual cycle on intertidal flats. Finding the answer requires testing the hypothesis that an activity reduces the birds' fitness by reducing their chances of surviving during the non-breeding season or while on migration and/or by lowering their chances of breeding successfully. To estimate the effect on bird numbers of any resulting change in the per capita rates of survival and reproduction requires quantitative predictions; that is, mathematical modelling. Often predictions will be required for conditions that lie outside the present-day environmental range when the current demographic functions may not apply. Individual-based models (IBM) can make predictions for novel conditions because: (a) they include aspects of species' natural history that are likely to affect individual fitness, and (b) model birds use the same fitness-maximising decision rules in responding to environmental change as the birds use in nature. This chapter reviews these two aspects of IBMs as applied to wading birds, the tests of model predictions and the many scenarios to which they have been applied.

Keywords: individual-based model, pure and applied research, model validation, shorebirds, habitat change, Oystercatcher

Correspondence: johngc66@gmail.com

Introduction

From the inception of the Estuarine and Coastal Sciences Association, there has been a need to investigate whether human activities in the intertidal zone disadvantage migratory shorebirds (mainly waders Charadrii and wildfowl Anatidae) that depend for much of their annual cycle on intertidal flats. A minimal list of activities includes reclamation, shellfishing, pollution, sea-level and temperature rise, hunting, water sports, angling, bait collection, birdwatching, dog-walking, barrage construction, invasive species and military training (e.g. Lambeck, Goss-Custard and Triplet 1996).

John D. Goss-Custard and Richard A. Stillman, 'Predicting the Effect of Environmental Change on Non-breeding Shorebirds with Individual-based Modelling' in: *Challenges in Estuarine and Coastal Science*. Pelagic Publishing (2022). © John D. Goss-Custard and Richard A. Stillman. DOI: 10.53061/INUA1114

Over the same period in the UK, society's priorities have changed. This is well illustrated by the conflict between shellfishing and conservation, notably in the Burry Inlet, South Wales (Davidson 1968). The severe winter of 1962–3 killed most cockles, resulting in a massive spatfall followed by a huge increase in mature cockles of fishable size one year later. Subsequently, stocks and harvests declined dramatically. Because each of the many thousands of Eurasian Oystercatchers *Haematopus ostralegus* in the inlet consumed up to 500 mature cockles per day, they were viewed as serious competitors of the fishery, a pest of shellfisheries (Davidson 1968): at that time, producing food and creating jobs trumped conservation. Thousands of birds were then shot, but with no obvious benefit to cockle stocks or shellfishery (Prater 1981).

It is most unlikely that such action would even be contemplated these days. The research paradigm on wading birds Charadrii described here – individual-based (or agent-based) modelling (IBM) – was instigated in the 1970s to find a way of resolving such conflicts by providing solutions that take the interests of all parties into maximum account, the most reliable basis of long-term sustainability. The approach was individual based because (a) it is individual animals that respond to environmental change, not populations (Lomnicki 1978, 1980, 1982), and (b) model birds use evolved fitness-maximising decision rules when responding to environmental change and so respond in ways that real birds are likely to do: accordingly, the paradigm is more likely to maintain its predictive power to new environments than the empirical relationships on which more traditional methods, such as demographic and habitat-association models, are based (Goss-Custard 1985; Goss-Custard et al. 1995; Grimm and Railsback 2005; Evans 2012; Stillman et al. 2015).

Although currently the paradigm has been applied in waterbirds in 53 case studies of 28 species at 32 sites in 9 countries (Brown and Stillman in press), we focus here on waders, and particularly on the 'Oystercatcher–shellfish' problem – where it is now possible not only to quantify the harvest that safeguards the birds but also to devise management practices that increase both the birds' food supply and the harvest. In the absence of such a capability, impact assessments had often been based on the precautionary principle, which led to activities being controlled or banned even if there was little evidence that they had detrimental effect on birds (Goss-Custard 2016). By the same token, damaging activities were also allowed to continue. For example, in the Wash (Atkinson et al. 2003) and Wadden Sea (van Roomen et al. 2005), high mortality rates of mollusc-eating birds occurred as a result of shellfish overfishing because the birds' requirements had been greatly underestimated. Using IBMs, the findings of Goss-Custard et al. (2004) led to a change in policy that greatly increased the shellfish stocks reserved for the birds after harvesting.

Natural history of non-breeding shorebirds

To maximise 'fitness' during the non-breeding season, shorebirds must survive until the breeding season in good enough condition to migrate and breed successfully. A species' natural history sets the context in which the birds have to act to achieve this goal. Research attention focused early on the probable importance of the birds' invertebrate food supply during the non-breeding season, as many human activities seemed likely to reduce the birds' ability to obtain their food requirements. Accordingly, research focused on two questions: do shorebirds regularly have difficulty in obtaining their food, and would any difficulty be increased by human activities (Prater 1972; Evans 1976; Goss-Custard 1977; Evans et al. 1979)?

Whether waders can collect their energy requirements during the intertidal exposure period when the flats are accessible depends on the birds' intake rate, usually measured as kJ/s. Intake rate is influenced by numerical prey density – which determines the rate of encounter with prey – and by prey size – which determines the energy value of each prey (Zwarts et al. 1996a; Goss-Custard et al. 2006c). Intake rate is also affected by the prey's behaviour. Bar-tailed Godwits *Limosa lapponica*, for instance, can only catch Lugworms *Arenicola marina* when the worms are within reach of the bill at the sediment surface. Apart from temperature, the frequency at which this occurs depends on the wetness of the sediment and, for example, the proximity of the wetted perimeter at the tide edge (Smith 1975). Landmark research in the Dutch Wadden on spatial and temporal variations in (a) the interplay between the behaviour of waders and that of their prey, and (b) the accessibility, detectability, ingestibility, digestibility and profitability of the prey has played a pivotal role in identifying the many, and subtle, factors and processes that determine the intake rate of waders and their ability to acquire their food requirements in the time available (Esselink and Zwarts 1989; Zwarts and Esselink 1989; Wanink and Zwarts 1985, 1996; Zwarts and Wanink 1989, 1991, 1993; Zwarts 1991; Zwarts and Blomert 1992; Zwarts, Blomert and Wanink 1992; Zwarts et al. 1996a, 1996b, Zwarts, Wanink and Ens 1996).

Considerable evidence confirms that mortality from starvation does occur in many wader species, at least in the northern hemisphere where most research has been carried out until recently. The mortality rate can be especially high during severely cold and/or windy weather when the birds' energy requirements are high and reduced prey activity decreases their catchability (Goss-Custard 1969; Davidson 1981; Davidson and Evans 1982). In Oystercatchers, mortality can be very high in cold winters (Swennen 1984; Camphuysen et al. 1996; Hulscher, Exo and Clark 1996; Zwarts et al. 1996b; Duriez et al. 2009, 2012; Schwemmer et al. 2014) and when shellfish are scarce (Camphuysen et al. 1996; Atkinson et al. 2003; Duriez et al. 2012), and especially high (>30% in adults) when these conditions coincide (Duriez et al. 2012).

Severe winters and reduced shellfish stocks do not necessarily lead to high mortality in Oystercatchers, however. Not all severe winters kill large numbers (Camphuysen et al. 1996; Duriez et al. 2009, 2012). Nor do large decreases in the food supply necessarily reduce Oystercatcher survival. Engineering works in the Oosterschelde, for instance, reduced the intertidal area by one-third with no increase in Oystercatcher mortality in mild winters (Duriez et al. 2009). In the Wash and Wadden Sea, Oystercatchers maintained high survival when cockles were scarce but only when mussels were abundant (Atkinson et al. 2003; Ens, Smaal and de Vlas 2004). Clearly, human activities can reduce shorebird fitness, but not always.

Estimating the quantitative contribution human activities might make to shorebird mortality is even more challenging when the immediate cause of death is not starvation but enemies. Many non-breeding waders are killed by birds of prey, and the hungry ones seem to be most at risk because they have to feed in dangerous places. While replete birds are resting at high tide roosts at the beginning and end of the exposure period, hungry waders often forage on the upper shore and/or in fields over high tide (Goss-Custard 1969; Heppleston 1971; Goss-Custard et al. 1977; Smart and Gill 2003). Foraging upshore is dangerous because the birds are close to shoreline vegetation from where many birds of prey launch their attacks (Page and Whitacre 1975; Whitfield 1985, 2003; Cresswell 1994; Cresswell and Whitfield 1994; Dierschke, 1998; Cresswell and Quinn 2004; Dekker and Ydenburg 2004; Mindermann, Lind and Cresswell 2006; van den Hout et al. 2014); indeed, shorebirds may avoid upshore areas at night because of the greater predation risk (Piersma et al. 2006). There is some evidence also that hungry Oystercatchers are most at risk of parasitic infection because the parasites may be more abundant upshore than downshore (Goater 1993; Goater, Goss-Custard and Kennedy 1995) and in terrestrial habitats than in the intertidal zone (Le Drean-Quenec'hdu and Goss-Custard 1999). So, human activities that cause shorebirds to extend their foraging in both time and space may increase mortality, even if the immediate cause of death is not starvation.

A vital aspect of wader natural history to take into account when predicting the impact of human activity on fitness is intra-specific competition for food. Competition can be direct and obvious, as when individuals fight over a food item or feeding site, or indirect, as when the anti-predator responses of the prey reduces their accessibility to waders (Goss-Custard 1970, 1976, 1977, 1980; Goss-Custard et al. 1982a; Ens and Goss-Custard 1984; Selman and Goss-Custard 1988; Yates, Stillman and Goss-Custard 2000; Rutten et al. 2010a, b; Folmer, Olff and Piersma 2011; Bijleveld, Folmer and Piersma 2012; Dokter et al. 2017). The direct removal of space – as in reclamation – is likely to increase the density of birds on the remaining feeding grounds. Reducing general food abundance – as in harvesting – is likely to reduce the area with prey at densities and of a size that the birds require to feed at the required rate. Both these effects will intensify interference competition, although usually only after bird density has reached a critical 'interference' threshold (Stillman et al. 1996; Stillman, Goss-Custard and Caldow 1997) and/or when prey are scarce (Triplet, Stillman and Goss-Custard 1999). Accordingly, mortality caused directly or indirectly by food shortage is likely to be density-dependent and there is, in fact, increasing evidence from the northern hemisphere that mortality during the non-breeding season in several wader species can be density-dependent (Durell et al. 2001, 2003; Goss-Custard et al. 2001; Whitfield 2003; Ryan, Green and Dodd 2016).

Both intra-specific and inter-specific competition for food is common in Oystercatchers eating shellfish. This bird's unusually muscular and heavy build evolved to exploit heavily-armoured but highly profitable, large shellfish (Hulscher 1996). Even so, it can take an Oystercatcher several minutes to break into a large one (Zwarts et al. 1996d). Other Oystercatchers and kleptoparasites, such as Carrion Crows *Corvus corone* and Herring Gulls *Larus argentatus*, often attack an Oystercatcher that may have spent up to several minutes breaking into a shellfish before stealing the now accessible flesh: kleptoparasites alone can steal up to one-third of those opened by Oystercatchers (Zwarts and Drent 1981; Wood, Stillman and Goss-Custard 2015).

This outline of the natural history of northern hemisphere waders during the non-breeding season reveals the critical periods when human activities are most likely to intensify any difficulties the birds may have and the means by which they could do so. The critical period for Oystercatchers is December to March, when most mortality occurs and the birds are building their body reserves to fuel spring migration (Goss-Custard et al. 1996; Zwarts et al. 1996b). Daily energy requirements are high at that time; those of the average adult Oystercatcher in the eastern Dutch Wadden Sea, for example, increase from 718 kJ in September to a peak of 865 kJ in January (Goss-Custard and Stillman 2020). At the same time, the flesh content of shellfish is at its lowest while the accessibility and food value of alternative intertidal prey upshore, such as the bivalve *Macoma balthica*, is low (Zwarts and Wanink 1993). For Oystercatchers, the months December to March are the appropriate period for evaluating the potential impacts of human activities on the birds.

Individual variation

It is individual animals that respond to a changed environment, not populations (Lomnicki 1978, 1980, 1982). Because individuals differ, many survive harsh conditions even though many others die. To predict by modelling an increased mortality caused by a human activity, individual variation must be taken into account (Goss-Custard 1985). Individuals undoubtedly vary in many aspects of physiology, morphology, susceptibility to enemies, but at present little is known about individual variations except for foraging, and particularly in Oystercatchers (Sutherland et al. 1996).

To date, two components of individual variation in foraging have been included in IBMs. Each line in Fig. 13.1 represents an 'interference function' that relates the intake rate of an individual bird to the density of competitors around it. As the intercept measures intake rate in the absence of interference competition, it reflects the bird's basic foraging ability, or foraging efficiency (FE). At competitor densities above the interference threshold, intake rate declines if the bird is affected by competition. The slope measures its susceptibility to interference (STI); that is, by how much intake rate is decreased by every unit increase in competitor density. FE and STI vary independently across individuals and depend on a bird's age, feeding technique and, in the case of STI alone, on its percentage success in aggressive encounters over food or 'social dominance' (Ens and Goss-Custard 1984; Goss-Custard and Durell 1987a, b, c, 1988; Rutten et al. 2010a, b). On the Exe estuary, a bird's FE and STI affect: (a) on which mussel bed it forages (Goss-Custard et al. 1982b); (b) whether it must forage for upshore mussels at the start and end of the exposure period (Goss-Custard et al. 2001); and (c) how frequently it feeds on non-shellfish prey when the mussel beds are inaccessible under water (Caldow et al. 1999). Individuals with a low FE and/or a high STI were more

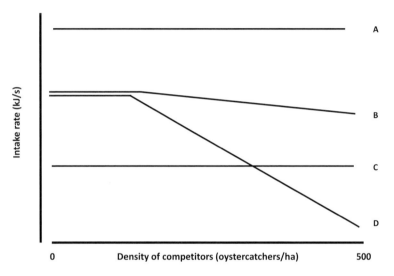

Figure 13.1 How the competitive ability of individual mussel-eating Oystercatchers was measured and is represented in the IBM. The interference functions of four individuals of differing foraging efficiency (FE) and susceptibility (STI) are shown. **Bird A:** In the absence of competitors, its intake rate – the interference-free intake rate – is high, so it has a high FE. As the density of competitors around it increases, its intake rate remains high because it is a dominant bird and wins 100% of the aggressive interactions with other Oystercatchers. **Bird B:** This is a bird of average FE but its intake rate decreases as competitor density rises above the disturbance threshold of c. 100 Oystercatchers/ha. Because it has high dominance (e.g. 75% wins), the slope of the decline is shallow, so this bird has a low STI. **Bird C:** This bird has a very low FE but is a dominant bird, so its intake rate is unaffected by competitor density, and its STI is therefore very low. **Bird D:** This bird has an average FE but, because of its low dominance (25% wins), its intake rate decreases rapidly as competitor density increases so it has a high STI.

The coefficients describing STI and FE were derived from fieldwork in which individually identified birds were observed regularly throughout one or two winters on a mussel bed marked out in 25 × 25 m squares, within which the number of Oystercatchers, and thus competitor density, was recorded at the start and end of each five-minute observation period during which the mussels consumed, their size and the social interactions of the subject bird were recorded, as detailed in Ens and Goss-Custard (1984) and Goss-Custard and Durell (1988).

In the model, the STI depends on a bird's dominance in a strictly linear hierarchy with values ascribed at random to each model individual over the range 0–100% (Stillman et al. 2000). The FE is also ascribed at random and independently of an individual's STI from a normal distribution with a SD of 15% of the mean. A discussion of most appropriate statistical model for representing interference is in van der Meer and Ens (1997).

likely to feed for longer, including in fields over high water, and so seem likely to be most at risk of starvation and from enemies when feeding conditions deteriorate.

Individual-based models

Shorebird IBMs are based on two theoretical frameworks, themselves derived from evolutionary thinking, which are thought to provide a reliable basis for predicting the choices made by individual shorebirds in changed feeding conditions resulting from human activity (Goss-Custard and Sutherland 1997). Game theory provides the conceptual framework for predicting how model birds would respond to changes in the spatial distribution of the food supply and competitor numbers (Sutherland and Parker 1985). Foraging theory (formerly optimal foraging theory) provides the decision rules by which individuals select prey species and size classes, the most widespread one used being the maximise intake rate rule. Based on field and laboratory research on Oystercatchers (Wanink and Zwarts 1985, 1996; Cayford and Goss-Custard 1990; Ens, Smaal and de Vlas 1996a, b; Meire 1996a; Johnstone and Norris 2000), some Oystercatcher–shellfish IBMs use foraging theoretic algorithms to predict the sizes of shellfish taken by Oystercatchers and their intake rate when harvesting changes the shellfish size distribution (Stillman et al. 2001).

Using a standard format for reporting IBMs (Grim et al. 2006), Stillman (2008) describes the IBM 'MORPH' that tracks the diet, foraging location, and body condition of each individual bird and whether it starves before the end of the non-breeding season. The food supply is in discrete patches. Patches can vary in the prey species, numerical density, size and energy content of the food items they contain. Patches vary in their height on the shore and therefore the time for which they are exposed and so accessible to shorebirds through spring and neap tidal cycles. Each bird must consume per 24 hours enough food to meet its energy demands that increase as ambient temperature decreases below 10 °C (Kersten and Piersma 1987) and – added recently – as body weight increases (Zwarts et al. 1996b). An individual forages in the locations, times of day and stages of the tidal cycle where its intake rate is highest. Although all individuals employ the same optimisation principle (intake rate maximisation) their choices differ as each individual's unique combination of FE and STI – its competitive ability – is different. The model is game theoretic, as each animal reacts to the decisions made by competitors in deciding when, where and on what to feed. When daily energy consumption exceeds daily expenditure, individuals accumulate energy reserves or maintain them if a maximum level has already been reached. When daily requirements exceed daily consumption, individuals draw on their reserves. If these fall to zero, it starves.

Any overwinter reduction in individual prey mass is included and their numerical density is reduced daily through depletion by the birds and by other agents, such as storms. In Oystercatcher–shellfish models, the harvest is deducted daily from the shellfish stocks. In some Oystercatcher–shellfish models, the gut processing rate of ingested flesh is represented so that model birds, like real ones, must feed over two tidal cycles if all their food is to come from shellfish (Kersten and Visser 1996; Zwarts et al. 1996d; Sitters 2000). In all shorebird IBMs, birds disturbed by people or raptors spend time and energy relocating to an undisturbed area (West et al. 2002; Goss-Custard et al. 2006b). If a feeding patch is removed, either permanently by habitat loss or temporarily by disturbance, the birds that previously fed there choose another patch. They do this using the normal rate-maximising and game theoretic decision rules. Because wintering shorebirds seem frequently to use this decision framework to decide where, when, and on what to feed (Goss-Custard 1985; Goss-Custard and Sutherland 1997), model birds probably react to changes in their feeding environment as birds in nature would respond. This is the fundamental reason why IBMs are believed to provide a sound basis for predicting to the novel environments resulting from, for example, habitat loss or shellfishing.

Calibration and validation

Although parameterised to the best of current knowledge, it is sometimes necessary to calibrate the IBM so that starvation rate predictions more closely match field estimates. Re-calibration is done by varying a parameter that probably remains constant in all scenarios but currently is not well estimated, such as night-time FE. An Oystercatcher–shellfish IBM for the Wash, east England, for example, was calibrated by varying the (unmeasured) intake rate of birds foraging upshore of the main cockle and mussel beds (Fig. 13.2). Encouragingly, the calibration coefficient exactly matched the mean of 15 field estimates made in four other systems over a period of 25 years (Fig. 13.2). This example also illustrates how foraging upshore of the main feeding areas when these are covered by the tide can be very important to shorebird survival. Even though upshore intake rates in Oystercatchers are small compared with the typical rates of >2 mg AFDM/s (ash-free dry mass) on the main downshore shellfish beds (Zwarts et al. 1996a), upshore foraging increases the time available for feeding per 24 hours (Goss-Custard and Stillman 2020).

MORPH has been applied to many issues and shorebird species and its two most important predictions have been regularly tested (Brown and Stillman in press). As measuring body condition just before birds depart on migration is difficult, the main predictions tested are the proportion of birds starving over the non-breeding season and the hours spent foraging per day by the average bird at a specified stage of the non-breeding season, this being a proxy measure of the starvation rate when this cannot be measured directly. The agreement so far between prediction and observation is encouraging (Fig. 13.3). Particularly encouraging are those cases in which MORPH predicted the mortality rate in

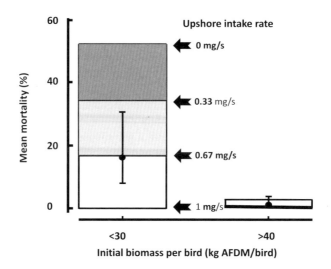

Figure 13.2 The mortality rate of shellfish-eating Oystercatchers on the Wash as predicted by the model MORPH in winters when shellfish are abundant (>40 kg AFDM) per Oystercatcher) and scarce (<30 kg AFDM per Oystercatcher) in relation to the assumed intake rate of the birds on the upshore flats. When shellfish are abundant, the upshore intake rate has little effect on the predicted mortality rate, which is always very small because food is so abundant in the main, downshore feeding areas. In winters of shellfish scarcity, and with no upshore foraging available, the predicted rate of starvation is >50%. The predicted rate decreases as the assumed intake rate on the upshore flats increases and is 0% when the birds obtain 1 mg AFDM/s by feeding on the upshore flats on the receding and advancing tides (Stillman et al. 2003). The predicted mortality matched the observed mortality with an upshore intake rate of 0.67 mg AFDM/s: this is identical to the mean of 15 field estimates (0.67, S.E. = ±0.08) obtained from four locations over 25 years (Goss-Custard, unpublished information).

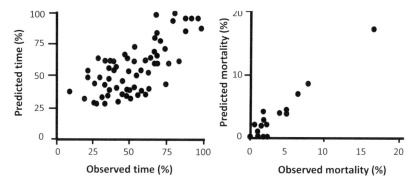

Figure 13.3 Field tests of the predictions of the model MORPH for the time spent feeding during a tidal exposure period (left), and for mortality from starvation within the non-breeding period (right). Species included for the time spent feeding panel were: Eurasian Oystercatcher *Haematopus ostralegus*, Pied Oystercatcher *Haematopus longirostris*, Little Stint *Calidris minuta*, Sanderling *C. alba*, Dunlin *C. alpina*, Curlew *Numenius arquata*, Ringed Plover *Charadrius hiaticula*, Turnstone *Arenaria interpres*, Knot *Calidris canutus*, Grey Plover *Pluvialis squatarola*, Bar-tailed Godwit, Black-tailed Godwit *L. limosa*, Avocet *Recurvirostra avosetta*, Shelduck *Tadorna tadorna*, Mute Swan *Cygnus olor*, Pacific Black Brant *Branta bernicla nigricans* and Pink-footed Goose *Anser brachyrhynchus*. Species included for the mortality rate panel were: Redshank, Eurasian Oystercatcher, Common Scoter *Melanitta nigra*, Mute Swan *Cygnus olor*, Brent Goose *B.b. bernicla*, Eider *Somateria mollissima*, Pacific Black Brant and Pink-footed Goose. Tests were carried out in several countries within Europe and America.

novel circumstances different from those for which the model had been parameterised, this being an essential requirement if a model's predictive reliability is to be accepted: (a) increased Oystercatcher density on the Exe estuary (Stillman et al. 2000); (b) habitat loss in Redshank *Tringa totanus* of the Severn estuary (Goss-Custard et al. 2006a), and (c) the decreasing mortality rate relative to that in adults in sub-adult Oystercatchers during severe weather in the Dutch Wadden Sea (Goss-Custard and Stillman 2020).

Applications

To date, MORPH has been applied on 53 occasions involving 28 species in 32 sites in 9 countries (Brown and Stillman in press). Here we focus on its application to the Oystercatcher–shellfish issue. During the 1990s, it was often assumed that Oystercatchers and other shellfish-dependent birds only required 70% of their aggregate consumption to remain after shellfish harvesting as the remainder could be provided by other prey, such as *M. balthica*. Their required ration, or ecological requirement (ER), was thus 0.7 of their physiological requirement (PR). However, foraging is a rate process and the rate of consumption must at least equal the rate of expenditure. As consumption rate is affected by many processes, the notion of 'ration' is inappropriate (Goss-Custard et al. 2002). Shellfish may be abundant but many of them may be inaccessible (Meire 1996a) or too dangerous (Norris 1999; Johnstone and Norris 2000) or time consuming (Zwarts et al. 1996a) for Oystercatchers. Consumption rate is also affected by the decrease in the flesh content of shellfish over the non-breeding season and by interference competition between birds. Consequently, the quantity of shellfish that must remain after harvesting to maintain the birds' normally low mortality rate can be several times greater than the quantity they actually consume: that is, the ratio ER/PR is >1. Using the IBM MORPH as illustrated in Fig. 13.4, this ratio – the ecological multiplier (EM) – can be 6–10 in mussel eaters, though less in cockle eaters (Goss-Custard et al. 2004; Stillman, Wood and Goss-Custard 2016). Accordingly, shellfishing authorities in the UK and the Netherlands nowadays require

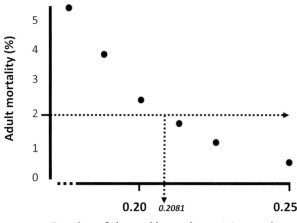

Figure 13.4 The stock of shellfish that should remain after harvesting if Oystercatchers are to maintain their usually low mortality rate over the non-breeding season is calculated using the IBM MORPH. In these simulations, the initial stock present on the first day of the non-breeding season is multiplied by a range of fractions to identify the fraction that coincides with a mortality rate of 2%, the target rate for adult mortality in this case. The solid circles show the mortality rate predicted by simulations with MORPH against the fraction of the initial stock assumed to be present, each data point being based on ten simulations. The intersection of the horizontal and vertical lines shows how the fraction of the initial stock that gives a mortality rate of 2% is derived from the quadratic equation fitted to the data points.

An example of how the simulations are used: the question is how much of the 1,000 tonnes AFDM present at the start of the non-breeding season could be harvested through autumn and winter without causing more than 2% of the 10,000 Oystercatchers to starve. The answer is: 208 tonnes (0.2081 × 1,000). This is the birds' ecological requirement (ER). The average Oystercatcher consumes over the non-breeding season a total of 10 kg AFDM. The physiological requirement (PR) of the whole overwintering population is therefore 100,000 kg AFDM (10,000 × 10), or 100 tonnes. The precautionary assumption is that the birds should obtain all their energy requirements from shellfish. The simulations show that, if the normal adult mortality rate of 2% is to be maintained, the stock of shellfish remaining after harvesting needs to be 2.08 (208/100) times the amount they will actually consume. This ratio ER/PR is called the ecological multiplier (EM).

much larger amounts of shellfish to be left for Oystercatchers after harvesting than was often previously the case. MORPH also shows how shellfishing can be managed to simultaneously increase the harvest and to improve the feeding conditions for Oystercatchers (Caldow et al. 2004; Goss-Custard, Bowgen and Stillman 2019). It can also show when an invasive non-native shellfish is or is not a threat to Oystercatchers (Caldow et al. 2007; Herbert et al. 2018) and whether a proposed measure to mitigate for a substantial change in the foraging environment of Oystercatchers is likely to be effective (Durell et al. 2005).

Shorebird IBMs are quite complex, which can make them difficult for non-specialists to use, but this difficulty can be circumvented (Stillman, Wood and Goss-Custard 2016). IBMs can produce simple rules of thumb for decision-makers that can easily be applied to their particular circumstances. Most progress on this approach has been made on the shorebird–shellfish issue. The values of the EM being obtained from IBMs for an increasing number of systems, the very simple Bird Food Model, can be used by non-specialists to calculate how much shellfish should be left after harvesting to ensure that the fitness, and thus numbers, of shorebirds that also depend on the shellfish are protected (Stillman, Wood and Goss-Custard 2016).

Future developments

As the natural histories of species define the strategies individuals use to maximise fitness, and to place the assumptions of shorebird IBMs on a sounder basis, several aspects of shorebirds' natural history require further investigation: (a) differences in foraging and competitive behaviour between night and day (Turpie and Hockey 1993; Piersma et al. 2006; Kuwae 2010; Santos et al. 2010; Sitters 2000; Dwyer et al. 2013); (b) reduction in FE and dominance in birds moving from their established home range to a new area (Verhulst et al. 2004; Rutten et al. 2010a, b; Duriez et al. 2012); (c) prevalence and fitness costs of parasitism, and by how much infestation varies between individual hosts (Borgsteede, Broek and Swennen 1988; Goater, Goss-Custard and Kennedy 1995); (d) risk of being predated in relation to birds' nutritional status (Page and Whitacre 1975; Cresswell 1994; Cresswell and Whitfield 1994; Dierschke, 1998; Whitfield 2003; Cresswell and Quinn 2004; Dekker and Ydenburg 2004; Mindermann, Lind and Cresswell 2006; Pomeroy et al. 2008; van den Hout et al. 2013, 2014); (e) foraging efficiencies of individuals with subtly-different foraging specialisms (van de Pol et al. 2009b) and whether any of these individualisms co-correlate; (f) magnitude and fitness consequences of individual variation in physiology and morphology; (g) magnitude and fitness consequences of individual differences in the response to disturbance (van der Kolk et al. 2021); (h) the food value of that fraction of the prey that are accessible to shorebirds (Zwarts and Wanink 1991) and, in Oystercatchers, (i) the trade-off between the profitability of large shellfish and the increased risk of bill damage (Rutten et al. 2006).

The structure of shorebird IBMs vary according to species natural history, to whether parameter estimates are available and to the issue being explored. For example, the time available for foraging was reduced during tidal surge barrier construction in the Netherlands and the birds' limited gut processing rate (Kersten and Visser 1996; Meire 1996b; Zwarts et al 1996d) could prevent birds from obtaining all their requirements in the limited time available. The gut processing rate of Oystercatchers therefore needed to be included in the model WEBTICS of that system (Rappoldt et al. 2003a, b; Rappoldt, Kersten and Ens 2006).

The timescale over which shorebird IBMs operate is likely to be extended. While current shorebird–shellfish IBMs provide estimates as to how much shellfish should remain after harvesting to maintain the current bird fitness of the current population of birds, the effect of harvest size on future shellfish abundance is one key long-term issue that has yet to be resolved (e.g. Stillman et al. 2001; Goss-Custard and Stillman 2019). The second is that shorebirds in current IBMs make decisions based on maximising current intake rate within a single non-breeding season whereas real birds must take longer-term fitness consequences into account (Durell 2007; van de Pol et al. 2009a). Incorporating such life-history decisions into IBMs is a major challenge that could have far-reaching effects on the way these models develop.

The numbers of shorebirds spending the non-breeding season in a particular coastal site will depend on many factors operating outside the site itself: in other non-breeding areas, on migration stop-over sites and on the breeding grounds. The breeding success of inland-nesting waders is widely affected by predation and changes in agricultural practice (Teunissen, Schekkerman and Willems 2005; Klok, Roodbergen and Hemerik 2009; Roodbergen et al. 2011). Whereas the annual survival of adult and sub-adult Oystercatchers breeding inland in Europe did not change over the four decades prior to 2006, the nesting success declined strongly across all meadow-breeding birds from approximately 40% to around 20% (Klok et al. 2009; Roodbergen et al. 2011). This decline in breeding success coincided in all species with about a 40% increase in the rate of predation of nests, the rate being particularly high in Oystercatchers and Curlews *Numenius arquata* (Roodbergen et al. 2011). Accordingly, a third long-term issue is how to incorporate the predictions of IBMs for the non-breeding season into year-round, multi-year dynamical population models of shorebirds (e.g. Pettifor et al. 2000; van de Pol et al. 2010), including any carry-over effects between seasons (Hotker 2002; Norris 2005; Calvert, Walde and Taylor 2009; Harrison et al. 2010; Duriez et al. 2012; Rakhimberdiev et al. 2015). It seems likely that, for this purpose, IBMs will be used to predict how a human activity changes the coefficients of demographic functions during the non-breeding season, such as the intercept and the slope of the density-dependent function between mortality rate and population size (Goss-Custard 2017).

Concluding remarks

A modelling approach was important so that predictions could be quantitative. An example of a useful prediction might be: 'If a particular proposal was implemented, the number of shorebirds would be reduced by 10% but if the proposed mitigation measure was implemented, shorebird numbers would remain at current levels.' If this could be done in such a way that the decision-makers had confidence in the predictions, it would save everyone a lot of time and expense, help conservation authorities to target their limited funds and also reduce the risk of vetoing worthwhile projects for no good reason and so risk losing public support for the very good, and popular, cause of shorebird conservation (Goss-Custard 2016, 2017). To be able to provide such quantitative predictions was why the shorebird IBM research programme was started on the Exe estuary in 1976, not long after the Estuarine and Coastal Sciences Association itself was established. The research focused initially on Oystercatchers eating mussels as this was technically a suitable system on which to develop and test the IBM approach, and, of course, because of the then recent shorebird–shellfishing controversy in the Burry Inlet.

The applied issues for which shorebird IBMs were developed also provided a great scientific opportunity. Experiments to test ecological predictions are often impractical, especially at large spatial scales. Comparing model predictions against the outcomes of human-induced environmental change provides an opportunity to test our quantitative understanding as well as our ability to make reliable predictions. Quantitative predictions provide better tests of our understanding than do non-quantitative ones expressed in words, of course, because there is a far greater chance that they will not be matched by outcomes in the real world: the close correspondence between the predicted (3.65%) and observed (3.17%) increase in the mortality rate of redshank displaced by barrage building in the Severn estuary is an example from shorebirds (Goss-Custard et al. 2006a). Accordingly, the traditional divide between 'pure' and 'applied' science is inappropriate. Successful tests of prediction are vital if our forecasts are to be respected by all parties involved in a dispute. After all, fair treatment to all is the essential basis for long-term sustainability (Redpath et al. 2013).

Acknowledgements

The senior author acknowledges the role Professor A. Lomnicki played in the development of the work described in this chapter, as his pioneering thinking on the role of individual differences in the regulation of animal populations was a fundamental influence on the development of shorebird IBMs.

References

Atkinson, P.W., Clark, N.A., Bell, M.C., Dare, P.J., Clark, J.A. and Ireland, P.L. (2003) Changes in commercially fished shellfish stocks and shorebird populations in the Wash, England. *Biological Conservation* 114: 127–41.

Bijleveld, A.I., Folmer, E.O. and Piersma, T. (2012) Experimental evidence for cryptic interference among socially foraging shorebirds. *Behavioural Ecology* 23: 806–14. https://doi.10.1093/beheco/ars034

Borgsteede, F.H.M., Broek, E. van den and Swennen, C. (1988) Helminth parasites of the digestive tract of the oystercatcher, *Haematopus ostralegus*, in the Wadden Sea, The Netherlands. *Netherlands Journal of Sea Research* 22: 171–4.

Brown, S. and Stillman, R.A. (in press) Evidence-based conservation in a changing world: lessons from waterbird individual-based models. *Ecosphere*. https://doi.org/10.31223/X5KP43

Caldow, R.W.G., Goss-Custard, J.D., Stillman, R.A., Durell, S.E.A. le V. dit, Swinfen, R. and Bregnballe, T. (1999) Individual variation in the competitive ability of interference-prone foragers: the relative importance of foraging efficiency and susceptibility to interference. *Journal of Animal Ecology* 68: 869–78.

Caldow, R.W.G., McGrorty, S., Stillman, R.A., Goss-Custard, J.D., Durell, S.E.A. le V. dit, West, A.D., Beadman, A.D., Kaiser, M.J., Mould, K. and Wilson, A. (2004) A behaviour-based modelling approach to predicting how best to reduce shorebird–shellfish conflicts. *Ecological Applications* 14: 1411–27.

Caldow, R.W.G., Stillman, R.A., Durell, S.E.A. le V. dit, West, A.D., McGrorty, S., Goss-Custard, J.D., Wood, P.J. and Humphreys, J. (2007) Benefits to shorebirds from invasion of a non-native shellfish. *Proceedings of the Royal Society London B* 274: 1449–55.

Calvert, A.M., Walde, S.J. and Taylor, P.D. (2009) Nonbreeding-season drivers of population dynamics in seasonal migrants: conservation parallels across taxa. *Avian Conservation and Ecology* 4: 5.

Camphuysen, C.J., Ens, B.J., Heg, D., Hulscher, J., van der Meer, J. and Smit, C.J. (1996) Oystercatcher *Haematopus ostralegus* winter-mortality in The Netherlands: the effect of severe weather and food supply. *Ardea* 84A: 469–92.

Cayford, J.T. and Goss-Custard, J.D. (1990) Seasonal changes in the size selection of mussels, *Mytilus edulis*, by oystercatchers, *Haematopus ostralegus*: an optimality approach. *Animal Behaviour* 40: 609–24.

Cresswell, W. (1994) Age-dependent choice of Redshank (*Tringa totanus*) feeding location – profitability or risk. *Journal of Animal Ecology* 63: 589–600.

Cresswell, W. and Quinn, J.L. (2004) Faced with a choice, sparrowhawks more often attack the more vulnerable prey group. *Oikos* 104: 71–6.

Cresswell, W. and Whitfield, D.P. (1994) The effects of raptor predation on wintering wader populations at the Tyninghame estuary, southeast Scotland. *Ibis* 136: 223–32.

Davidson, N.C. (1981) Survival of shorebirds (*Charadrii*) during severe weather: the role of nutritional reserves. In N.V. Jones and W.J. Wolff (eds) *Feeding and Survival Strategies of Estuarine Organisms*, pp. 231–49. New York: Plenum Press.

Davidson, N.C. and Evans, P.R. (1982) Mortality of redshanks and oystercatchers from starvation during severe weather. *Bird Study* 29: 183–8.

Davidson, P.E. (1968) The oystercatcher – a pest of shellfisheries. In R.K. Murton and E.N. Wright (eds) *The Problems of Birds as Pests*, pp. 141–55. London: Academic Press.

Dekker, D. and Ydenburg, R. (2004) Raptor predation on wintering dunlins in relation to the tidal cycle. *The Condor* 106: 415–19.

Dierschke, V. (1998) High profit at high risk for juvenile dunlins *Calidris alpina* stopping over at Helgoland (German Bight). *Ardea* 85: 59–69.

Dokter, A.M., van Loon, E.E., Rappoldt, C., Osterbeek, K., Baptist, M.J., Bouten, W. and Ens, B.J. (2017) Balancing food and density-dependence in the spatial distribution of an interference-prone forager. *Oikos* 126: 1184–96.

Duriez, O., Saether, S.A., Ens, B.J., Choquet, R., Pradel, R., Lambeck, R.H.D. and Klaassen, M. (2009) Estimating survival and movements using both live and dead recoveries: a case study of oystercatchers confronted with habitat change. *Journal of Applied Ecology* 46: 144–53.

Duriez, O., Ens, B.J., Choquet, R., Pradel, R. and Klaassen, M. (2012) Comparing the seasonal survival of resident and migratory oystercatchers: carry-over effects of habitat quality and weather conditions. *Oikos* 121: 862–73.

Durell, S.E.A. le V. dit (2007) Differential survival in adult Eurasian oystercatchers *Haematopus ostralegus*. *Journal of Avian Biology* 38: 530–5.

Durell, S.E.A. le V. dit., Goss-Custard, J.D., Stillman, R.A. and West, R.A. (2001) The effect of weather and density-dependence on oystercatcher *Haematopus ostralegus* winter mortality. *Ibis* 143: 498–9.

Durell, S.E.A. le V. dit, Goss-Custard, J.D., Clarke, R.T. and McGrorty, S. (2003) Density-dependent mortality in wintering Eurasian Oystercatchers *Haematopus ostralegus*. *Ibis* 145: 496–8.

Durell, S.E.A. le V. dit, Stillman, R.A., Triplet, P., Aulert, C., Ono dit Biot, D., Bouchet, A., Duhamel, S., Mayot, S. and Goss-Custard, J.D. (2005) Modelling the efficacy of proposed mitigation areas for shorebirds on the Seine estuary, France. *Biological Conservation* 123: 67–77.

Dwyer, R.G., Bearhop, S., Campbell, H.A. and Bryant, D.M. (2013) Shedding light on light: benefits of anthropogenic illumination to a nocturnally foraging shorebird. *Journal of Animal Ecology* 82: 478–85.

Ens, B.J. and Goss-Custard, J.D. (1984) Interference among oystercatchers, *Haematopus ostralegus*, feeding on mussels, *Mytilus edulis*, on the Exe estuary. *Journal of Animal Ecology* 53: 217–31.

Ens, B.J., Dirksen, S., Smit, C.J. and Bunskoeke, E.J. (1996) Seasonal changes in size selection and intake rate of oystercatchers *Haematopus ostralegus* feeding on the bivalves *Mytilus edulis* and *Cerastoderma edule*. *Ardea* 84A: 159–76.

Ens, B.J., Smaal, A.C. and de Vlas, J. (2004) The effects of shellfish fishery on the ecosystems of the Dutch Wadden Sea and Oosterschelde: final report on the second phase of the scientific evaluation of the Dutch shellfishery policy (EVA II). Alterra-rapport 1011; RIVO-rapport C056/04; RIKZ-rapport/2004.031. Available at: https://research.wur.nl/en/publications/the-effects-of-shellfish-fishery-on-the-ecosystems-of-the-dutch-w (Accessed: 7 September 2021).

Esselink, P. and Zwarts, L. (1989) Seasonal trend in burrow depth and tidal variation in feeding activity of *Nereis diversicolor*. *Marine Ecology Progress Series* 56: 243–54.

Evans, M.R. (2012) Modelling ecological systems in a changing world. *Philosophical Transactions of the Royal Society B: Biological Sciences* 367: 181–90.

Evans, P.R. (1976) Energy balance and optimal foraging strategies in shorebirds: some implications for their distribution and movements in the non-breeding season. *Ardea* 64: 117–39.

Evans, P.R., Herdson, D.M., Knights, P.J. and Pienkowski, M.W. (1979) Short-term effects of reclamation of parts of Seal Sand, Teesmouth, on wintering waders and Shelduck. I. Shorebird diets, invertebrate densities, and the impact of predation on the invertebrates. *Oecologia* 41: 183–206.

Folmer, E.O., Olff, H. and Piersma, T. (2011) The spatial distribution of flocking foragers: disentangling the effects of food availability, interference and conspecific attraction by means of spatial autoregressive modelling. *Oikos* 121: 551–61. https://doi.org/10.1111/j.1600-0706.2011.19739.x

Goater, C.P. (1993) Population biology of *Meiogymnophallus minutus* (Trematoda: Gymnophallidae) in cockles *Cerastoderma edule* from the Exe estuary, England. *Journal of the Marine Biological Association* 73: 163–77.

Goater, C.P., Goss-Custard, J.D. and Kennedy, C.R. (1995) Population biology of two species of helminths in oystercarchers, *Haematopus ostralegus*. *Canadian Journal of Zoology* 75: 296–300.

Goss-Custard, J.D. (1969) The winter feeding ecology of the redshank, *Tringa totanus*. *Ibis* 111: 338–56.

Goss-Custard, J.D. (1970) Feeding dispersion in some overwintering wading birds. In J.H. Crook (ed.) *Social Behaviour in Birds and Mammals*, pp. 3–35. London: Academic Press.

Goss-Custard, J.D. (1976) Variation in the dispersion of redshank, *Tringa totanus*, on their winter-feeding grounds. *Ibis* 118: 257–63.

Goss-Custard, J.D. (1977) The ecology of the Wash. III. Density-related behaviour and the possible effects of a loss of feeding grounds on wading birds (Charadrii). *Journal of Applied Ecology* 14: 721–39.

Goss-Custard, J.D. (1980) Competition for food and interference among waders. *Ardea* 68: 31–52.

Goss-Custard, J.D. (1985) Foraging behaviour of wading birds and the carrying capacity of estuaries. In R.M. Sibly and R.H. Smith (eds) *Behavioural Ecology: Ecological Consequences of Adaptive Behaviour*, pp. 169–88. Oxford: Blackwells.

Goss-Custard, J.D. (2016) Mud, birds and poppycock. *Bulletin of the British Ecological Society* 47: 26–9.

Goss-Custard, J.D. (2017) *Birds and People: Resolving the conflict on estuaries*. Amazon UK.

Goss-Custard, J.D. and Sutherland, W.J. (1997) Individual behaviour, populations and conservation. In J.R. Krebs and N.B. Davies (eds) *Behavioural Ecology: an Evolutionary Approach*. 4th edition, pp. 373–95. Oxford: Blackwell Science.

Goss-Custard, J.D., Bowgen, K.M. and Stillman, R.A. (2019) Increasing the harvest for mussels *Mytilus edulis* without harming oystercatchers *Haematopus ostralegus*. *Marine Ecology Progress Series* 612: 101–10.

Goss-Custard, J.D. and Durell, S.E.A. le V. dit (1987a) Age-related effects in oystercatchers, *Haematopus ostralegus*, feeding on mussels, *Mytilus edulis*. 1. Foraging efficiency and interference. *Journal of Animal Ecology* 56: 521–36.

Goss-Custard, J.D. and Durell, S.E.A. le V. dit (1987b) Age-related effects in oystercatchers, *Haematopus ostralegus*, feeding on mussels, *Mytilus edulis*. 2. Aggression. *Journal of Animal Ecology* 56: 537–48.

Goss-Custard, J.D. and Durell, S.E.A. le V. dit (1987c) Age-related effects in oystercatchers, *Haematopus ostralegus*, feeding on mussels, *Mytilus edulis*. 3. The effect of interference on overall intake rate. *Journal of Animal Ecology* 56: 549–58.

Goss-Custard, J.D. and Durell, S.E.A. le V. dit (1988) The effect of dominance and feeding method on the intake rates of oystercatchers, *Haematopus ostralegus*, feeding on mussels, *Mytilus edulis*. *Journal of Animal Ecology* 57: 827–44.

Goss-Custard, J.D. and Stillman, R.A. (2020) How manual cockle-raking may affect availability of cockles *Cerastoderma edule* for oystercatchers *Haematopus ostralegus* in the Dutch Wadden Sea. BU Global Environmental Solutions report BUG2842 to Province of Fryslân, Bournemouth. Available at: https://rijkewaddenzee.nl/wp-content/uploads/2021/01/BUG2842-Province-of-Fryslan-Wadden-Sea-Final-2020-08-21.pdf (Accessed: 7 September 2021).

Goss-Custard, J.D., Durell, S.E.A. le V. dit and Ens, B.J. (1982a) Individual differences in aggressiveness and food stealing among wintering oystercatchers,

Haematopus ostralegus L. *Animal Behaviour* 30: 917–28.

Goss-Custard, J.D., Durell, S.E.A. le V. dit, McGrorty, S., Reading, C.J. (1982b) Use of mussel, *Mytilus edulis* beds by oystercatchers *Haematopus ostralegus* according to age and population size. *Journal of Animal Ecology* 51: 543–54.

Goss-Custard, J.D., Jenyon, R.A., Jones, R.E., Newbery, P.E. and Williams, R. le B. (1977) The ecology of the Wash. II. Seasonal variation in the feeding conditions of wading birds (Charadrii). *Journal of Applied Ecology* 14: 701–19.

Goss-Custard, J.D., Caldow, R.W.G., Clarke, R.T., Durell S.E.A. le V. dit, Urfi, A.J. and West, A.D. (1995) Consequences of habitat loss and change to populations of wintering migratory birds: predicting the local and global effects from studies of individuals. *Ibis* 137: S56–66.

Goss-Custard, J.D., Durell, S.E.A. le V. dit, Goater, Hulscher, J.B., Lambeck, R.H.D., Meininger, P.L. and Urfi, J. (1996) How oystercatchers survive the winter. In J.D. Goss-Custard (ed.), *The Oystercatcher: From Individuals to Populations*, pp. 133–54. Oxford: Oxford University Press.

Goss-Custard, J.D., Stillman, R.A., West, A.D., Caldow, R.W.G. and McGrorty, S. (2002) Carrying capacity in overwintering migratory birds. *Biological Conservation* 105: 27–41.

Goss-Custard, J.D., Stillman, R.A., West, A.D., Caldow, R.W.G., Triplet, P., Durell, S.E.A. le V. dit and McGrorty, S. (2004) When enough is not enough: shorebirds and shellfish. *Proceedings of the Royal Society London B* 271: 233–7.

Goss-Custard, J.D., Burton, N.H.K., Clark, N.A., Ferns, P.N., McGrorty, S., Reading, C.J., Rehfisch, M.M., Stillman, R.A., Townend, I., West, A.D. and Worrall, D.H. (2006a) Test of a behaviour-based individual-based model: response of shorebird mortality to habitat loss. *Ecological Applications* 16: 2205–22.

Goss-Custard, J.D., Triplet, P., Sueur, F. and West, A.D. (2006b) Critical thresholds of disturbance by people and raptors in foraging wading birds. *Biological Conservation* 127: 88–97.

Goss-Custard, J.D., West, A.D., Yates, M.G. and 31 other authors (2006c) Intake rates and the functional response in shorebirds (Charadriiformes) eating macro-invertebrates. *Biological Reviews* 81: 501–29.

Goss-Custard, J.D., West, A.D., Stillman, R.A., Durell, S.E.A. le V. dit, Caldow, R.W.G., McGrorty, S. and Nagarajan, R. (2001) Density-dependent starvation in a vertebrate without significant depletion. *Journal of Animal Ecology* 70: 955–65.

Grimm, V. and Railsback, S.F. (2005) *Individual-based modeling and ecology.* Princeton, NJ: Princeton University Press.

Grimm, V., Berger, U., Bastiansen, F., Eliassen, S., Ginot, V., Giske, J., Goss-Custard, J.D. and 21 others (2006) A standard protocol for describing individual-based and agent-based models. *Ecological Modelling* 198: 115–26.

Harrison, X.A., Blount, J.D., Inger, R., Norris, D.R. and Bearhoop, S. (2010) Carry-over effects as drivers of fitness differences in animals. *Journal of Animal Ecology* 80: 1–43. https://doi.org/10.1111/j.1365 -2656.2010.01740.x

Herbert, R.J.H., Davies, C.J., Bowgen, K.M., Hatton, J. and Stillman, R.A. (2018) The importance of non-native Pacific oyster reefs as supplementary feeding areas for coastal birds on estuary mudflats. *Aquatic Conservation: Marine and Freshwater Ecosystem* 28: 1294–1307.

Heppleston, P.B. (1971) The feeding ecology of oystercatchers (*Haematopus ostralegus* L.) in winter in Northern Scotland. *Journal of Animal Ecology* 40: 651–72.

Hotker, H. (2002) Arrival of pied avocets *Recurvirostra avosetta* at the breeding sites: effects of winter quarters and consequences for reproductive success. *Ardea* 90: 379–87.

Hulscher, J.B. (1996) Food and feeding behaviour. In J.D. Goss-Custard (ed.) *The Oystercatcher: From Individuals to Populations*, pp. 7–29. Oxford: Oxford University Press.

Hulscher, J.B., Exo, K-M. and Clark, N.A. (1996) Why do oystercatchers migrate? In J.D. Goss-Custard (ed.) *The Oystercatcher: From Individuals to Populations*, pp. 155–85. Oxford: Oxford University Press.

Johnstone, I. and Norris, K. (2000) Not all oystercatchers *Haematopus ostralegus* select the most profitable common cockles *Cerastoderma edule*: a difference between feeding methods. *Ardea* 88: 137–53.

Kersten, M. and Piersma, T. (1987) High levels of energy expenditure in shorebirds: metabolic adaptations to an energetically expensive way of life. *Ardea* 75: 175–87.

Kersten, M. and Visser, W. (1996) The rate of food processing in the oystercatcher: food intake and energy expenditure constrained by a digestive bottleneck. *Functional Ecology* 10: 440–8.

Klok, C., Roodbergen, M. and Hemerik, L. (2009) Diagnosing declining grassland wader populations using simple matrix models. *Animal Biology* 50: 127–44.

Kuwae, T. (2007) Diurnal and nocturnal feeding rate in Kentish plovers *Charadrius alexandricus* on an intertidal flat as recorded by telescope video systems. *Marine Biology* 151: 663–73.

Lambeck, R.H.D., Goss-Custard, J.D. and Triplet, P. (1996) Oystercatchers and man in the coastal zone. In J.D. Goss-Custard (ed.), *The Oystercatcher: From Individuals to Populations*, pp. 289–326. Oxford: Oxford University Press.

Le Drean-Quenec'hdu, S. and Goss-Custard, J.D. (1999) Repartition spatiale des Huitriers-pie (Limicoles) en hivernage et charge parasitaire intestinal. Spatial distribution of Oystercatchers and gut parasite load. *Canadian Journal of Zoology* 77: 1117–27.

Lomnicki, A. (1978) Individual differences between animals and the natural regulation of their numbers. *Journal of Animal Ecology* 47: 461–75.

Lomnicki, A. (1980) Regulation of population density due to individual differences and patchy environment. *Oikos* 35: 185–93.

Lomnicki, A. (1982) Individual heterogeneity and population regulation. In King's College Sociobiology Group (eds) *Current Problems in Sociobiology*, pp. 153–67. Cambridge: Cambridge University Press.

Meire, P.M. (1996a) Using optimal foraging theory to determine the density of mussels *Mytilus edulis* that can be harvested by hammering oystercatchers *Haematopus ostralegus*. *Ardea* 84A: 141–52.

Meire, P.M. (1996b) Feeding behaviour of oystercatchers *Haematopus ostralegus* during a period of tidal manipulation. *Ardea* 84A: 509–24.

Mindermann, J., Lind, J. and Cresswell, W. (2006) Behaviourally mediated indirect effects: interference competition increases predation mortality in foraging redshanks. *Journal of Animal Ecology* 75: 713–23.

Norris, D.R. (2005) Carry-over effects and habitat quality in migratory populations. *Oikos* 109: 178–86.

Norris, K. (1999) A trade-off between energy intake and exposure to parasites in oystercatchers feeding on a bivalve mollusc. *Proceedings of the Royal Society London B* 266: 1703–9.

Page, G. and Whitacre, D.F. (1975) Raptor predation on wintering shorebirds. *Condor* 77: 73–83.

Pettifor, R.A., Caldow, R.W.G., Rowcliffe, J.M., Goss-Custard, J.D., Black, J.M., Hodder, K.H., Houston, A.I., Lang, A. and Webb, J. (2000) Spatially explicit, individual-based behaviour models of the annual cycle of two migratory goose populations – model development, theoretical insights and applications. *Journal of Applied Ecology* 37 (Supplement 1): 103–35.

Piersma, T., Gill, R.E., de Goeij, P., Dekinga, A., Shepherd, M.L., Ruthrauff, D. and Tibbitts, L. (2006) Shorebird avoidance of nearshore feeding and roosting areas at night correlates with presence of a nocturnal avian predator. *Wader Study Group Bulletin* 109: 73–6.

Pomeroy, A.C., Acevedo Seaman, D.A., Butler, R.W., Elner, R.W., Williams, T.D. and Ydenberg, R.C. (2008). Feeding–danger trade-offs underlie stopover site selection by migrants. *Avian Conservation and Ecology – Écologie et conservation des oiseaux* 3 (1): 7. https://doi.org/10.5751/ACE-00240-030107

Prater, A.J. (1972) The ecology of Morecambe Bay. III. The food and feeding habits of knot (*Calidris canutus*) in Morecambe Bay. *Journal of Applied Ecology* 9: 179–94.

Prater, A.J. (1981) *Estuary Birds of Britain and Ireland*. Carlton: Poyser.

Rakhimberdiev, E., van den Hout, P.J., Brugge, M., Spaans, B. and Piersma, T. (2015) Seasonal mortality and sequential density dependence in a migratory bird. *Journal of Avian Biology* 46: 1–10.

Rappoldt, C., Ens, B.J., Kersten, M.A.J.M. and Dijkman, E.M. (2003a) Wader energy balance & tidal cycle simulator WEBTICS: Technical Documentation Version 1.1. Alterra-rapport 869.

Rappoldt, C., Ens, B.J., Berrevoets, C.M., Geutrts van Kessel, A.J.M., Bult, T.P. and Dijkman, E.M. (2003b) Scholeksters en hun voedsel in de Oosterschelde; rapport voor deelproject D2 thema 1 van EV II, de tweede fase van het evaluatieonderzoek naar de effecten van schelpdiervisserij op natuurwaarden in de Waddenzee en Oosterschelde 1999–2003. Wageningen, Alterra-Rapport 883. Available at: https://library.wur.nl/WebQuery/wurpubs/327146 (Accessed: 7 September 2021).

Rappoldt, C., Kersten, M. and Ens, B.J. (2006) Schleksters en de droogvalduur van kokkels in de Oosterschelde. EcoCurves rapport 2; SOVON-onderzoeksrapport 2006/12. Available at: https://www.sovon.nl/nl/content/scholeksters-en-de-droogvalduur-van-kokkels-de-oosterschelde (Accessed: 7 September 2021).

Redpath, S.M., Young, J., Evely, A., Adams, W.M., Sutherland, W.J., Whitehouse, A., Amar, A., Lambert, R.A., Linnell, J.D.C., Watt, A. and Gutierrez, R.J. (2013) Understanding and managing conservation conflicts. *Trends in Ecology and Evolution* 28: 100–9.

Roodbergen, M., van der Werf, B. and Hotker, H. (2011) Revealing the contributions of reproduction and survival to the Europe-wide decline in meadow birds: review and meta-analysis. *Journal of Ornithology* 153: 53–74. https://doi.org/10.1007/s10336-011-0733-y

Rutten, A.L., Oosterbeek, K., Ens, B.J. and Verhulst, S. (2006) Optimal foraging on perilous prey: risk of bill damage reduces optimal prey size in oystercatchers. *Behavioral Ecology* 17 (2): 297–302. https://doi.org/10.1093/beheco/arj029

Rutten, A.L., Oosterbeek, K., van der Meer, J., Verhulst, S. and Ens, B.J. (2010a) Experimental evidence for interference competition in oystercatchers, *Haematopus ostralegus*. 1. Free-living birds. *Behavioral Ecology* 21 (6): 1251–60. https://doi.org/10.1093/beheco/arq129

Rutten, A.L., Oosterbeek, K., Verhulst, S., Dingemanse, N.J. and Ens, B.J. (2010b) Experimental evidence for interference competition in oystercatchers, *Haematopus ostralegus*. II. Captive birds. *Behavioral Ecology* 21 (6): 1261–70. https://doi.org/10.1093/beheco/arq130

Ryan, L.J., Green, J.A. and Dodd, S.G. (2016) Weather conditions and conspecific density influence survival of overwintering dunlin *Calidris alpina* in North Wales. *Bird Study* 63: 1–9.

Santos, C.D., Miranda, A.C., Granadeiro, J.P., Lourenco, P.M., Saraiva, S. and Palmeirim, J.M. (2010) Effects of artificial illumination on the nocturnal foraging of waders. *Acta Oecologica* 36: 166–73.

Schwemmer, P., Halterlein, B., Getter, O., Gunther, K., Gorman, V.M. and Garthe, S. (2014)

Weather-related winter mortality of European oystercatchers (*Haematopus ostralegus*) in the North-western Wadden Sea. *Waterbirds* 37: 319–30.

Selman, J. and Goss-Custard, J.D. (1988) Interference between foraging redshank, *Tringa totanus*. *Animal Behaviour* 36: 1542–5.

Sitters, H.P. (2000) The role of night-feeding in shore-birds in an estuarine environment with specific reference to mussel-eating oystercatchers. D.Phil. thesis, Oxford University.

Smart, J. and Gill, J.A. (2003) Non-intertidal habitat use by shorebirds: a reflection of inadequate inter-tidal resources? *Biological Conservation* 111: 359–69.

Smith, P.C. (1975) A study of the winter-feeding ecology and behaviour of the bar-tailed godwit (*Limosa lapponica*). PhD thesis, University of Durham. Available at: http://etheses.dur.ac.uk/8179/ (Accessed: 7 September 2021).

Stillman, R.A. (2008) MORPH: an individual-based model to predict the effect of environmental change on foraging animal populations. *Ecological Modelling* 216: 265–76.

Stillman, R.A. and Goss-Custard, J.D. (2010) Individual-based ecology of coastal birds. *Biological Reviews* 85: 413–34. https://doi.org/10.1111/j.1469-185X.2009.00106.x

Stillman, R.A., Goss-Custard, J.D. and Caldow, R.W.G. (1997) Modelling interference from basic foraging behaviour. *Journal of Animal Ecology* 66: 692–703.

Stillman, R.A., Wood, K.A. and Goss-Custard, J.D. (2016) Deriving simple predictions from complex models to support environmental decision-making. *Ecological Modelling* 32: 134–41.

Stillman, R.A., Goss-Custard, J.D., Clarke, R.T. and Durell, S.E.A. le V. dit (1996) Shape of the inter-ference function in a foraging vertebrate. *Journal of Animal Ecology* 65: 813–24.

Stillman, R.A., Goss-Custard, J.D., West, A.D., Durell, S.E.A. le V. dit, Caldow, R.W.G., McGrorty, S. and Clarke, R.T. (2000) Predicting to novel environ-ments: tests and sensitivity of a behaviour-based population model. *Journal of Animal Ecology* 37: 564–88.

Stillman, R.A., Goss-Custard, J.D., West, A.D., McGrorty, S., Caldow, R.W.G., Durell, S.E.A. le V. dit, Norris, K.J., Johnstone, I.G., Ens, B.J., van der Meer, J. and Triplet, P. (2001) Predicting oystercatcher mortality and population size under different regimes of shellfishery management. *Journal of Applied Ecology* 38: 857–68.

Stillman, R.A., Railsback, S.F., Giske, J., Berger, U. and Grimm, V. (2015) Making predictions in a changing world: the benefits of individual-based ecology. *Bioscience* 65: 140–50. https://doi.org/10.1093/biosci/biu192

Sutherland, W.J. and Parker, G.A. (1985) Distribution of unequal competitors. In R.M. Sibly and R.H. Smith (eds) *Behavioural Ecology: Ecological Conse-quences of Adaptive Behaviour*, pp. 255–74. Oxford: Blackwell Scientific Publications.

Sutherland, W.J., Ens, B.J., Goss-Custard, J.D. and Hulscher, J.B. (1996) Specialisation. In J.D. Goss-Custard (ed.), *The Oystercatcher: From Individuals to Populations*, pp. 56–76. Oxford: Oxford Univer-sity Press.

Swennen, C. (1984) Differences in quality of roosting flocks of Oystercatchers In P.R. Evans, J.D. Goss-Custard and W.G. Hale (eds) *Coastal Waders and Wildfowl in Winter*, pp. 177–89. Cambridge: Cambridge University Press.

Teunissen, W.A., Schekkerman, H. and Willems, F. (2005) Predatie bij weidevogels. Op zoek naar de mogelijke effecten van predatie op de weidevo-gelstand. SOVON-ondeerzoeksrapport 2005/11; Alterra-rapport 1292. Available at: https://www.sovon.nl/sites/default/files/doc/Rapporten/Predatie%20bij%20weidevogels_rap2005_11.pdf (Accessed: 7 September 2021).

Triplet, P., Stillman, R.A. and Goss-Custard, J.D. (1999) Prey abundance and the strength of inter-ference in a foraging shorebird. *Journal of Animal Ecology* 68: 254–65.

Turpie, J.K. and Hockey, P.A.R. (1993) Comparative diurnal and nocturnal foraging behaviour and energy intake of premigratory grey plovers *Pluvi-alis squatarola* and whimbrels *Numenius phaeopus* in South Africa. *Ibis* 135: 156–65.

van der Meer, J. and Ens, B.J. (1997) Models of inter-ference and their consequences for the spatial distribution of ideal free predators. *Journal of Animal Ecology* 66: 846–858.

van den Hout, P.J., van Gils, J.A., Robin, F., van der Geest, M., Dekinga, A. and Piersma, T. (2014) Interference from adults forces young red knots to forage for longer and in dangerous places. *Animal Behaviour* 88: 137–46.

van de Pol, M., Brouwer, L., Ens, B.J., Oosterbeek, K. and Tinbergen, J.M. (2009a) Fluctuating selection and the maintenance of individual and sex-specific diet specialization in free-living oystercatchers. *Evolution* 64: 836–51.

van der Kolk, H.-J., Ens, B.J., Frauendorf, M., Jonge-jans, E., Oosterbeek, K., Bouten, W. and van de Pol, M. (2021) Why time-limited individuals can make populations more vulnerable to disturbance. *Oikos* 130: 637–51. https://doi.org/10.1111/oik.08031

van de Pol, M., Ens, B.J., Oosterbeek, K., Brouwer, L., Verhulst, S., Tinbergen, J.M., Rutten, A.L. and de Jong, M. (2009b) Oystercatchers' bill shapes as a proxy for diet specialisation: more differentiation than meets the eye. *Ardea* 97: 335–47.

van de Pol., M., Vindenes, Y., Saether, B-E., Engen, S., Ens, B.J., Oosterbeekj, K. and Tinbergen, J.M. (2010) Effect of climate change and variability on population dynamics in a long-lived shorebird. *Ecology* 9: 1192–1204.

van Roomen, M., van Turnhout, C., van Winden, E., Koks, B., Goedhart, P., Leopold, M. and Smit, C. (2005) Trends in benthivorous waterbirds in the Dutch Wadden Sea 1975–2002: Large differences between shellfish-eaters and worm-eaters. *Limosa* 78: 21–38.

Verhulst, S., Oosterbeek, K., Rutten, A.I. and Ens, B.J. (2004) Shellfish fishery severely reduces condition and survival of oystercatchers despite creation of large marine protected areas. *Ecology and Society* 9 (1): 17. Available at: http://www.ecologyandsociety.org/vol9/iss1/art17/ (Accessed: 7 September 2021).

Wanink, J. and Zwarts, L. (1985) Does an optimally foraging oystercatcher obey the functional response? *Oecologia* 67: 98–106.

Wanink, J. and Zwarts, L. (1996) Can food specialisation by individual oystercatchers *Haematopus ostralegus* be explained by differences in prey specific handling efficiencies? *Ardea* 84A: 177–98.

West, A.D., Goss-Custard, J.D., Stillman, R.A., Caldow, R.W.G., Durell, S.E.A le V. dit and McGrorty, S. (2002) Predicting the impacts of disturbance on wintering waders using a behaviour-based individuals model. *Biological Conservation* 106: 319–28.

Whitfield, D.P. (1985) Raptor predation on wintering waders in southeast Scotland. *Ibis* 127: 544–58.

Whitfield, D.P. (2003) Predation by Eurasian Sparrowhawks produces density-dependent mortality of wintering Redshanks. *Journal of Animal Ecology* 72: 27–35.

Wood, K.A., Stillman, R.A. and Goss-Custard, J.D. (2015) The effect of kleptoparasite and host numbers on the risk of food-stealing in an avian assemblage. *Journal of Avian Biology* 46: 589–96.

Yates, M.G., Stillman, R.A. and Goss-Custard, J.D. (2000) Contrasting interference functions and foraging dispersion in two species of shorebirds Charadrii. *Journal of Animal Ecology* 69: 314–22.

Zwarts, L. (1991) Seasonal variation in body weight of the bivalves *Macoma balthica, Scrobicularia plana, Mya arenaria* and *Cerastoderma edule* in the Dutch Wadden Sea. *Netherlands Journal of Sea Research* 28: 231–45. https://doi.org/10.1016/0077 -7579(91)90021-R

Zwarts, L. and Blomert, A-M. (1992) Why knot *Calidris canutus* take medium-sized *Macoma balthica* when six prey species are available. *Marine Ecology Progress Series* 83: 113–28.

Zwarts, L. and Drent, R.H. (1981) Prey depletion and the regulation of predator density: Oystercatchers (*Haematopus ostralegus*) feeding on mussels (*Mytilus edulis*). In N.V. Jones and W.J. Wolff (eds) *Feeding and Survival Strategies of Estuarine Organisms*, pp. 193–216. New York: Plenum Press.

Zwarts, L. and Esselink, P. (1989) Versatility of male curlews *Numenius arquata* preying upon *Nereis diversicolor*: deploying contrasting capture models dependent on prey availability. *Marine Ecology Progress Series* 56: 255–69.

Zwarts, L. and Wanink, J. (1989) Siphon size and burying depth in deposit- and suspension-feeding benthic bivalves. *Marine Biology* 100: 227–40.

Zwarts, L. and Wanink, J.H. (1991) The macrobenthos fraction accessible to waders may represent marginal prey. *Oecologia* 87: 581–7. https://doi.org/ 10.1007/BF00320424

Zwarts, L. and Wanink, J.H. (1993) How the food supply harvestable by waders in the Wadden Sea depends on the variation in energy density, body weight, biomass, burying depth and behaviour if tidal-flat invertebrates. *Netherlands Journal of Sea Research* 31: 441–76. https://doi.org/10.1016/0077 -7579(93)90059-2

Zwarts, L., Blomert, A.-M. and Wanink, J.H. (1992) Annual and seasonal variation in the food supply harvestable by knot *Calidris canutus* staging in the Wadden Sea in late summer. *Marine Ecology Progress Series* 83: 129–39. https://doi.org/10.3354/ meps083129

Zwarts, L., Wanink, J.H. and Ens, B.J. (1996) Predicting seasonal and annual fluctuations in the local exploitation of different prey by oystercatchers *Haematopus ostralegus*: a ten-year study in the Wadden Sea. *Ardea* 1984A: 401–40.

Zwarts, L., Ens, B.J., Goss-Custard, J.D., Hulscher, J.B. and Durell, S.E.A. le V. dit (1996a) Causes of variation in prey profitability and its consequences for the intake rate of the Oystercatchers *Haematopus ostralegus*. *Ardea* 84: 229–68.

Zwarts, L., Hulscher, J.B., Koopman, K., Piersma, T. and Zegers, P.M. (1996b) Seasonal and annual variation in body weight, nutrient stores and mortality of Oystercatchers *Haematopus ostralegus*. *Ardea* 84A: 327–56.

Zwarts, L., Cayford, J.T., Hulscher, J.B., Kersten, M., Meire, P.M. and Triplet, P. (1996d) Prey size selection and intake rate. In J.D. Goss-Custard (ed.) *The Oystercatcher: From Individuals to Populations*, pp. 30–55. Oxford: Oxford University Press.

CHAPTER 14

The Role of Estuaries and Coastal Areas as Nurseries for Fish: Concepts, Methodological Challenges and Future Perspectives

HENRIQUE CABRAL

Abstract

Estuaries and coastal zones have long been recognised as important nursery areas for fish, since they offer high food availability, low predation pressure and good conditions for rapid growth. However, the nursery concept has been changing, as have the methodological approaches used to evaluate this function. The value of nursery areas has historically been assessed according to the abundance of juvenile fish, their growth, their survival and the effective contribution (in numbers of fish) to adult habitats. A more complex and dynamic perspective of nursery areas has now been proposed, accounting for ontogenetic shifts between habitat patches, and recognising a seascape nursery concept. Both natural and anthropogenic factors make the relative importance of this habitat highly variable, and a more dynamic view is proposed here. There is no consensus on the main advantages of fish nursery areas, and evidence from a wide range of studies and even meta-analyses has failed to build a unanimously accepted ecological framework. Methodologies for assessing the nursery value of coastal areas have major limitations that limit the possibility for comparative analyses, and further developments and innovative tools are required. Available knowledge is extremely biased, focusing mainly on Europe and North America, and several topics related to patterns and processes driving juvenile fish ecology in nursery areas need further research. Despite existing marine protected area networks often integrating nursery areas, their effective and sustainable management and conservation is required in order to cope with increasing anthropogenic pressure in coastal ecosystems worldwide.

Keywords: fish nurseries, juvenile fish, essential juvenile habitat, seascape nursery, estuarine management and conservation

Correspondence: henrique.cabral@inrae.fr

The nursery concept: origin and evolution

The nursery role concept dates back to the early 1900s (Gunther 1967; Beck et al. 2001), when it was applied to fish and invertebrate species that present a complex lifecycle, with adults occurring in the continental shelf or deeper coastal waters and juveniles concentrating in

Henrique Cabral, 'The Role of Estuaries and Coastal Areas as Nurseries for Fish: Concepts, Methodological Challenges and Future Perspectives' in: *Challenges in Estuarine and Coastal Science*. Pelagic Publishing (2022). © Henrique Cabral. DOI: 10.53061/ESFX8762 shallow areas, particularly in estuaries. The concept was rapidly generalised to a large number of species and marine ecosystems worldwide (Beck et al. 2001), and the term nursery (or equivalent expressions) is used to acknowledge the occurrence of juveniles, often in abundance, in certain areas. This broader concept coincides with the term 'juvenile habitat'. Adults and juveniles of a large number of marine species share roughly the same habitats, although to a limited extent spatial segregation is often observed, with juveniles occurring in shallower and more protected areas (e.g. Gillanders et al. 2003; Prista et al. 2003; França et al. 2004; San Martín, Cubillos and Saavedra 2011; Vallisneri et al. 2014). Juvenile habitats are still unknown for a wide diversity of fish species, especially deep-water or mesopelagic species (e.g. Parin 1986; Priede 2019).

The nursery concept gained major relevance in the theoretical framework of population dynamics in fisheries science models of fish stocks. Huse (2015) synthesised its introduction and evolution, outlining the fact that it was integrated in Hjort's hypotheses on recruitment variability (the interannual variability in the drift of eggs and larvae away from nursery areas influences recruitment) (Hjort 1926). Further developments were made by Harden Jones (1968), who proposed the so-called migration triangle (Fig. 14.1) and established almost as a rule that most temperate continental shelf fish undergo migrations between spawning, wintering and feeding in spatially distinct areas. Iles and Sinclair (1982) and Sinclair (1988) postulated that fish populations tend to spawn at specific areas and periods according to predictable ocean circulation features that limit larval dispersal and favour larval concentration in particular areas (the member–vagrant hypothesis). Those authors have also highlighted the relevance of density-dependent processes.

In estuarine systems, the application of the functional guild approach to the study of fish assemblages (e.g. McHugh 1967; Haedrich 1983; Elliott, O'Reilly and Taylor 1990; Elliott and Dewailly, 1995) has long recognised a group of species that are only represented by juveniles and for which estuaries may be considered as nursery areas (e.g. belonging to guilds such as marine migrant, marine estuarine-opportunist or marine estuarine-dependent (Elliott and Dewailly 1995; Elliott et al. 2007; Potter et al. 2015).

A milestone in the evolution of the nursery concept, as outlined by Beck et al. (2001), was to consider that habitats or sub-areas within an ecosystem do not have the same value for

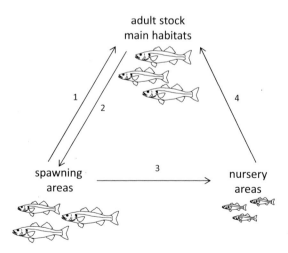

Figure 14.1 Main movements of marine fish using coastal ecosystems as nurseries (usually known as the migration triangle), depicting concentration of adults in spawning areas and movements between these areas and other adult habitats (1 and 2), larval drift and migration from spawning areas towards coastal nurseries (3) and movements of juveniles from nurseries towards adult habitats (4) (adapted from Harden Jones 1968)

fish juveniles. This dimension stressed the need to compare juvenile abundance, and later survival and condition, in different habitats or areas within an ecosystem, often estuaries or marine coastal areas. Although a large number of studies have addressed this issue (e.g. Orth, Heck Jr and van Montfrans 1984; Heck, Nadeau and Thomas 1997; Able 1999; Minello, 1999; Lefcheck et al. 2019), comparisons have often been limited to a few habitats, owing to the difficulty in using the same sampling techniques or obtaining comparable estimates of fish density, and they have been restricted to a small number of geographical areas. An overlooked aspect has been the opportunistic nature of some of this habitat use. Several fish species that use estuarine habitats as nursery areas may also use adjoining marine shallow coastal areas (e.g. Sole *Solea solea*, Seabass *Dicentrarchus labrax* and Plaice *Pleuronectes platessa*, among many others), but overall assessments or comparative analyses of nurseries importance for juveniles, including their densities, growth or survival, are still very scarce. The opportunistic use of nursery areas has also nourished a major and lengthy debate regarding the concept of estuarine dependency (e.g. McHugh 1967; Haedrich 1983; Potter et al. 1990; Able 2005; Elliott et al. 2007; Potter et al. 2015), with it being outlined in more recent reviews that fish species strictly dependent on estuaries as nursery areas are relatively rare, and that a continuum in terms of dependency is perhaps most appropriate for describing the relationship of fish species with estuaries (Whitfield 2020).

Beck et al. (2001) have indeed changed the nursery concept, transforming it with their nursery role hypothesis, which formulates that a habitat is a nursery for juveniles of a particular species if its contribution per unit area to the production of individuals that recruit to adult populations is greater, on average, than production from other habitats

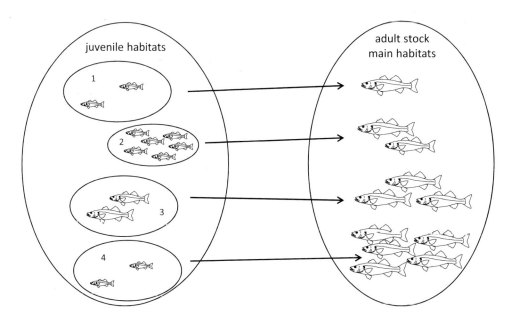

Figure 14.2 Schematic representation of the factors operating in juvenile and nursery habitats, according to Beck et al. (2001). Juveniles occur in discrete nursery habitats/areas and their density and contribution to adult stock may be different: 1 – a low juvenile density habitat with a modest contribution to adult stock; 2 – a high juvenile density habitat contributing moderately to adult stock; 3 – a low juvenile density habitat with a high contribution to adult stock, both in numbers and in condition of fish (larger juveniles); 4 – a low juvenile density habitat with a strong contribution to adult stock. According to Beck et al. (2001), nursey habitats should contribute disproportionally in numbers (or in quality) per unit area, and thus only habitats 2 and 4 should be considered as nursery habitats (1 and 2 being simply juvenile habitats) (adapted from Beck et al. 2001)

in which juveniles occur (Fig. 14.2). The major innovative inputs of this nursery role hypothesis are: (a) the development of a framework that could be testable; (b) the need to compare all habitats that juveniles use, and the fact that only a subset of juvenile habitats are nursery habitats; (c) the necessity to estimate movement of individuals from juvenile to adult habitats as a measure of assessing their contributions; and (d) the proposal that the total biomass of individuals recruiting to adult populations is the best single measure of the contribution from juvenile habitats, since it integrates density, growth and survival (key aspects for the comparison of the relative value of different putative nursery areas). To put into practice this new concept, major methodological challenges had to be addressed, and these are discussed in the following sections.

The methodological framework proposed by Beck et al. (2001) underestimates the value of juvenile habitats that have a large surface area but a low density of organisms, since the importance is expressed by unit area. Dahlgren et al. (2006) suggested that the assessment of the value of nursery areas should take into account the total contribution of individuals to the population, and proposed the concept of effective juvenile habitat to refer to habitats that make a greater than average overall contribution to adult populations. Sheaves, Baker and Johnston (2006) identified some weaknesses in this concept, especially the fact that it does not account for spatial complexity and connectivity or essential resources supporting juveniles, and limits a nursery value assessment to its numeric contribution to adult stock. These authors stressed that connectivity between habitat mosaics should be integrated in the nursery concept.

The most recent developments have been made by Nagelkerken et al. (2015) who, based on the paradigm that juveniles make ontogenetic shifts across habitats (Adams et al. 2006), introduced the concept of 'seascape nurseries' – a nursery as a spatially explicit seascape consisting of multiple mosaics of habitat patches that are functionally connected (Fig. 14.3). Fish abundance and biomass may indicate the core habitats or areas, but migration pathways connecting such core areas at larger spatial and temporal scales, through ontogenetic habitat shifts or inshore–offshore migrations, should also be identified and incorporated. This proposal was considered to be more realistic and useful for management and conservation purposes than previous formulations. Although a methodological framework has been proposed by these authors to address nursery value assessment, how complexity and dynamics should be integrated was not fully explored.

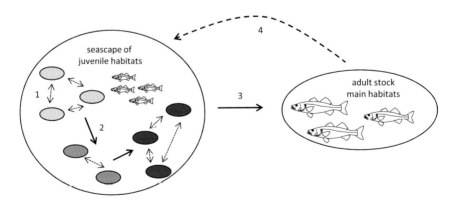

Figure 14.3 Representation of the seascape nursery concept proposed by Nagelkerken et al. (2015). The seascape nursery consists of several habitat mosaics (represented by grey ellipses, each pattern representing a habitat type) connected through periodic (1, dashed arrows) and/or ontogenetic movements (2, solid arrows). After the nursery period, juveniles migrate towards adult main habitats (3). Spawning occurs offshore, and eggs, larvae and early juveniles migrate towards juvenile habitats (4) (adapted from Nagelkerken et al. 2015)

Spatial and temporal dimensions of the nursery concept: towards a more dynamic perspective of the seascape nurseries concept

Spatial and temporal dimensions of the nursery concept are critical, but they have been often disregarded. As outlined by Beck et al. (2001) the spatial segregation between juveniles and adults is a core issue for the nursery role hypothesis. Indeed, for the vast majority of fish species there is some overlap between juveniles and adult habitats. For instance, for coastal fish such as wrasses (Labridae), blennies (Blenniidae), gobies (Gobiidae) or even sea bream (Sparidae), juveniles occupy fairly similar habitats to adults, although often occurring in shallower and more protected microhabitats (e.g. algal tufts, rock crevices, tidal pools) and progressively sharing the same areas as adults as they grow (e.g. Hofrichter and Patzner 2000; Gonçalves et al. 2002; Pais et al. 2013; Ventura, Jona Lasinio and Ardizzone 2015). These areas may represent essential juvenile habitats, but they should not be considered as nurseries according to the Beck et al. (2001) definition, since the nursery role hypothesis implies a disjunction between juvenile and adult habitats, and a movement after the juvenile phase to adult habitats (often associated with reproduction) (Fig. 14.4).

Another critical aspect to consider is that, for the majority of fish species, a nursery area does not correspond to a fixed area or to a specific habitat. Juveniles search actively for a favourable environmental context (see the next section) that is not spatially fixed. Furthermore, for certain species, it is common that estuarine or coastal nurseries in different areas present different conditions and even different locations along the main estuarine longitudinal gradient (e.g. Cabral et al. 2007). The spatial extent of nursery areas may also vary according to juvenile contingent, expanding or contracting in line with juvenile numbers. With regard to habitats, this dimension has been the basis of the nursery role hypothesis. Beck et al. (2001) suggested that juvenile habitats should be comparatively analysed in order to evaluate their contribution to the population (using the four main criteria described above). These authors stressed that nursery habitats are a subset of juvenile habitats and that a comparison of all habitats would be needed to discover which should be considered to be nurseries. Several studies were developed with that purpose, and they have highlighted the preponderance of certain habitats compared with others (e.g. Lefcheck et al. 2019). However, as largely argued by Nagelkerken et al. (2015), habitat use by juveniles is typically not limited to a very precise habitat; rather, they move between a wide diversity of habitats according to diet, tidal, seasonal and/or ontogenetic patterns (e.g. Gonçalves et al. 2002; Sheaves 2005; Nagelkerken 2007; Hammerschlag, Heithaus and Serafy 2010; Boström et al. 2011; Baker et al. 2013; Igulu et al. 2013; Olds et al. 2013). A certain habitat may have a higher relevance in terms of fish density, growth, survival

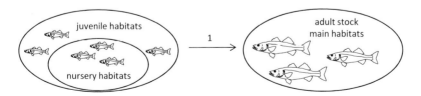

Figure 14.4 Relationship between juvenile, nursery and adult habitats, according to Beck et al. (2001). Nursery habitats are a portion of all juvenile habitats, for which their contribution to adult stocks are disproportionately higher per unit area. Even if a partial overlap between juvenile and adult habitats may exist, there must be a unidirectional movement (1) to non-juvenile habitats (usually related to reproduction or ontogenetic habitat shifts) for a species to be considered to have a nursery habitat (adapted from Beck et al. 2001)

or contribution to the overall population, but if it is used intermittently (e.g. intertidal mudflats, saltmarsh creeks) or in a sequential way it cannot be considered more valuable than the others, thus emphasising the seascape nurseries approach (cf. Sheaves, 2005; Nagelkerken et al. 2015). Unfortunately, our knowledge regarding fine-scale habitat use patterns for most fish species using coastal areas as nurseries is extremely scarce, and this clearly constrains these assessments.

The temporal dimension of the nursery role hypothesis also has implications for several aspects of habitat use, such as seasonal patterns (in particular nursery colonisation timing or the timing of juvenile migration towards offshore waters), which are extremely variable according to latitude and type of coastal system, among other factors, even for the same species. Probably the two most overlooked aspects of the temporal dimension in the nursery role hypothesis are the interannual variability in connectivity patterns among spawning areas and nursery areas, and consequently the colonisation of nurseries by larvae and juveniles, and changes in environmental context (especially habitat characteristics) of juvenile habitats over time. Results from several studies addressing connectivity patterns between spawning areas and nursery grounds have outlined differences over years that result from natural variability (especially that of oceanographic processes during fish larval drift) and even human-induced variability (e.g. related to changes in coastal oceanography owing to the increase in the emission of greenhouse gases, and their consequences for ocean temperature, pH, salinity, currents, etc.) (e.g. Hufnagl et al. 2013; Tanner et al. 2017; Lacroix, Barbut and Volckaert 2018; Cabral et al. 2021). Therefore, the value of a nursery may vary significantly depending on the success of the colonisation process in a certain year, and its relative importance may also vary accordingly.

Coastal areas, and estuaries in particular, are ecosystems that are strongly impacted by human activities and, consequently, their habitats change at different rates owing to both natural and anthropogenic variability. Changes in rainfall or in temperature may alter habitat suitability for fish juveniles and affect an environment's value as a nursery. Similarly, pollution, dredging and fishing, among many other human pressures, may decrease the value of certain habitats as nurseries, since they may affect density, growth, survival or migration to offshore areas. Thus, the value of individual habitats in a wider seascape may change drastically according to different timescales, driven both by natural or anthropogenic variabilities and processes.

Although complexity and dynamics have been recognised as major attributes of nursery habitats (e.g. Sheaves et al. 2015), these aspects have not previously been integrated in the seascape nursery concept. Appreciation of changes in nurseries and their relative importance over time informs the more dynamic seascape nursery concept that is proposed here (Fig. 14.5). This concept derives from the perspective proposed by Nagelkerken et al. (2015), which assumes that nursery habitats are extremely dynamic over time, because of natural and human-induced variability, which introduces complexity and uncertainty to any evaluations that are made.

Incorporating spatial and temporal scales in the nursery hypothesis introduces a much higher level of complexity in assessment and testing, owing to the need for recurring evaluation. A more holistic perspective is important, applying wide spatial and temporal scales, integrating habitat patches that, independently of a particular period or year, are used as nurseries. Efforts to assess habitat patch mosaics that are functionally connected and correspond to nursery areas are of major importance for conservation purposes. Research into seasonal and interannual variability in nursery habitat use patterns is also of great value, as it brings a better understanding of why, how and when fish use these areas, allowing the formation of ecological insights that allow a more unified theory.

According to this conceptual framework, the habitat used by fish juveniles is dynamic and variable both spatially and temporally, integrating a perspective of ontogenetic habitat

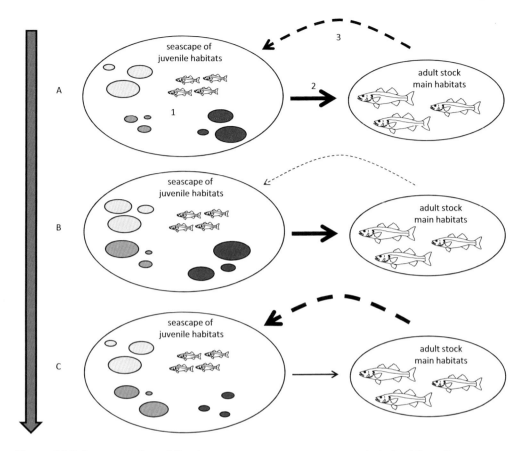

Figure 14.5 Representation of the dynamic seascape nursery concept, derived from the seascape nursery perspective proposed by Nagelkerken et al. (2015). Nurseries consist of several habitat mosaics (represented by grey ellipses, each pattern representing a habitat type) connected through periodic and/or ontogenetic movements (1). After the nursery period juveniles migrate towards main adult habitats (2). Spawning occurs offshore, and eggs, larvae and early juveniles migrate towards juvenile habitats (3). The value of nursery habitats is extremely dynamic through time (A, B, C ...), owing to both natural and human-induced variability, which introduces complexity and uncertainty in evaluations. In this example, period A represents moderate recruitment to nursery areas and also a moderate contribution from seascape nursery habitats to adult stock; in period B, a low recruitment to nursery areas is represented (e.g. corresponding to a high mortality of eggs and larvae during pelagic planktonic phase or unfavourable oceanographic conditions), but a moderate contribution is ensured (e.g. due to low juvenile mortality within seascape nursery habitats); finally, in period C, a strong recruitment to nursery areas is represented, but a low contribution to adult stocks occurs (e.g. owing to a high juvenile mortality within nursery habitats).

shifts (Nagelkerken et al. 2015), but also an interannual habitat quality and quantity dimension, owing to natural and anthropogenic factors (e.g. Sheaves 2016; Sheaves, Baker and Johnston 2006; Sheaves et al. 2015). This more dynamic perspective of the seascape nurseries concept has two domains that should be further explored: one addresses the management and conservation of interconnected habitats, independently of their relative value at a particular moment, and the other addresses research questions centred on the variability of habitat use by fish juveniles, insights into which are detailed in the last section of this chapter.

Why fish juveniles concentrate in nursery areas

Three interrelated conditions have usually been pointed out as the main drivers for juvenile concentration in coastal nurseries: high food availability, low predation pressure and appropriate abiotic conditions for rapid growth (e.g. Haedrich 1983; Miller, Crowder and Moser 1985; Beck et al. 2001; Lefcheck et al. 2019). These benefits imply some costs (Miller, Crowder and Moser 1985), namely a high mortality in early life stages (eggs and larvae), active migration-associated costs in the early juvenile phase, and metabolic costs and/or adaptations to an extremely variable environment, especially in terms of osmoregulation. Some of these aspects remain overlooked, and evidence from recent studies has suggested that they should be further investigated (e.g. the food limitation hypothesis in estuarine and coastal nurseries, Chevillot et al. 2018; Tableau et al. 2019; extremely low numbers of juveniles in certain estuarine nurseries owing to human pressures, Rochette et al. 2010).

Primary and secondary production is usually higher in estuaries and coastal zones than in other marine areas, especially in benthic organisms, which are typically the most abundant prey of fish species that use these areas as nursery grounds (e.g. Cabral 2000). Quantitative comparative studies (for nursery versus non-nursery areas) are scarce, but some have shown that food availability is higher within estuarine nursery areas than in adjacent non-nursery grounds, although the abundance of prey among nursery areas is extremely variable (e.g. Wouters and Cabral 2009). Nonetheless, the consumption of benthic prey has rarely been estimated, especially considering other groups than fish. Rosa et al. (2008) performed an exclusion experiment in an estuary to compare the seasonal variation of one key prey species (the polychaete Ragworm *Hediste diversicolor*) in sediment plots available to both bird and nekton predators, just to nekton, and without predators. The authors found that abundance of this prey in the plots protected from all predators was eight times greater than in those without any protection. They suggested that predation is a key factor for the population dynamics of prey, and that food availability can be a limiting factor in estuarine systems.

Le Pape and Bonhommeau (2015) reviewed the literature regarding food resources within nursery areas in order to evaluate whether or not food constitutes a limiting factor for juvenile growth and mortality. They collected evidence from different types of study regarding flatfish species that use coastal habitats as nursery grounds (from laboratory environments to in situ; experimental versus observational studies; individual versus population analyses), and outlined that, while some authors mentioned that food might be a limiting factor (e.g. Van der Veer and Witte 1993; Nash and Geffen 2000; Craig et al. 2007; Nash et al. 2007), others suggested that the carrying capacity of nursery areas is not reached (e.g. Rogers 1994; Shi, Gunderson and Sullivan 1997; Van der Veer et al. 2000; Vinagre and Cabral 2008). Le Pape and Bonhommeau (2015) proposed the food limitation hypothesis, stating that juveniles concentrate in estuaries and coastal productive areas where their fitness is enhanced through better feeding conditions and optimal growth, but that food limitation is of major importance in determining nursery habitat capacity. Further evidence of this hypothesis has been inferred by different methodological approaches (e.g. Tableau et al. 2016, 2019; Saulnier et al. 2020).

Low predation pressure in nursery areas is another generally accepted favourable condition for juvenile survival. The main support for this idea is the lack or scarcity of piscivorous fish in these areas (Sheaves 2001; Whitfield 2020). However, the occurrence of high trophic-level predators may vary according to geographical area or could even be a recent phenomenon. In some well-studied European estuarine systems a decline in top predators has been reported (e.g. some elasmobranchs such as the Smoothhounds *Mustelus* spp., Meagre *Argyrosomus regius*; see Costa 1982). In addition, predator groups such as birds or mammals have been underestimated. Moreover, non-indigenous piscivorous species have been reported in estuaries (e.g. Wels Catfish *Silurus glanis* in several European estuaries; Ferreira et al. 2019). It is also clear that it is not only top predators that may have a high impact on juvenile fish mortality. A large number of studies have documented high predation rates in nursery areas by crustaceans, such as shrimps (e.g. Brown Shrimp *Crangon crangon*), which are extremely abundant in coastal areas and may greatly affect juvenile cohort strength (e.g. Pihl and Van der Veer 1992; Sheaves 2001; Taylor 2005). Methods for estimating predation pressure on juveniles are also critical to provide insights here. Most of the studies that have estimated survival or mortality rates have used either predator or prey abundance at a particular moment or through time (e.g. Taylor 2005). This methodological approach may generate biased estimations by not accounting for predation avoidance behaviour (e.g. burrowing behaviour for flatfishes, selecting feeding areas with low numbers of predators, such as intertidal areas) or even under- or over-estimation of mortality rates owing to ontogenetic habitat shifts, including migration towards offshore areas (e.g. Vasconcelos et al. 2014).

Favourable conditions for rapid fish growth within nursery areas have rarely been tested and are mainly attributed to higher temperatures in spring and summer periods in estuaries and shallow coastal areas, when compared with deeper shelf waters. In fact, though, higher temperatures may not always support higher growth rates, or may promote less suitable conditions owing to interrelationships with other factors, such as hypoxic or anoxic zones (e.g. Schmidt et al. 2019). A large number of studies have compared fish juvenile growth in different nursery areas, but not nurseries versus non-nursery areas. These have often pointed out a pattern that may be attributed to latitude (higher growth rates in lower latitudes compared with northern areas, reflecting a latitudinal gradient in water temperature and not habitat-specific growth (e.g. Vinagre et al. 2008)). Some authors have reviewed the literature regarding fish juvenile growth rates in different habitats in which juveniles occur, and have concluded that certain habitats, namely seagrass beds, support an increased growth of juveniles (e.g. Heck, Hays and Orth 2003, McDevitt-Irwin et al. 2016). In a recent review, Lefcheck et al. (2019) conducted a meta-analysis using 160 studies on the role of structured habitats in promoting juvenile performance, and concluded that the pattern relative to growth was not particularly clear, and neither was the latitudinal trend in growth rate. The study of fish growth has some major methodological constraints, so it is difficult to validate the statement that fish grow more rapidly in some habitats than in others. The majority of studies have used fish otolith readings or cohort progression analyses using fish obtained from a relatively large area, and do not allow identification of a particular habitat within nurseries (e.g. Cabral and Costa, 2001; Cabral 2003). Several indicators that may be considered proxies of growth, especially fish condition, have also been comparatively analysed among different nursery areas (e.g. Vasconcelos et al. 2009; Schloesser and Fabrizio 2019), but trends and global patterns are difficult to discern in such evidence.

Methodologies for assessing the value of nursery areas

The methodological approaches that have been used for assessing nursery values typically cover one of the four dimensions according to Beck et al. (2001): fish abundance (densities or biomass), growth, survival and contribution to adult population. Integrative and holistic assessments are extremely scarce. The estimation of fish densities within juvenile habitats has been one of the most used methods for the evaluation of the habitat value as nursery. Beck et al. (2001) have outlined that nursery habitats should produce more adult recruits per unit area than other juvenile habitats, while Dahlgren et al. (2006) have stressed that the identification of nurseries should take into consideration their total contribution to the adult population, and thus that the area covered by nurseries and overall juvenile numbers should be determined. Despite this controversy, the majority of studies evaluating nursery function have been based on a comparison of fish densities (number of juveniles per unit of area) between different habitats (e.g. Lefcheck et al. 2019). Standard methods of fish sampling in estuaries and coastal zones have been used and compared (e.g. beam trawl, beach seines, drop or lift nets, among others; Elliott and Hemingway 2002), and these have focused on a wide diversity of habitats (e.g. mudflats, seagrass beds, saltmarsh or mangrove creeks, oyster reefs, sandy bottoms) in coastal areas worldwide (e.g. Heck and Crowder 1991; Able 1999; Lefcheck et al. 2019). Two major constraints, not usually considered in these comparisons, are the differential efficiency of fishing gears according to habitat or ontogenetic stages, and the difficulty in comparing estimates obtained using different fishing techniques. Furthermore, most of the studies neglect diel, short-term or ontogenetic movements between habitats, which may produce biased perspectives. For example, an intertidal mudflat may be extremely important as a feeding ground and may host a high density of juveniles, compared with other habitats, but it is an intermittent nursery ground, since fish move away from these areas during low tide. Fish densities are an important criterion for assessing the nursery value of a habitat, but they may not be directly related to growth and survival and thus may not reflect the real contribution to the population (e.g. Beck et al. 2001).

The study of fish growth also has some major methodological constraints that make it difficult to conclude that fish grow more rapidly in some habitats than in others. The majority of the studies used fish otolith readings or cohort progression analyses, with fish obtained from a relatively large area and not allowing the identification of a particular habitat within nurseries (e.g. Cabral 2003; Able et al. 2013). Studies of a wider geographical scope often outlined differences in growth patterns, which reflected latitudinal gradients in water temperature and not habitat-specific growth (as detailed above; e.g. Vinagre et al. 2008). Several indicators that may be considered proxies of growth, especially fish condition, have also been comparatively analysed among different nursery areas (e.g. Vasconcelos et al. 2009), but again trends and global patterns are difficult to reliably discern. Some methodological tools, such as stable isotopes or otolith geochemistry, have been concomitantly used in order to ensure a certain level of site fidelity (allowing the assumption that growth estimates were relative to fish using a particular habitat/area), but in most of the cases this relation is not easily established. Another major source of bias may be due to extended or late recruitment of contingents of juveniles to nursery areas (smaller numbers than previously), or the early migration of juveniles towards offshore areas (predominantly larger juveniles, who presumably grew faster) (Le Pape and Bonhommeau 2015). Survival estimates may be impacted by all the preceding constraints and sources of bias, with estimates of a decrease in abundance along with the occurrence period within nurseries being the traditional methodological approach used (Le Pape and Bonhommeau 2015). Several studies have focused instead on the mortality rates within nursery areas, using mainly predation or anthropogenic pressures (especially fisheries) as the main drivers (e.g. Pihl 1990; Cabral et al. 2002).

Finally, the quantification of the contribution of different nurseries to adult populations is not straightforward, since fish movement patterns towards offshore areas are poorly understood for many species (e.g. Gillanders 2002, Hamer, Jenkins and Gillanders 2005, Rooker et al. 2010). Most artificial tagging techniques, despite considerable technical advances, are not viable for small fish. Stable isotope analyses are extremely dependent on diet composition, which may not necessarily reflect a location or may be too complex to provide useful information to estimate connectivity patterns (fishes occupying higher trophic levels; e.g. meagre juveniles, Meagre *Argyrosomus regius*, within estuarine environments; Lagardère and Mariani 2006). Genetic markers provide relevant information whenever a population discontinuity is found, but are less useful in reduced connectivity scenarios, since a low level of mixing between populational sub-units usually corresponds to a lack of genetic differentiation (e.g. Calvès et al. 2013; Reis-Santos et al. 2018). Other natural tags, such as the chemical composition of otoliths, have been used successfully in this context, allowing discrimination of chemical elemental signatures in otoliths to assign or suggest nursery origin (e.g. Elsdon et al. 2008; Gillanders, 2005, 2009; Thorrold et al. 2001; Vasconcelos et al. 2008; Wells, Rooker and Itano 2012). By estimating the proportion of adults that have a particular signature, and thus have spent the juvenile period in a certain nursery, it may be possible to evaluate the contribution of the different nursery areas to the population, and ultimately to compare their value as nurseries. But aside from some technical or analytical aspects that are not straightforward, it has also been documented that these chemical signatures are not consistent over time, which implies that a long-term assessment is necessary (e.g. Reis-Santos et al. 2013).

Major knowledge gaps and future perspectives

Nursery-role concepts have been evolving and have nourished an interesting and useful scientific debate. Despite the extensive literature supporting the understanding of this topic, there are still major knowledge gaps worthy of further research, some of which are pointed out below without the assumption that they are exhaustive. Species and geographical areas that have been studied constitute a small and biased group, with African, Asian and South American coastal fish assemblages and systems clearly underrepresented in the literature. Inter-habitat use patterns, especially processes driving nursery colonisation, habitat selection or preference, and optimal/sub-optimal environmental conditions, are topics that even for the most studied species need further research. The high variability and patchy nature of juvenile fish density within nursery areas may suggest a strong individual behavioural component, reflecting an intra-species high variability in habitat use patterns (e.g. Cabral et al. 2007; Nagelkerken et al. 2015), which is very difficult to estimate and model without methodological approaches that allow individual tagging and monitoring. It would also be helpful if nursery concept theories could benefit from accumulated knowledge on fish ecophysiology, in particular cost–benefit analyses or energy-oriented modelling (e.g. Dynamic Energy Budget model, Mounier et al. 2020).

Effective indicators of fish condition and health are also needed to better understand the value of nurseries in terms of their contribution to the export of 'high-quality' juveniles (without diseases, pollutants, parasites, etc.) to offshore habitats where adults concentrate. Furthermore, while the quantification of human pressures has been undertaken in several studies, their relationships with the nursery value of coastal habitats have only rarely been established. One dimension that clearly needs to be investigated is the impact of climate change on coastal nurseries. The vast majority of the literature on this topic has tackled larval drift processes prior to nursery colonisation and outlined changes in connectivity and recruitment success (e.g. Britten, Dowd and Worm 2016; Lacroix, Barbut and Volckaert 2018; Young et al. 2018; Cabral et al. 2021). However, studies on patterns and processes, and their trends within nursery areas are very scarce, despite their relevance in terms of scientific and management perspectives.

A final comment should be made concerning the use of available scientific knowledge for management and conservation purposes. The seascape nursery concept, or even the more dynamic view that is introduced here, may be discouraging for managers or for conservation purposes, since it implies a high variability in the ranking of the most important habitats for fish juveniles, extremely dependent on ontogenetic stages and environmental context (including human pressures). Nonetheless, the focus for managers and conservation stakeholders should be to establish an interconnected habitat spatial plan that constitutes the major management or conservation unit. The sustainable management and conservation of coastal fish nurseries will surely benefit coastal ecosystem health and the services these ecosystems provide.

References

Able, K.W. (1999) Measures of juvenile fish habitat quality: examples from a national estuarine research reserve. In L.R. Benaka (ed.) *Fish Habitat: Essential Fish Habitat and Rehabilitation*, pp. 134–47. Bethesda, MD: American Fisheries Society. https://doi.org/10.47886/9781888569124.ch11

Able, K.W. (2005) A re-examination of fish estuarine dependence: evidence for connectivity between estuarine and ocean habitats. *Estuarine, Coastal and Shelf Science* 64: 5–17. https://doi.org/10.1016/j.ecss.2005.02.002

Able, K.W., Wuenschel, M.J., Grothues, T.M., Vasslides, J.M. and Rowe, P.M. (2013) Do surf zones in New Jersey provide 'nursery' habitat for southern fishes? *Environmental Biology of Fishes* 96: 661–75. https://doi.org/10.1007/s10641-012-0056-8

Adams, A.J., Dahlgren, C.P., Kellison, G.T., Kendall, M.S., Layman, C.A., Ley, J.A., Nagelkerken, I. and Serafy, J.E. (2006) Nursery function of tropical back-reef systems. *Marine Ecology Progress Series* 318: 287–301. https://doi.org/10.3354/meps318287

Baker, R., Fry, B., Rozas, L.P. and Minello, T.J. (2013) Hydrodynamic regulation of salt marsh contributions to aquatic food webs. *Marine Ecology Progress Series* 490: 37–52. https://doi.org/10.3354/meps10442

Beck, M.W., Heck Jr., K.L., Able, K.W., Childers, D.L., Eggleston, D.B., Gillanders, B.M., Halpern, B., Hays, C.G., Hoshino, K., Minello, T.J., Orth, R.J., Sheridan, P.F. and Weinstein, M.P. (2001) The identification, conservation, and management of estuarine and marine nurseries for fish and invertebrates. *Bioscience* 51: 633–41. https://doi.org/10.1641/0006-3568(2001)051[0633:ticamo]2.0.co;2

Boström, C., Pittman, S.J., Simenstad, C. and Kneib, R.T. (2011) Seascape ecology of coastal biogenic habitats: advances, gaps, and challenges. *Marine Ecology Progress Series* 427: 191–217. https://doi.org/10.3354/meps09051

Britten, G.L., Dowd, M. and Worm, B. (2016) Changing recruitment capacity in global fish stocks. *Proceedings of the National Academy of Sciences of the United States of America* 113: 134–9. https://doi.org/10.1073/pnas.1504709112

Cabral, H. (2000) Comparative feeding ecology of sympatric *Solea* and *S. senegalensis*, within the nursery areas of the Tagus estuary, Portugal. *Journal of Fish Biology* 57: 1550–62. https://doi.org/10.1006/jfbi.2000.140

Cabral, H. (2003) Differences in growth rates of juvenile *Solea* and *Solea senegalensis* in the Tagus estuary, Portugal. *Journal of the Marine Biological Association UK* 83: 861–8. https://doi.org/10.1017/S0025315403007902h

Cabral, H. and Costa, M.J. (2001) Abundance, feeding ecology and growth of 0-group sea bass, *Dicentrarchus labrax*, within the nursery areas of the Tagus estuary. *Journal of the Marine Biological Association UK* 81: 679–82. https://doi.org/10.1017/S0025315401004362

Cabral, H.N., Teixeira, C.M., Gamito, R. and Costa, M.J. (2002) Importance of discards of a beam trawl fishery as input of organic matter into nursery areas within the Tagus estuary. *Hydrobiologia* 475/476: 449–55. https://doi.org/10.1007/978-94-017-2464-7_34

Cabral, H., Drouineau, H., Teles-Machado, A., Pierre, M., Lepage, M., Lobry, J., Reis-Santos, P. and Tanner, S. (2021) Contrasting impacts of climate change on connectivity and larval recruitment to estuarine nursery areas. *Progress in Oceanography* 196: 102608. https://doi.org/10.1016/j.pocean.2021.102608

Cabral, H.N., Vasconcelos, R., Vinagre, C., França, S., Fonseca, V., Maia, A., Reis-Santos, P., Lopes, M., Ruano, M., Campos, J., Freitas, V., Santos, P.T. and Costa, M.J. (2007) Relative importance of estuarine flatfish nurseries along the Portuguese coast. *Journal of Sea Research* 57: 209–17. https://doi.org/10.1016/j.seares.2006.08.007

Calvès, I., Lavergne, E., Meistertzheim, A.L., Charrier, G., Cabral, H., Guinand, B., Quiniou, L. and Laroche, J. (2013) Genetic structure of European flounder *Platichthys flesus*: effects of both the southern limit of the species' range and chemical stress. *Marine Ecology Progress Series* 472: 257–73. https://doi.org/10.3354/meps09797

Chevillot, X., Tecchio, S., Chaalali, A., Lassalle, G., Selleslagh, J., Castelnaud, G., David, V., Bachelet, G., Niquil, N., Sautour, B. and Lobry, J. (2019) Global changes jeopardize the trophic carrying capacity and functioning of estuarine ecosystems. *Ecosystems* 22: 473–95. https://doi.org/10.1007/s10021-018-0282-9

Costa, M.J. (1982) Contribution à l'étude de l'écologie des poissons de l'estuaire du Tage (Portugal). PhD thesis. Université Paris VII, Paris.

Craig, J.K., Rice, J.A., Crowder, L.B. and Nadeau, D.A. (2007) Density dependent growth and mortality in an estuary-dependent fish: an experimental approach with juvenile spot *Leiostomus xanthurus*. *Marine Ecology Progress Series* 343: 251–62. https://doi.org/10.3354/meps06864

Dahlgren, C.P., Kellison, G.T., Adams, A.J., Gillanders, B.M., Kendall, M.S., Layman, C.A., Ley, J.A., Nagelkerken, I. and Serafy, J.E. (2006) Marine nurseries and effective juvenile habitats: concepts and applications. *Marine Ecology Progress Series* 312: 291–5. https://doi.org/10.3354/meps312291

Elliott, M. and Dewailly, F. (1995) Structure and components of European estuarine fish assemblages. *Netherlands Journal of Aquatic Ecology* 29: 397–417. https://doi.org/10.1007/bf02084239

Elliott, M. and Hemingway, K. (2002) Field methods. In M. Elliott and K. Hemingway (eds) *Fish in Estuaries*, pp. 410–509. Oxford: Blackwell Science. https://doi.org/10.1002/9780470995228.ch8

Elliott, M., O'Reilly, M.G. and Taylor, C.J.L. (1990) The Forth estuary: a nursery and overwintering area for North Sea fishes. *Hydrobiologia* 195: 89–103. https://doi.org/10.1007/bf00026816

Elliott, M., Whitfield, A.K., Potter, I.C., Blaber, S.J.M., Cyrus, D.P., Nordlie, F.G. and Harrison, T.D. (2007) The guild approach to categorizing estuarine fish assemblages: a global review. *Fish and Fisheries* 8: 241–68. https://doi.org/10.1111/j.1467-2679.2007.00253.x

Elsdon, T.S., Wells, B.K., Campana, S.E., Gillanders, B.M., Jones, C.M., Limburg, K.E., Secor, D.H., Thorrold, S.R. and Walther, B.D. (2008) Otolith chemistry to describe movements and life-history parameters of fishes: hypotheses, assumptions, limitations and inferences. *Oceanography and Marine Biology: An Annual Review* 46: 297–330. https://doi.org/10.1201/9781420065756.ch7

Ferreira, M., Gago, J. and Ribeiro, F. (2019) Diet of European catfish in a newly invaded region. *Fishes* 4: 58. https://doi.org/10.3390/fishes4040058

França, S., Vinagre, C., Costa, M.J. and Cabral, H.N. (2004) Use of the coastal areas adjacent to the Douro estuary as a nursery area for pouting, *Trisopterus luscus* Linnaeus, 1758. *Journal of Applied Ichthyology* 20: 99–104. https://doi.org/10.1046/j.1439-0426.2003.00525.x

Gillanders, B.M. (2005) Using elemental chemistry of fish otoliths to determine connectivity between estuarine and coastal habitats. *Estuarine, Coastal and Shelf Science* 64: 47–57. https://doi.org/10.1016/j.ecss.2005.02.005

Gillanders, B.M. (2009) Tools for studying biological marine ecosystem interactions – natural and artificial tags. In I. Nagelkerken (ed.) *Ecological Connectivity among Tropical Coastal Ecosystems*, pp. 457–92. Amsterdam: Springer. https://doi.org/10.1007/978-90-481-2406-0_13

Gillanders, B.M., Able, K.W., Brown, J.A., Eggleston, D.B. and Sheridan, P.F. (2003) Evidence of connectivity between juvenile and adult habitats for mobile marine fauna: an important component of nurseries. *Marine Ecology Progress Series* 247: 281–95. https://doi.org/10.3354/meps247281

Gillanders, B.M. (2002) Connectivity between juvenile and adult fish populations: do adults remain near their recruitment estuaries? *Marine Ecology Progress Series* 240: 215–23. https://doi.org/10.3354/meps240215

Gonçalves, E.J., Barbosa, M., Cabral, H.N. and Henriques, M. (2002) Ontogenetic shifts in patterns of microhabitat utilization in the small-headed clingfish, *Apletodon dentatus* (Gobiesocidae). *Environmental Biology of Fishes* 63: 333–9. https://doi.org/10.1023/a:1014302319622

Gunter, G. (1967) Some relationships of estuaries to the fisheries of the Gulf of Mexico. In G.H. Lauff (ed.) *Estuaries*, pp. 621–38. Washington: American Association for the Advancement of Science.

Haedrich, R.L. (1983) Estuarine fishes. In B. Ketchun (ed.) *Estuaries and Enclosed Seas*, pp. 183–207. Amsterdam: Elsevier.

Hamer, P.A., Jenkins, G.P. and Gillanders, B.M. (2005) Chemical tags in otoliths indicate the importance of local and distant settlement areas to populations of a temperate sparid, *Pagrus auratus*. *Canadian Journal of Fisheries and Aquatic Sciences* 62: 623–30. https://doi.org/10.1139/f04-221

Hammerschlag, N., Heithaus, M.R. and Serafy, J.E. (2010) Influence of predation risk and food supply on nocturnal fish foraging distributions along a mangrove–seagrass ecotone. *Marine Ecology Progress Series* 414: 223–35. https://doi.org/10.3354/meps08731

Harden Jones, F.R. (1968) *Fish Migration*. London: Edward Arnold Ltd.

Heck Jr., K.L. and Crowder, L.B. (1991) Habitat structure and predator–prey interactions in vegetated aquatic systems. In S.S. Bell, E.D. McCoy and H.R. Mushinsky (eds) *Habitat Structure: The Physical Arrangement of Objects in Space*, pp. 282–99. New York: Chapman and Hall. https://doi.org/10.1007/978-94-011-3076-9_14

Heck Jr., K.L., Hays, G. and Orth, R.J. (2003) Critical evaluation of the nursery role hypothesis for seagrass meadows. *Marine Ecology Progress Series* 253: 123–36. https://doi.org/10.3354/meps253123

Heck Jr, K.L., Nadeau, D.A. and Thomas, R. (1997) The nursery role of seagrass beds. *Gulf of Mexico Science* 1: 50–4. https://doi.org/10.18785/goms.1501.08

Hjort, J. (1926) Fluctuations in the year classes of important food fishes. *ICES Journal of Marine Science* 1: 5–38. https://doi.org/10.1093/icesjms/1.1.5

Hofrichter, R. and Patzner, R.A. (2000) Habitat and microhabitat of Mediterranean clingfishes (Teleostei: Gobiesociformes: Gobiesocidae). *PSZN Marine Ecology* 21: 41–53. https://doi.org/10.1046/j.1439-0485.2000.00689.x

Hufnagl, M., Peck, M.A., Nash, R.D.M., Pohlmann, T. and Rijnsdorp, A.D. (2013) Changes in potential North Sea spawning grounds of plaice (*Pleuronectes platessa* L.) based on early life stage connectivity to nursery habitats. *Journal of Sea Research* 84: 26–39. https://doi.org/10.1016/j.seares.2012.10.007

Huse, G. and Marshall, C.T. (2016) A spatial approach to understanding herring population dynamics. *Canadian Journal of Fisheries and Aquatic Sciences* 73: 177–88. https://doi.org/10.1139/cjfas-2015-0095

Igulu, M.M., Nagelkerken, I., van der Velde, G. and Mgaya, Y.D. (2013) Mangrove fish production is largely fuelled by external food sources: a stable isotope analysis of fishes at the individual, species, and community levels from across the globe. *Ecosystems* 16: 1336–52. https://doi.org/10.1007/s10021-013-9687-7

Iles, T.D. and Sinclair, M. (1982) Atlantic herring: stock discreteness and abundance. *Science* 215: 627–33. https://doi.org/10.1126/science.215.4533.627

Lacroix, G., Barbut, L. and Volckaert, F.A.M. (2018) Complex effect of projected sea temperature and wind change on flatfish dispersal. *Global Change Biology* 24: 85–100. https://doi.org/10.1111/gcb. 13915

Lagardère, J.P. and Mariani, A. (2006) Spawning sounds in meagre *Argyrosomus regius* recorded in the Gironde estuary, France. *Journal of Fish Biology* 69: 1697–1708. https://doi.org/10.1111/j.1095-8649. 2006.01237.x

Le Pape, O. and Bonhommeau, S. (2015) The food limitation hypothesis for juvenile marine fish. *Fish and Fisheries* 16: 373–98. https://doi.org/10. 1111/faf.12063

Lefcheck, J.S., Hughes, B.B., Johnson, A.J., Pfirrmann, B.W., Rasher, D.B., Smyth, A.R., Williams, B.L., Beck, M.W. and Orth, R.J. (2019) Are coastal habitats important nurseries? A meta-analysis. *Conservation Letters* 12: e12645. https://doi.org/10. 1111/conl.12645

McDevitt-Irwin, J.M., Iacarella, J.C. and Baum, J.K. (2016) Reassessing the nursery role of seagrass habitats from temperate to tropical regions: a meta-analysis. *Marine Ecology Progress Series* 557: 133–43. https://doi.org/10.3354/meps11848

McHugh, J.L. (1967) Estuarine nekton. *American Association for the Advancement of Science* 83: 581–620.

Miller, J.M., Crowder, L.B. and Moser, M.L. (1985) Migration and utilization of estuarine nurseries by juvenile fishes: an evolutionary perspective. *Contributions in Marine Science* 27: 338–52.

Minello, T. (1999) Nekton densities in shallow estuarine habitats of Texas and Louisiana and the identification of essential fish habitat. In L.R. Benaka (ed.) *Fish Habitat: Essential Fish Habitat and Rehabilitation*, pp. 43–75. Bethesda, MD: American Fisheries Society.

Mounier, F., Loizeau, V., Pecquerie, L., Drouineau, H., Labadie, P., Budzinski, H. and Lobry, J. (2020) Dietary bioaccumulation of persistent organic pollutants in the common sole *Solea* in the context of global change. Part 2: Sensitivity of juvenile growth and contamination to toxicokinetic parameters uncertainty and environmental conditions variability in estuaries. *Ecological Modelling* 431: 109196. https://doi.org/10.1016/j.ecolmodel.2020.109196

Nagelkerken, I., Sheaves, M., Baker, R. and Connolly, R.M. (2015) The seascape nursery: a novel spatial approach to identify and manage nurseries for coastal marine fauna. *Fish and Fisheries* 16: 362–71. https://doi.org/10.1111/faf.12057

Nagelkerken, I. (2007) Are non-estuarine mangroves connected to coral reefs through fish migration? *Bulletin of Marine Science* 80: 595–607.

Nash, R.D M. and Geffen, A.J. (2000) The influence of nursery grounds processes in the determination of year-class strength in juvenile plaice *Pleuronectes platessa* L. in Port Erin Bay, Irish Sea. *Journal of Sea Research* 44: 101–10. https://doi.org/10.1016/S1385 -1101(00)00044-7

Nash, R.D.M., Geffen, A.J., Burrows, M.T. and Gibson, R.N. (2007) Dynamics of shallow-water juvenile flatfish nursery grounds: application of the

shelf-thinning rule. *Marine Ecology Progress Series* 344: 231–44. https://doi.org/10.3354/meps06933

Olds, A.D., Albert, S., Maxwell, P.S., Pitt, K.A. and Connolly, R.M. (2013) Mangrove-reef connectivity promotes the effectiveness of marine reserves across the western Pacific. *Global Ecology and Biogeography* 22: 1040–1049. https://doi.org/10.1111/geb.12072

Orth, R.J., Heck Jr., K.L. and van Montfrans, J. (1984) Faunal communities in seagrass beds: a review of the influence of plant structure and prey characteristics on predator–prey relationships. *Estuaries* 7: 339–50. https://doi.org/10.2307/1351618

Pais, M.P., Henriques, S., Batista, M.I., Costa, M.J. and Cabral, H. (2013) Seeking functional homogeneity: a framework for definition and classification of fish assemblage types to support assessment tools on temperate reefs. *Ecological Indicators* 34: 231–45. https://doi.org/10.1016/j.ecolind.2013.05.006

Parin, N.V. (1986) Trichiuridae. In P.J.P. Whitehead, M.-L. Bauchot, J.-C. Hureau, J. Nielsen and E. Tortonese (eds) *Fishes of the North-Eastern Atlantic and the Mediterranean*, Vol. 2, pp. 976–80. Paris: UNESCO.

Pihl, L. and Van der Veer, H.W. (1992) Importance of exposure and habitat structure for the population density of 0-group plaice, *Pleuronectes platessa* L., in coastal nursery areas. *Netherlands Journal of Sea Research* 29: 145–52. https://doi.org/10.1016/0077 -7579(92)90015-7

Pihl, L. (1990) Year-class strength regulation in plaice (*Pleuronectes platessa*) on the Swedish west coast. *Hydrobiologia* 195: 79–88. https://doi.org/10.1007/BF00026815

Potter, I.C., Beckley, L.E., Whitfield, A.K. and Lenanton, R.C. (1990) Comparisons between the roles played by estuaries in the life cycles of fishes in temperate Western Australia and Southern Africa. *Environmental Biology of Fishes* 28: 143–78. https://doi.org/10.1007/bf00751033

Potter, I.C., Tweedley, J.R., Elliott, M. and Whitfield, A. (2015) The ways in which fish use estuaries: a refinement and expansion of the guild approach. *Fish and Fisheries* 16: 230–9. https://doi.org/10.1111/faf.12050

Priede, I.G. (2019) Deep-sea fishes literature review. JNCC Report No. 619. JNCC, Peterborough.

Prista, N., Vasconcelos, R.P., Costa, M.J. and Cabral, H. (2003) The demersal fish assemblage of the coastal area adjacent to the Tagus estuary (Portugal): relationships with environmental conditions. *Oceanologica Acta*, 26: 525–36. https://doi.org/10.1016/S0399-1784(03)00047-1

Reis-Santos, P., Tanner, S.E., Vasconcelos, R.P., Elsdon, T.S., Cabral, H.N. and Gillanders, B.M. (2013) Connectivity between estuarine and coastal fish populations: contributions of estuaries are not consistent over time. *Marine Ecology Progress Series* 491, 177–186. https://doi.org/10.3354/meps10458

Reis-Santos, P., Tanner, S.E., Aboim, M.A., Vasconcelos, R.P., Laroche, J., Charrier, G., Perez, M.,

Presa, P., Gillanders, B.M. and Cabral, H.N. (2018) Reconciling differences in natural tags to infer demographic and genetic connectivity in marine fish populations. *Scientific Reports* 8: 10343. https://doi.org/10.1038/s41598-018-28701-6

Rochette, S., Rivot, E., Morin, J., Mackinson, S., Riou, P. and Le Pape, O. (2010) Effect of nursery habitat degradation on flatfish population: Application to *Solea* in the Eastern Channel (Western Europe). *Journal of Sea Research* 64: 34–44. https://doi.org/10.1016/j.seares.2009.08.003

Rogers, S.I. (1994) Population density and growth rate of juvenile sole *Solea* (L.). *Netherlands Journal of Sea Research* 32: 353–60. https://doi.org/10.1016/0077-7579(94)90012-4

Rooker, J.R., Stunz, G.W., Holt, S.A. and Minello, T.J. (2010) Population connectivity of red drum in the northern Gulf of Mexico. *Marine Ecology Progress Series* 407: 187–96. https://doi.org/10.3354/meps08605

Rosa, S., Granadeiro, J.P., Vinagre, C., França, S., Cabral, H.N. and Palmeirim, J.M. (2008) Impact of predation on the polychaete *Hediste diversicolor* in estuarine intertidal flats. *Estuarine, Coastal and Shelf Science* 78: 655–64. https://doi.org/10.1016/j.ecss.2008.02.001

San Martín, M.A., Cubillos, L.A. and Saavedra, J.C. (2011) The spatio-temporal distribution of juvenile hake (*Merluccius gayi gayi*) off central southern Chile (1997–2006). *Aquatic Living Resources* 24: 161–8. https://doi.org/10.1051/alr/2011120

Saulnier, E., Le Bris, H., Tableau, A., Dauvin, J.C. and Brind'Amour, A. (2020) Food limitation of juvenile marine fish in a coastal and estuarine nursery. *Estuarine, Coastal and Shelf Science* 241: 106670. https://doi.org/10.1016/j.ecss.2020.106670

Schloesser, R.W. and Fabrizio, M.C. (2019) Nursery habitat quality assessed by the condition of juvenile fishes: Not all estuarine areas are equal. *Estuaries and Coasts* 42: 548–66. https://doi.org/10.1007/s12237-018-0468-6

Schmidt, S., Diallo, I.I., Derriennic, H., Fallou, H. and Lepage, M. (2019) Exploring the susceptibility of turbid estuaries to hypoxia as a prerequisite to designing a pertinent monitoring strategy of dissolved oxygen. *Frontiers in Marine Science* 6: 352. https://doi.org/10.3389/fmars.2019.00352

Sheaves, M. (2001) Are there really few piscivorous fishes in shallow estuarine habitats? *Marine Ecology Progress Series* 222: 279–90. https://doi.org/10.3354/meps222279

Sheaves, M. (2005) Nature and consequences of biological connectivity in mangrove systems. *Marine Ecology Progress Series* 302: 293–305. https://doi.org/10.3354/meps302293

Sheaves, M. (2016) Simple processes drive unpredictable differences in estuarine fish assemblages: Baselines for understanding site-specific ecological and anthropogenic impacts. *Estuarine, Coastal and Shelf Science* 170: 61–9. https://doi.org/10.1016/j.ecss.2015.12.025

Sheaves, M., Baker, R. and Johnston, R. (2006) Marine nurseries and effective juvenile habitats: an alternative view. *Marine Ecology Progress Series* 318: 303–6. https://doi.org/10.3354/meps318303

Sheaves, M., Baker, R., Nagelkerken, I. and Connolly, R.M. (2015) True value of estuarine and coastal nurseries for fish: incorporating complexity and dynamics. *Estuaries and Coasts* 38: 401–14. https://doi.org/10.1007/s12237-014-9846-x

Shi, Y., Gunderson, D.R. and Sullivan, P.J. (1997) Growth and survival of 0 super (+) English sole, *Pleuronectes vetulus*, in estuaries and adjacent nearshore waters off Washington. *Fishery Bulletin* 95: 161–73.

Sinclair, M. (1988) *Marine Populations*. Seattle: University of Washington Press.

Tableau, A., Brind'Amour, A., Woillez, M. and Le Bris, H. (2016) Influence of food availability on the spatial distribution of juvenile fish within soft sediment nursery habitats. *Journal of Sea Research* 111: 76–87. https://doi.org/10.1016/j.seares.2015.12.004

Tableau, A., Le Bris, H., Saulnier, E., Le Pape, O. and Brind'Amour, A. (2019) Novel approach for testing the food limitation hypothesis in estuarine and coastal fish nurseries. *Marine Ecology Progress Series* 629: 117–31. https://doi.org/10.3354/meps13090

Tanner, S.E., Teles-Machado, A., Martinho, F., Peliz, A., Cabral, H.N., 2017. Modelling larval dispersal dynamics of common sole (*Solea solea*) along the western Iberian coast. *Progress in Oceanography* 156: 78–90. https://doi.org/10.1016/j.pocean.2017.06.005

Taylor, D.L. (2005) Predation on post-settlement winter flounder *Pseudopleuronectes americanus* by sand shrimp *Crangon septemspinosa* in NW Atlantic estuaries. *Marine Ecology Progress Series* 289: 245–62. https://doi.org/10.3354/meps289245

Thorrold, S.R., Latkoczy, C., Swart, P.K. and Jones, C.M. (2001) Natal homing in marine fish metapopulation. *Science* 291: 297–9. https://doi.org/10.1126/science.291.5502.297

Vallisneri, M., Tommasini, S., Stagioni, M., Manfredi, C., Isajlović, I. and Montanini, S. (2014) Distribution and some biological parameters of the red gurnard, *Chelidonichthys cuculus* (Actinopterygii, Scorpaeniformes, Triglidae) in the north-central Adriatic Sea. *Acta Ichthyologica et Piscatoria* 44: 173–80. https://doi.org/10.3750/AIP2014.44.3.01

van der Veer, H.W. and Witte, J.I.J. (1993) The 'maximum growth/optimal food condition' hypothesis: a test for 0-group plaice *Pleuronectes platessa* in the Dutch Wadden Sea. *Marine Ecology Progress Series* 10: 81–90. https://doi.org/10.3354/meps101081

Van der Veer, H., Bergham, R., Miller, J. and Rijnsdorp, A. (2000) Recruitment in flatfish, with special emphasis on North Atlantic species: Progress made by the Flatfish Symposia. *ICES Journal of Marine Science* 57: 202–15. https://doi.org/10.1006/jmsc.1999.0523

Vasconcelos, R.P., Reis-Santos, P., Fonseca, V., Ruano, M., Tanner, S., Costa, M.J. and Cabral, H. (2009) Juvenile fish condition in estuarine nurseries along the Portuguese coast. *Estuarine, Coastal and Shelf Science* 82: 128–38. https://doi.org/10.1016/j.ecss.2009.01.002

Vasconcelos, R.P., Reis-Santos, P., Tanner, S., Maia, A., Latkoczy, C., Günther, D., Costa, M.J. and Cabral, H. (2008) Evidence of estuarine nursery origin of five coastal fish species along the Portuguese coast through otolith elemental fingerprints. *Estuarine, Coastal and Shelf Science* 79: 317–27. https://doi.org/10.1016/j.ecss.2008.04.006

Vasconcelos, R.P., Eggleston, D.B., Le Pape, O. and Tulp, I. (2014) Patterns and processes of habitat-specific demographic variability in exploited marine species. *ICES Journal of Marine Science* 71: 638–47. https://doi.org/10.1093/icesjms/fst136

Ventura, D., Jona Lasinio, G. and Ardizzone, G. (2015) Temporal partitioning of microhabitat use among four juvenile fish species of the genus *Diplodus* (Pisces: Perciformes, Sparidae). *Marine Ecology* 36: 1013–32. https://doi.org/10.1111/maec.12198

Vinagre, C. and Cabral, H. (2008) Prey consumption by juvenile soles, *Solea* and *Solea senegalensis*, in the Tagus estuary, Portugal. *Estuarine, Coastal and Shelf Science* 78: 45–50. https://doi.org/10.1016/j.ecss.2007.11.009

Vinagre, C., Amara, R., Maia, A. and Cabral, H.N. (2008) Latitudinal comparison of spawning season and growth of 0-group sole, *Solea* (L.). *Estuarine, Coastal and Shelf Science* 78: 521–8. https://doi.org/10.1016/j.ecss.2008.01.012

Wells, R.J.D., Rooker, J.R. and Itano, D.G. (2012) Nursery origin of yellowfin tuna in the Hawaiian Islands. *Marine Ecology Progress Series* 461: 187–96. https://doi.org/10.3354/meps09833

Whitfield, A.K. (2020) Fish species in estuaries – from partial association to complete dependency. *Journal of Fish Biology* 97: 1262–4. https://doi.org/10.1111/jfb.14476

Wouters, N. and Cabral, H.N. (2009) Are flatfish nursery grounds richer in benthic prey? *Estuarine, Coastal and Shelf Science* 83: 613–20. https://doi.org/10.1016/j.ecss.2009.05.011

Young, E.F., Tysklind, N., Meredith, M.P., de Bruyn, M., Belchier, M., Murphy, E.J. and Carvalho, G.R. (2018) Stepping stones to isolation: Impacts of a changing climate on the connectivity of fragmented fish populations. *Evolutionary Applications* 11: 978–94. https://doi.org/10.1111/eva.12613

CHAPTER 15

Marine Conservation: Smoke and Mirrors in the Coastal Zone

JOHN HUMPHREYS

Abstract

This chapter elucidates certain aspects of marine conservation policy that provide governments with opportunities to use a rhetoric of achievement that is inconsistent with reality. To this end, British government obligations are critically tracked from the high-level global targets to which it is a signatory, through its own self-assessment and claims of world-leading status, to a reality exemplified by an estuarine marine protected area on the south coast of England. The chapter illustrates how proxy targets, combined with a fragmented system of environmental legislation and statutory instruments, and consequently non-strategic regulatory responses, effectively obscure failure and enable embellishment of the reality of environmental degradation in coastal seas. It appears that amenability to regulation trumps environmental risk in determining the priorities of government interventions, and that amenability to regulation is a function of politics and influence. Consequently, the strong are weakly regulated while the weak are strongly regulated, and an emphasis on measures of uncertain utility leaves relatively undisturbed the major underlying practices standing in the way of environmental protection and remediation. Alternative approaches are suggested that, if properly formulated, could rectify some of these deficits and improve the efficacy of future marine conservation policy and practice.

Keywords: marine conservation, marine protected areas, UK marine strategy, Poole Harbour

Correspondence: jhc@jhc.co

Introduction

The British government has established a powerful public relations message around its 'Blue Belt' Marine Protected Area (MPA) programme. Government ministers claim that Britain is 'leading the world' in marine protection, based on the percentage sea area within its Exclusive Economic Zone (EEZ) which falls within a designated MPA. On this foundation the UK is promoting a new 30% global MPA coverage target in international environmental arenas such as the G7 (Defra 2019a and b).

This chapter examines the relationship between the rhetoric and the reality, arguing that British government grandstanding on marine protection is far removed from the reality implied. More particularly, it shows how a fragmented collection of obligations, legislation and policy instruments, at international, national and local levels, enables such claims in the context of a degraded environment. To this end, the chapter first considers global policy on MPAs before determining the veracity of government claims with reference to its obligations and examining a particular case on the ground.

The scope of the chapter therefore reaches from top to bottom: from international policy to a particular estuarine MPA on the south coast of England. In so doing, it attempts to analyse weaknesses in conservation policy and distortions in practice, in such a way as to allow reflection on the future for marine conservation.

Smoke and mirrors 1: international targets

Global policy on MPAs is established through the Convention on Biological Diversity (CBD) to which the British government is a party. In 2010, the international community under the auspices of the CBD met in Japan to establish a series of Aichi Targets for 2020. Target 11 required that 10% of the sea be conserved within MPAs. Intrinsic to the area target were four defining attributes of an MPA that served as qualifying clauses determining the requirements for an MPA to count towards the area target. In 2020, at the end of the target period, the CBD published its final report on the Aichi Targets, based on submissions from member states (CBD 2020).

For the purpose of assessing the outcome, the CBD Secretariat effectively converted a single holistic target with qualifying (essentially defining) clauses into six separately assessed elements. Aichi Target 11 is stated below with numbers added by the author to indicate the separated elements, but with 1 omitted since it relates exclusively to the terrestrial environment:

> By 2020 at least … 10% of coastal and marine areas (2), especially areas of particular importance for biodiversity and ecosystem services (3) are conserved through effectively and equitably managed (4), ecologically representative (5) and well-connected systems of protected areas … and integrated into the wider … seascape (6). (CBD 2020, with numbers added by the author)

Of these, element 2 is reported as achieved, whereas elements 3–6 are not. On this basis, the CBD claimed 'partial achievement' (Humphreys 2021).

The CBD report was itself based on reports from its signatory states. In 2019 Britain reported that it was 'on track to achieve Target 11' as follows:

> UK's protected area network currently covers … 24% of its sea area … The … network has been designated following principles to help identify they are ecologically representative and well connected. Over 60% of sites within the UK protected area network are compliant with global management effectiveness criteria … Nevertheless … work to fully implement marine protected area management measures and monitor their effectiveness is ongoing. (JNCC 2019)

For Britain, although not always distinguishing between terrestrial and marine protected areas, it appears success was claimed on elements 2, 3, 5 and 6 of Target 11 but, in the marine environment, despite the ambiguous prose, not 4, since management measures are not fully implemented.

Globally and in British seas, the 10% quantitative area target appears to be achieved, while three (Britain) or none (the world) of the other elements are. Should we be reassured that at least 10% of the global marine environment or that 24% (by now more) of the UK's EEZ was reported as falling within a designated marine protected area?

The post hoc finessing of Target 11 so that each element is separately evaluated enables disingenuous claims of achievement. In fact, the individual elements are not sequential. Elements 4–6 qualify element 2. When other elements have not been achieved, the achievement of element 2 is strictly meaningless, and the MPA should not be counted as such. What has been 'achieved' are MPA designations: lines in the sea identifying the boundaries of ostensibly conserved areas, which may or may not be important, managed, representative or well connected.

Assessing qualifying clauses as separate elements is of particular significance if element 4 is not achieved, since the management of an MPA is the one essential process that distinguishes it from any other part of the sea. An unmanaged protected area is essentially an unprotected area. Management is also the element that requires ongoing and often local resourcing, in contrast to other elements that are part of the initial designation process. In 2019, the British government announced 41 new Marine Conservation Zones with no new funding for management (Defra 2019a).

Reality exemplified

To elucidate the reality behind the British government's self-declared 'world leading' status and ostensible 'achievement' of the CBD area target, it is instructive to move from the general to the specific. Although an increasing number of very large MPAs are designated in the high seas, the greater number are located in inshore areas where anthropogenic impacts are most clearly understood and the practice of government regulation is most observable. Estuaries in particular are revealing in these respects, as their environmental protections often involve a complex collection of statutory instruments whose efficacy in practice can be evaluated. Poole Harbour on the central south coast of England is one such estuary.

Poole Harbour is an MPA with several overlapping statutory designations, most significantly as a Site of Special Scientific Interest since 1991, and a Special Protection Area since 1999. It is also a listed Ramsar Convention site (Humphreys and May 2005). These various statutory MPA designations are further overlayed with two general aquatic conservation instruments that are distinct from the MPA designations: The Water Environment Regulations (WER, transposed from the EU Water Framework Directive), which although primarily applied to inland waters in fact extends into coastal waters; the UK Marine Strategy (UKMS, also transposed from an EU directive), which extends to the mean high-water level of spring tides. In contrast to the MPA designations that specify particular features for protection, both WER and UKMS are examples of an ecosystem approach to conservation, with the holistic integrated goal of achieving Good Environmental Status (GES).

It might be supposed that such a site, in contributing to the claimed British MPA achievement, would demonstrate the benefits of effective environmental regulation. Yet the harbour also has a history of water quality issues, which technically have no direct bearing on its MPA status. These include high dissolved nitrate levels originating mainly from agriculture in the catchment, and sewage effluent released directly or from upstream sources (Wardlaw 2005). Consequently, in addition to its protected status the harbour has also been classified as Polluted Waters (Eutrophic) and a Sensitive Area (Eutrophic) under European directives (Humphreys and Hall 2022), and in unfavourable condition owing to eutrophication (Bowles 2020). The two contrasting styles of designation for Poole Harbour: heavily protected yet heavily polluted, are not uncommon in British estuaries, many of which contribute to the government's MPA area target. It is noteworthy that if Poole Harbour MPA is considered to fulfil all the CBD's qualifying requirements for an MPA, it would be considered an 'achievement' despite the failure to achieve GES.

However, the problem is not just nitrates. Antiquated foul water infrastructure combined with inadequate treatment facilities result in rainfall events in which the capacity of sewage treatment works to process and/or hold effluent is overwhelmed. As a result, untreated 'raw' sewage is released directly into many estuaries and their catchments.

In 2019, privatised water utility companies in England discharged raw sewage from such 'storm overflows' for a total of 1.5 million hours over 204,000 release episodes, and in 2020, the reported number of raw sewage discharges was over 420,000 (Humphreys and Hall 2022).

In Poole Harbour, over 700 such overflows from seven outfalls occurred between 19 December 2019 and 17 February 2020. Although some were of short duration, many were not. On some days, a single outflow was releasing raw sewage into the harbour for over 22 hours. These overflows are legal in so far as they remain within the limits expressed in permits from the government regulator, the Environment Agency (EA), whose approach to raw sewage overflow episodes involves a 'lighter touch' regulatory regime involving 'operator self-monitoring' (Humphreys and Hall 2022).

Smoke and mirrors 2: UK marine strategy

It is also illuminating to contrast the reality of Poole Harbour with the British government's general self-assessment in relation to the aim of GES under the UK Marine Strategy. GES is defined in terms of 'ecologically diverse and dynamic ocean and seas which are clean, healthy and productive, within their intrinsic conditions, and the use of the marine environment is at a level that is sustainable, thus safeguarding the potential for uses and activities by current and future generations' (Defra 2019c).

The concept of GES allows for a level of exploitation for human benefit and therefore a degree of anthropogenically induced change, but only within environmental limits; that is, 'intrinsic conditions'. In a sense it implies a sort of 'carrying capacity' for humans somewhat analogous to that sometimes attributed to an estuary's ability to support protected bird populations.

Reporting on progress in 2019, the government announced: 'We have largely achieved GES for eutrophication' (Defra 2019c). Leaving aside what Poole Harbour may tell us about the meaning of the phrase 'largely achieved', this statement reveals a finessing of apparent 'achievement' from actual failure that is reminiscent of that claimed for MPAs. Again, an integrated target, in this case for GES, is broken down into separate elements that are then individually assessed in such a way as to claim some sort of achievement.

The point about GES was that it represented a whole ecosystem approach to conservation as an improvement over a features approach in which specified species or habitats are conserved while potentially allowing continued degradation at system level. In this context, the idea that 'We have largely achieved GES for eutrophication' is a contradiction in terms. The problem is not that the achievement of GES should not be assessed through a combination of measures, but rather that the improvements in one element can be regarded in any sense a GES achievement in the absence of the others. In fact, GES, in common with the achievement of Aichi Target 11, can only exist when all elements are in place.

Gaming and grandstanding

These problems with both Aichi Target 11 and GES are not merely semantic; they facilitate a scientifically incoherent but politically advantageous game.

However good the theory and intentions behind Aichi Target 11, and the introduction of GES in marine conservation policy, in their application both have been finessed or 'gamed' in such a way to claim achievement where none strictly exists. The simple reality is that, in these cases, Aichi Target 11 and GES have not been achieved. In both, an original conception has been retrospectively distorted in the reporting, in such a way as appears to evade proper recognition of failure.

The gaming by governments of targets is well established in the literature. Gaming is characterised by minimising qualitative standards in order to achieve quantitative targets – a phenomenon perfectly demonstrated by the achievement of the one quantitative element of Aichi Target 11 (10% coverage), on the basis of failing on qualitative elements.

Evading the recognition of failure is itself problematic, but at worst it also provides political benefits in terms of opportunities for government grandstanding on the results. As technical documents are represented in press releases and speeches, failures are finessed away and 'achievements' become amplified. This is particularly evident with MPAs: lines in the sea become enough. The 2019 designation of 41 new MPAs enabled the British government to claim that 'The UK is already leading the rest of the world by protecting over 30% of our ocean' (Defra 2019a). Through such means, the government assumes 'world leader' status, encouraging a new CBD target of 30% that it claims to have already achieved, while actually it is technically unable to demonstrate its achievement even of the 10% target.

Smoke and mirrors 3: greater attention to lesser risks

Salmon and sea trout migrate through the Poole Harbour estuary from and to their upstream spawning grounds. The Atlantic Salmon *Salmo salar* has been in decline since the industrialisation of the eighteenth century, with severe impacts caused by human activities, notably water pollution (WWF 2001). Salmon thrive in well-oxygenated water with low levels of organic and inorganic pollution. In the UK, 86% of English rivers are not of good ecological status (EA 2020). Noting that salmon stocks in many rivers across England had failed to meet their minimum safe levels, the government regulator, the EA, adopted several formal commitments in 2015. These included Commitment 1.2 'Optimize survival of adult salmon and smolts in estuarine and coastal environments', and Commitment 5.5 'Assess priorities for action in addressing water pollution impacts on salmon' (EA 2015).

In seeking to fulfil Commitment 1.2, EA has turned its attention to Poole Harbour, in which netting for marine fish is long established. These fisheries are actively managed, sustainable and do not target salmonids. For the EA, the issue is bycatch. Their concern is that nets set to target marine species may intercept migrating salmonids and therefore (with minor exceptions) 'are of the opinion that closing all sea fish netting exploitation in estuaries by byelaw … represents the only logical and pragmatic means to manage this issue in the longer term' (EA 2017).

Some insight into the sentiment underlying this assertion is provided by an EA opinion relating to nearby Southampton Water in which they equate the loss of a single adult female salmon to be sufficient to conclude (using the technical language of conservation legislation) 'that no adverse effect on site integrity could not be demonstrated'. This single female salmon determination has been raised in relation to netting in Poole Harbour.

So, the commitment to optimise the survival of adult salmon in estuaries implies the cessation of all marine fish netting, in order that a single salmon should not be accidentally caught. In contrast, the long-standing need to improve chronic pollution requires no direct regulatory action, only the belated formulation of a plan: '5.5 Assess priorities for action in addressing water pollution impacts on salmon'. How can we explain this?

In contrasting regulator responses to water pollution and salmonids in Poole Harbour, it is not necessary to imply that estuarine water quality is currently the highest risk causal factor in the continuing fall in the North Atlantic stock. There are, for example, reasons to believe that the current sustained decline may be connected to oceanic water temperature trends (e.g. Nicola et al. 2018). However, water quality is a major general environmental problem with well-understood detriment across many habitats and species, including salmonids. In contrast, already regulated marine artisanal netting fisheries are not causing ecosystem-wide impacts. Moreover, the extent to which marine fish netting bycatch would have any material effect on salmonid stocks is questionable, both in terms of the lack of secure quantification of bycatch and the materiality of any such bycatch on a salmon population already compromised by many other challenges, including chronic water pollution.

In essence, it is doubtful whether cessation of netting for other species will have any measurable beneficial effect in terms of salmon stock recovery. The response to this situation is to cite the precautionary principle, which in this context is used to justify attention to netting, effectively to compensate for either an inability (marine survival) or apparent disinclination (water quality) to act on more evidenced causes of decline. The result is a weaker regulatory response to a higher risk (water quality) and a stronger regulatory response to a lesser risk (netting bycatch). It appears that regulatory attention is non-strategic and as such determined as much by amenability to regulation as it is to environmental risk. For reasons outlined below, water quality is not amenable to regulation but artisanal netting for other species is.

Cui bono; who benefits? In terms of avoiding the substantial direct costs of reducing water pollution, the polluters do: agricultural landowners and commercial water companies. Whereas landowners and water companies are relatively organised, economically powerful and politically well represented, inshore fishers are in contrast politically weak and economically significant only in the local context of coastal communities. Moreover, solutions to raw sewage releases will require an injection of capital funding for infrastructure projects. However, the system is such that privatised water companies keen to protect their profits, will effectively expect compensation in the form of increases in charges to consumers, something another government regulator, The Water Services Regulation Authority (Ofwat – which sets limits on charges to consumers) is reluctant to facilitate. In this context, the EA's response to pollution is weak.

Here we have an explanation for the inverse relationship between the environmental risk and the regulatory response: The strong are weakly regulated while the weak are strongly regulated, which in turn suggests a government manifestly reluctant to push too hard against economic interests to which they are for various reasons politically committed – but at the same time needing to be seen to be doing something.

It would be naive to suppose that economics and politics would not be factors in the determination of environmental outcomes. But the case of salmonids sheds light on the current state of statutory conservation. The citing of a single fish as the basis for control or cessation of an activity is revealing, not in itself, but in so far as it is not applied to the larger risk of water pollution. Neither does the argument appear to be applied to upstream recreational salmon fisheries, for which the regulator (EA) is also directly responsible, and which, even in the context of a catch-and-return policy, must result in a less than 'optimised' level of fish mortality.

Solutions?

The modern MPA movement has its roots in the terrestrial National Park movement that since 1960 has become structurally integrated (as the World Commission on Protected Areas) into a United Nations environmental nexus of which the Convention on Biological Diversity is a part. As marine conservation interest grew, the National Park movement was quick to step in with a readymade strategy (Humphreys and Clark 2020). However, there are many differences between terrestrial and marine ecosystems, not least in terms of the nature of the ambient medium, which makes it more difficult in the marine sphere to manage the environment in spatially defined units. Poole Harbour illustrates this difficulty for coastal seas and estuaries, where MPA management is overwhelmed by inadequate practices elsewhere. Moreover, a degraded marine environment is much less publicly visible than a degraded landscape.

Nevertheless, the momentum for increasing MPA coverage as a solution to marine protection continues to grow. Since the beginning of the century academic books on marine conservation have focused on the principles and practice of MPAs (e.g. Roff and Zacharias 2001). The UK now claims to be on track to safeguard '50% of our precious marine habitats' (Defra 2019d), and along with environmental non-governmental organisations is now promoting a new global 30% by 2030 MPA area target ('30 by 30'), on the back of two failures at 10% (Humphreys and Clark 2020).

While there is strong evidence for the utility of MPAs in particular and specific circumstances, it is questionable whether, in their current formulation, they can provide a consistently effective strategy even for locally delineated coastal protection, and there is little evidential support for the general protection of the oceans through increasing global area targets. Moreover, alternative approaches are available.

In the 50 years since the foundation of the Estuarine and Coastal Sciences Association, if we omit fisheries conservation, two distinct policy trajectories for marine conservation can be discerned, both of which are manifest in Poole Harbour: the first scientifically informed and progressive, the second essentially reactive, involving arbitrary boundaries and proxy targets.

The first of these is based on a growing scientific understanding of environmental complexity, connectivity and ecological function. Within this trajectory, progressive improvements include the move from features approaches, through which species or habitats are protected out of context from the system within which they are integral, to habitat approaches that recognise aspects of this wider context, to whole ecosystem approaches, of which GES is, in principle, an example.

The second trajectory, initiated as an extension of a terrestrial National Park movement into the marine sphere, has involved the development of policy as a reaction to conservation failure. Examples include: the shifting goals of MPAs from local protections to a whole ocean strategy with arbitrary area targets (Humphreys 2021) to compensate for perceived failures of fisheries regulation (Humphreys and Clark 2020); the tightening of the definition of MPAs after recognition that governments where counting unmanaged but designated areas (Humphreys and Clark 2020); and lately the growing emphasis on Highly Protected Marine Areas (e.g. Benyon 2020), from which all or most human activity is excluded.

The reactive (MPA) trajectory is problematic in the sense of combining the twin agendas of greatly increased sea area with increased exclusion. However well-intentioned this dual direction may be, in reality there must be an inverse relationship between the two aims: the level of economic exclusion must at some point become more limited. The logic of the 10, 30, (and emerging) 50% trend must increasingly be reconciled with sustainable exploitation. In contrast, ecosystem approaches explicitly accommodate human economic activity – for example in the UK Marine Strategy's concept of GES, as defined above (Defra 2019c).

Marine protection policies that tend to polarise rather than reconcile the twin social imperatives of ecological conservation and economic benefit are unlikely to be a solution. However difficult it may be to achieve such reconciliation, policies that promote just one side of the ecology/economic equation represent an avoidance of the underlying issues, however virtuous they may seem.

From this some general characteristics of an improved approach to marine conservation emerge in which a whole ecosystem approach is applied to 100% of the marine realm, and within this MPAs are used for their original purpose: the strict protection of representative and rare areas. This goal, somewhat analogous to GES for the whole sea, if formulated in such a way as to preclude gaming, implies a more strategic approach in which degraded environments cannot contribute to the ostensible achievement of environmental targets, and in which risks are prioritised in such a way that the highest receive most resources and attention.

Such a global 100% GES-type policy would have the additional advantage of forcing real effort into the reconciliation of ecological and economic imperatives so that the intrinsic limits or ecological 'carrying capacity' for human use is not exceeded. It could also provide a single framework for cooperation, mutual recognition and common cause between the currently somewhat estranged disciplines and practices of fisheries protection and marine conservation. It would also be more aligned to the general globally adopted principle of sustainable development (Humphreys and Herbert 2018).

Such a shift in global direction, in so far as it recognises a continuing role for MPAs, might not be impossible for the CBD to achieve. Furthermore, by demanding assessment based on ecological function rather than arbitrary targets and proxy measures, it would at least reduce the opportunities available to disingenuous governments who find MPAs a particularly productive arena for gaming and grandstanding.

While this could be a step towards more effective marine conservation policy and practice, progress will continue to be constrained without more fundamental economic system change as recommended by Das Gupta (2021) in the context of the climate crisis. In any event, unless the current poorly formulated, fragmented, ineffectively regulated and piecemeal approach is improved, marine conservation will remain to some extent a public relations exercise; and the underlying economic interests and practices of environmental degradation will remain largely unrestrained.

Acknowledgement

This chapter draws significantly on three earlier published contributions in which can be found elaborations of the arguments and more complete sources: Humphreys and Clark (2020); Humphreys and Hall (2021); Humphreys (2021). The co-authors and publishers of these earlier works are gratefully acknowledged. The opinions expressed above are only those of the author.

References

Benyon, R. (2020) The Benyon review into highly protected marine areas. Commissioned by the British Government.

Bowles, F. (2020) Nitrogen pollution in coastal Marine Protected Areas: a river catchment partnership to plan and deliver targets in a UK estuarine Special Protection Area. In J. Humphreys and R.W.E. Clark (eds) *Marine Protected Areas: Science, Policy and Management*. Amsterdam: Elsevier. https://doi.org/10.1016/B978-0-08-102698-4.00038-1

CBD (2020) Global Diversity Outlook 5 Convention on Biological Diversity Secretariat, Montreal. https://www.cbd.int/gbo/gbo5/publication/gbo-5-en.pdf

Dasgupta, P. (2021) *The Economics of Biodiversity*. The Dasgupta Review. London: HM Treasury.

Defra (2019a) England's marine life protected with blue belt expansion. Press release, Gov.UK, 31 May.

Defra (2019b) Britannia protects the waves. Press release Gov.UK, 26 August.

Defra (2019c) Marine Strategy part one: UK updated assessment and good environmental status, October.

Defra (2019d) UK creates global alliance to help protect the world's oceans. Press release, Gov.UK, 24 September.

EA (2014) Protection of freshwater and migratory species. Letter from DEFRA to IFCA chief executive officers, 13 May.

EA (2015) Review of protection measures for Atlantic salmon and sea trout in inshore waters environment. K. Sumner, Evidence Directorate, October.

EA (2016) Risks posed to migratory salmonid fish species by sea fish netting in Poole and Christchurch Harbours. EA South West, December.

EA (2017) Environment Agency evidence for nearshore netting. Letter from EA to Southern IFCA, 10 February.

EA (2020) Environment Agency catchment data explorer: South West; Dorset; Poole Harbour rivers. Updated 17 September 2020.

Humphreys, J. (2021) Marine Protected Areas: one step forward, four steps back. *The Marine Biologist* 17: 15–16, Special Edition: The UN Decade of Ocean Science for Sustainable Development.

Humphreys, J. and Clark, R.W.E. (2020) A critical history of marine protected areas. In J. Humphreys and R.W.E. Clark (eds) *Marine Protected Areas: Science, Policy and Management*. Amsterdam: Elsevier. https://doi.org/10.1016/B978-0-08-102698-4.00001-0

Humphreys, J. and Hall, A. (2022) Environmental regulation in an industrialised estuary. In J. Humphreys and A. Hall (eds) *Harbour Ecology: Environment and Development in Poole Harbour*. Pelagic Publishing. In Press.

Humphreys, J. and Herbert, R.J.H. (2018) Marine protected areas: science policy and management. *Estuarine, Coastal and Shelf Science* 215: 215–18. https://doi.org/10.1016/j.ecss.2018.10.014

Humphreys, J. and May, V. (eds) (2005) *The Ecology of Poole Harbour*. Amsterdam: Elsevier Proceedings in Marine Science 7.

JNCC (2019). Sixth national report to the United Nations Convention on Biological Diversity: United Kingdom of Great Britain and Northern Ireland. Overview of the UK assessments of progress for the Aichi targets. JNCC, Peterborough.

Nicola, G.G., Elvira, B., Jonsson, B., Ayllon, D. and Almodóvar, A. (2018) Local and global climatic drivers of Atlantic salmon decline in Southern Europe. *Fisheries Research* 198: 78–85. https://doi.org/10.1016/j.fishres.2017.10.012

Roff, J. and Zacharias, M. (2001) *Marine Conservation Ecology*. London: Earthscan.

Wardlaw, J. (2005) Water quality and pollution monitoring in Poole Harbour. In J. Humphreys and V. May (eds) (2005) *The Ecology of Poole Harbour*, Proceedings in Marine Science 7. Amsterdam: Elsevier.

WWF (2001) The status of Wild Atlantic salmon. A river by river assessment. May. https://wwfeu.awsassets.panda.org/downloads/salmon2.pdf

Lessons from the Past Half-Century: Challenges, Opportunities and Priorities for Future Estuarine, Coastal and Marine Management

MICHAEL ELLIOTT and ALAN WHITFIELD

Abstract

Estuaries and coasts worldwide continue to face many environmental challenges that need to be addressed by environmental managers and their science support; these challenges have developed in recent centuries and show no signs of abating. Despite that, there is now half a century of advances in science and management and an increasing capacity to tackle complex problems using more sophisticated methods. However, by adopting a more holistic approach to environmental management, and techniques such as ecoengineering and nature-based solutions, we have opportunities to enhance the natural features and at the same time deliver the ecosystem services and the resulting goods and benefits on which society relies. Tackling these issues will remain a priority for estuarine and coastal managers but also needs collaboration amongst natural and social scientists. In particular, natural scientists need to develop an understanding of how these changes impact not only species distributions and physiology but also ecosystem functioning. Collaboration with social scientists to support science-based governance issues that force management to address societal wishes and determine what is possible in the coming decades remains a high priority. These aspects are illustrated using a southern hemisphere case study of the St Lucia estuarine system, which has been subjected to extreme global changes over the past half-century, and a northern hemisphere case study, of the Humber Estuary, UK, which shows how an increasing whole-estuary management approach over the past half-century can yield both natural and societal benefits.

Keywords: ecological functioning, ecoengineering, science, socio-ecological system

Correspondence: Mike.Elliott@hull.ac.uk

Introduction

Estuaries and coasts worldwide continue to be affected by the 'triple whammy' of increased industrialisation and urbanisation, increased use of resources (e.g. space, water, energy and foodstuffs) and a decreased resistance to negative climate change effects and other major stressors (Hewitt, Ellis and Thrush 2016; Al-Naswari, Hamylton and Jones 2018;

Michael Elliott and Alan Whitfield, 'Lessons from the Past Half-Century: Challenges, Opportunities and Priorities for Future Estuarine, Coastal and Marine Management' in: *Challenges in Estuarine and Coastal Science*. Pelagic Publishing (2022). © Michael Elliott and Alan Whitfield. DOI: 10.53061/PLYN4189

Defeo and Elliott 2021). Hence, the challenges facing estuarine managers, and the scientists supporting them, include: recovery from historical pollution by domestic, industrial and agricultural sources; adaptation to historical loss of area by land claim and the changed/changing shape of estuaries; accommodating continuing endogenic managed pressures from new industries and ports in coastal areas around the world, and accommodating exogenic unmanaged pressures such as climate change impacts and isostatic rebound (Elliott et al. 2014; Andersen et al. 2020). However, by adopting ecoengineering and nature-based solutions (e.g. Elliott et al. 2016), we have opportunities to ensure that we enhance the natural features (e.g. ecological structure and functioning) and at the same time deliver the ecosystem services and the resulting goods and benefits on which society relies. Tackling these will remain the priorities both for estuarine and coastal managers and also natural and social scientists.

Based on experience worldwide, this review indicates the science needed to support the prevailing governance (the plethora of policies, administrative bodies and legislative instruments) that enables management to address societal wishes, and to look ahead in terms of possibilities for estuarine management. This holistic science-based management is illustrated using two examples, firstly the St Lucia estuarine system in South Africa and secondly the Humber Estuary in the UK. The review also highlights that the responses of aquatic biota to global change are non-linear and that thresholds for many species have already been attained and are about to be exceeded, leading to loss of coastal biodiversity and decreased ecological functionality of particular systems.

Now that the Estuarine and Coastal Sciences Association is in its sixth decade, this gives the opportunity to assess the changes in estuarine, coastal and marine science and management over that period, to reflect on the fact that we know more, are better prepared and have different philosophies in treating the environment but that there are still major challenges. There is little doubt that the risks and hazards facing coasts and estuaries are increasing rapidly with each passing decade (Elliott et al. 2019), but there are also opportunities to harmonise the effects of human habitation with protecting nature and reducing the full impact of our 'footprint' on natural ecosystem functioning (Elliott, Borja and Cormier 2020a). This is our personal view: not only needing science to increase the fundamental understanding of the structure and functioning of these systems, but also the role of science for successful and sustainable management.

This chapter aims to briefly present some of the main features of each decade since the 1960s and to look forward from now and into future decades. We start from a science and management continuum, from ecological structure to ecological functioning, to ecosystem services and then, with the addition of human capital and complementary assets, to the societal goods and benefits (Fig. 16.1). Notably, the science base shows that ecosystem structure and functioning are degraded through human activities and pressures, and the ecosystem services and societal goods and benefits are decreased by the same activities and pressures. With increased management and associated governance, the ecosystem structure and functioning can recover, with ecosystem services, societal goods and benefits being enhanced (Fig. 16.1). The role of natural and social sciences is central to this, with the move to multi- and trans-disciplinary sciences being particularly important. In addition, the recent advent of translational science to ensure that our findings are used in management and governance is highlighted. In particular, the sequence of change in the scientific topics studied in estuaries and marine systems is also reflected by the types of environmental analysis and monitoring, again reflecting the change from single to multidisciplinary systems (Borja and Elliott 2021).

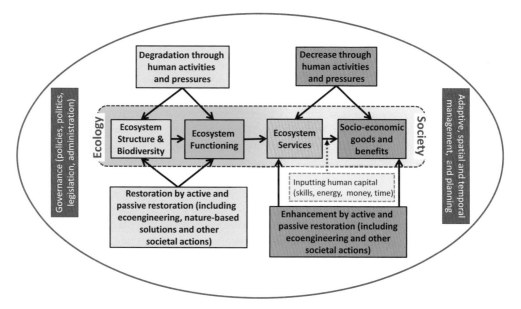

Figure 16.1 Underlying concepts of the socio-ecological system (expanded from Elliott et al. 2022; with acknowledgement to Dr Herman Hummel)

Where we have come from

The history of estuarine, coastal and marine science and management shows the trajectory starting from the decade when marine pollution issues first became of major concern, the 1960s, to the present. The text presented here has been greatly expanded from a summary table in Sheppard et al. (2021).

1960s

During this decade, the environmental quality emphasis was on chemical pollutants such as metals and pesticides, the latter not least because of the importance of Rachel Carson's seminal book *Silent Spring* and its emphasis on DDT (Carson 1962), but also with the advent of widely available machines such as atomic absorption spectrophotometers for detecting elements. There were also gross biota community assessments, understanding whole community changes, especially on the benthos and fishes, and particularly in small areas.

The advent of large tankers brought the first large marine oil spills, such as the *Torrey Canyon* in 1967 off the south-west coast of the UK; this led to the realisation that some clean-up technologies caused as much, if not more, damage than the oil itself. There were some large-scale fisheries agreements in the northern hemisphere such as the International Council for the Exploration of the Seas and the London Fisheries Convention, resulting from the realisation that even then certain fish stocks were under pressure and declining rapidly.

1970s

Biological analyses improved, with more studies being undertaken in community ecology – fundamental relationships such as the Pearson-Rosenberg model for benthic changes became widely known as the result of seminal work (Pearson and Rosenberg 1978). At an organism level, studies determining discharge limits and the toxicology of such events became part of water quality assessments. The role of estuaries as nursery areas for many species of fishes and invertebrates also came to the fore, highlighting the importance of shallow plant habitats such as seagrass beds, salt marshes and mangrove forests as fish nursery grounds. With the advent of better analytical equipment, it was easier to measure small contaminant levels of substances such as pesticides and fingerprint oils with the start of gas chromatography–mass spectrometry.

Estuarine and coastal management saw the advent of the terms Environmental Quality Objective (EQO) and Environmental Quality Standard (EQS). The UK Royal Commission on Environmental Pollution derived the three EQOs for estuaries – the ability of these systems to allow passage for migratory fish at all stages of the tide, that the channel bottom has the ability to support fish, and that the needs of all estuarine uses and users are addressed. These basic objectives are still valid, although perhaps they have become more elaborate.

There was an underlying philosophy of dealing with land-based discharges by better dilution for waste ('dilution is the solution to pollution'), especially by using marine outfall systems. The levels of materials discharged from pipes were then set according to either the EQS method (i.e. what level of discharge can a receiving body of water tolerate) or the Uniform Emission Standard (UES) method, where the same quantities and qualities are discharged irrespective of the receiving waters. The impacts of discharges were reduced as better dispersion modelling was applied to improved outfall designs. Paradoxically, this often involved separating the effluent from solid waste – and then discharging the liquid from pipes while putting the solid sewage waste into boats and dumping it offshore (the 'out of sight, out of mind' approach) as well as marine incineration of the waste (thereby polluting the air and sea while protecting the land).

This decade also saw terms such as Best Available Techniques Not Entailing Excessive Cost (BATNEEC) and Best Practical Environmental Option (BPEO) being developed – although the less politically correct version was Cheapest Available Technology Not Incurring Prosecution (CATNIP). Major threats to estuaries and restricted water bodies in developed countries were from organic enrichment and eutrophication, whereas in developing countries catchment degradation and sedimentation were major issues that persist today (Nedwell et al. 2002; Thrush et al. 2004). There was also a sectoral approach to management, with each sector (e.g. fisheries, recreation, navigation) treated independently and catchment degradation and waste disposal from the land being treated differently from dredging or waste disposal at sea.

The decade saw the advent of international agreements such as the UN Environment Programme (UNEP) Regional Seas pollution and dumping at sea conventions and a greater awareness of microbes and radioactivity as pollutants. Monitoring systems were developed, including the standardised 'Mussel Watch', bathing beach monitoring, and the advent of better legislation such as the US Clean Water Act and the UK Control of Pollution Acts (Borja and Elliott 2021). These steps were proactive in aiming to stop severe estuarine and marine pollution before it occurred.

Various biodiversity agreements and fisheries management protocols were introduced during this decade as a result of overexploitation of certain fish and invertebrate species. Marine Protected Areas and Marine Conservation Zones were declared by some governments, a trend that expanded over subsequent decades as marine and estuarine biota came under increasing exploitation pressure. The Ramsar Convention on Wetlands of International Importance especially as Waterfowl Habitat was signed by many governments in 1971, which then resulted in certain estuaries and coastal wetlands worldwide being given an elevated conservation status. The designation of Special Areas of Conservation and Special Protected Areas (for wetland birds) as the result of European Directives for habitats and species (1992) and wild birds (originally 1979) further promoted estuarine and coastal conservation.

1980s

This decade saw further studies on estuarine, coastal and marine ecological structure and functioning, and started the discussions on what was to become the biodiversity–ecosystem function debate (BEF) (Strong et al. 2015). There was an increase in both our ability to detect the presence of contaminants and monitor their biological effects. Methods for POPs (persistent organic pollutants) were derived as were methods for detecting sub-lethal biological effects on physiology and biochemistry, such as Imposex in dogwhelks and golf-ball shapes in oysters, both as the result of the vessel antifouling paints such as tri-butyl tin. Syntheses of marine environmental effects were given in the GEEP (Group of Experts on Environmental Pollution) workshops (Bayne, Clarke and Gray 1988) and the seminal and immensely foresighted work of McIntyre and Pearce (1980).

There was better onshore treatment of waste, especially in catchments, although estuaries and the coastal waters were still being regarded as 'high natural dispersion areas' with a (perceived) greater assimilative capacity to receive organic pollution than rivers and lakes. However, nutrient pollution and eutrophication continued to increase exponentially, with the growing human populations and new threats such as marine aggregate extraction and coastal erosion/flooding becoming a major problem in some localities.

While authorities in some countries had required industries and developers since the 1960s to carry out Environmental Impact Assessments, these and their resulting Environmental Statements became more detailed, formal and involving stakeholders in a legally required consultation. This was embodied in the European EIA Directive (originally 1985) and similar legislation worldwide.

Building on the previous decade, this period saw the advent of schemes to determine the overall environmental classification of estuaries and coasts such as the Scottish ADRIS classification scheme (of classes A–D, high, moderate, poor and bad), based on the level of contaminants, the status of the benthic and fish communities, and the aesthetic quality) (McLusky and Elliott 2004). This approach eventually became the EU Water Framework Directive, agreed to in the 2000s. It is ironic that the UK and Sweden helped draft this major directive, and then the UK left the EU under Brexit because some of the British public and their politicians did not like legislation 'from Brussels'!

1990s

This decade saw world summits such as the Rio UNCED (UN Convention on Environment and Development), where we were exhorted to 'think global and act local', and the Convention on Biological Diversity (CBD) – both of which brought the Ecosystem Approach into common usage and gave prominence to global marine problems, including the holed ozone layer and climate change. The Regional Seas conventions (created in the 1970s) increased their remit to give the holistic approach a boost, for example for the north-east Atlantic by merging the Paris Convention on land-based discharges and the Oslo Convention on dumping at sea (to create the Convention for the Protection of the Marine Environment of the North-East Atlantic, known as OSPAR).

The understanding of contaminants and pollutants increased with greater studies on endocrine-disrupting substances and their lethal and sublethal effects to the physiology, biochemistry, genetics, pathology, behaviour and population ecology of aquatic biota (e.g. Ribeiro et al. 2008), thus continuing from the earlier studies on Imposex. There was a large development of numerical indicators, especially for benthic communities, with diversity indicators making increasing use of packages such as PRIMER for analysing community structure. There was also an increasing awareness of the importance of non-indigenous species and their role as 'biological pollutants' (Elliott 2003). Environmental Management Systems were introduced and methods for the protection of habitats and species proposed, with integrated studies across the disciplines also beginning to gain momentum.

There were the first Quality Status Reports, such as required by the North Sea Task Force, and Integrated Coastal Zone Management led to habitat protection and more studies on restoration and recreation. There were the first offshore wind-power developments in Denmark as a precursor to the later, huge developments elsewhere in Europe. The increasing awareness of the effects of catchments and cities on estuaries required land-based nutrient controls which were successful in estuarine restoration; for example, the Thames Estuary in the UK regained its fish community.

In fisheries, there was increased single-species modelling and fisheries yield modelling but heavily exploited stocks continued to decline. In estuaries, there was an increasing use of fish assemblages, from the cellular to the community level, to provide information on the changing 'health' of these systems, and how these indicators could be used to monitor the decline or recovery of estuaries associated with anthropogenic influences (Whitfield and Elliott 2002; McLusky and Elliott 2004).

2000s

By this decade it was recognised that estuarine and coastal problems had to be controlled by a better understanding of connectivity with the catchment, thus leading to holistic River Basin Management Plans (McLusky and Elliott 2004). For example, the European WFD (Water Framework Directive) emphasised the use of indices for chemical and ecological quality, the latter using the biological quality elements phytoplankton, macroalgae, macrophytes, zoobenthos and fish; it also gave rise to the use of 'transitional waters' as a catch-all term for estuaries, fjords, rias, lagoons and so on (McLusky and Elliott 2007; Whitfield and Elliott 2011). Similar legislation in other parts of the world (e.g. the Water Reserve determinations for both the river and estuary in South Africa) introduced a holistic approach to estuarine and catchment management plans (van Niekerk et al. 2013).

This whole-ecosystem analysis considered estuaries and lagoons both as sources of material to the coast and as sinks. Catchment management monitoring, the growing threat of biological pollutants (i.e. non-native species) and declining estuary–coast connectivity, all attracted research and management attention (Borja and Elliott 2021). Marine governance and management improved and increased with the 2008 EU Marine Strategy Framework Directive and the US Oceans Act 2000. The former employs 11 descriptors to determine whether seas reach Good Environmental Status (Borja et al. 2010).

Schemes for ecosystem restoration were becoming commonplace, and monitoring increasingly detected whether integrated and multi-metric indicators were met. In fisheries, there was the advent of end-to-end modelling such as ATLANTIS, which aimed to link ecological processes, population dynamics and fisheries yields (Peck et al. 2018). On a wider scale, Particularly Sensitive Sea Areas were designated, and there was a greater understanding of the causes and consequences of the separate and joint effects of climate change. An example of site-specific effects is provided by Andersen et al. (2020) who examined multiple stressors in the Baltic/North Sea region. They found that nutrients and climate change were primary stressors across the entire region, with nutrients and physical conditions being most prevalent in estuaries and fisheries, contaminants and noise having a higher impact in offshore waters.

2010s

This decade encouraged integrated science and monitoring, although economic austerity restricted funding (Borja and Elliott 2013, 2021), which increased emphasis on rapid ways of collecting data such as remote scanning (satellite and unmanned techniques, landers, gliders, LiDAR, etc.). However, there also seemed to be an encouragement to reuse and model (and remodel) existing data rather than collect new data.

An increasing number of papers showed joined-up thinking from science to policy, legislation and administration. There were more proposals for solutions to global problems and collations of information such as the UN World Oceans Assessment I at the start of the decade and preparing for the second (II) iteration at the end (United Nations 2021). There was an increasing number of papers on micro- and mega-plastics and ocean litter, even if many of the papers merely showed the degree of contamination rather than the impact of pollution on the biota, and there were the first indications of noise and light as contaminants/pollutants (Borja and Elliott 2019). Changes in beauty and health trends led to pharmaceutical and personal care pollutants as a new class of contaminants entering estuaries and the coastal zone – hence the development of nanotoxicology, including the realisation that microbeads in such products were a further source of microplastics.

The merger of natural and social science approaches led to papers employing, or at least suggesting the use of systems analysis in marine and coastal management (Yoskowitz and Russell 2015; Elliott, Borja and Cormier 2020b). This aimed to lead to holistic governance and marine strategies, for example in Europe, to complement the Marine Strategy Framework Directive, which sought to ensure Good Environmental Status with the Maritime Spatial Planning Directive. There were also changes to marine uses, with the increased realisation that the fish stocks were being over-exploited and so there would be a need for increasing dependence on aquaculture. However, it was soon realised that using wild-caught fish for aquaculture feeds was not a sustainable approach. Overall, there was more emphasis on the economic aspects of estuaries, coasts and the sea, with several studies determining ecosystem services and societal goods and benefits (e.g. Turner and Schaafsma 2015).

Following the realisation that all problems in estuaries, coasts and seas are related to a set of natural and anthropogenic risks and hazards (Elliott et al. 2019), warranting a risk assessment and management approach, there was greater use of appropriate techniques that had long been used by industry (Cormier, Elliott and Rice 2019). However, towards the end of this decade, the same techniques were used for opportunity assessment and management (Elliott et al. 2020). Much of the emphasis then focused on environmental stabilisation, biodiversity, sustainability and marine conservation management, and the realisation that losing habitats was bad for the economy and society (human health) as well as ecology. This had been shown particularly by the 2007 tsunami in Indonesia where mangrove removal for aquaculture exacerbated the negative effects of tidal waves on coastal towns. This increased the need for a wider approach to habitat management, including the emphasis on ecohydrology as a way of linking the physical and ecological aspects (Wolanski and Elliott 2015). Habitat creation, recreation and restoration was also needed to reverse centuries of degradation, and so terms such as 'ecoengineering', 'working with nature' and 'nature-based solutions' became widely used (Elliott et al. 2016).

What we know and how we think

At the start of the 2010s, it became apparent that we could start synthesising many of the features of estuaries and looking for common themes, not least with the publication of the major 12-volume Treatise on Estuarine and Coastal Sciences (Wolanski and McLusky 2011). For example, the ideas of scale and connectivity in estuarine structure and functioning were emphasised by Elliott and Whitfield (2011) in their attempt to summarise estuarine structure and functioning in a set of paradigms (Table 16.1).

Two of the above paradigms (Table 16.1) include the effects of exogenic unmanaged pressures and endogenic managed pressures, both of which impact on the health of estuaries

Table 16.1 Estuarine paradigms (modified from Elliott and Whitfield 2011)

1	An estuary is an ecosystem in its own right but cannot function indefinitely in isolation and depends largely on other ecosystems, possibly more so than do other ecosystems.
2	As ecosystems, estuaries are more influenced by scale than any other aquatic system; their essence is in the connectivity across the various spatial and temporal scales and within the water body they are characterised by one or more ecotones.
3	Hydrogeomorphology is the key to understanding estuarine functioning but these systems are always influenced by salinity (and the resulting mixing and density/buoyancy currents) as a primary environmental driver.
4	Although estuaries behave as sources and sinks for nutrients and organic matter, in most systems allochthonous organic inputs dominate over autochthonous organic production.
5	Estuaries are physico-chemically more variable than other aquatic systems but estuarine communities are less diverse taxonomically, whereas the individuals are more physiologically adapted to environmental variability than equivalent organisms in other aquatic systems.
6	Estuaries are systems with low diversity/high biomass/high abundance and their ecological components show a diversity minimum in the oligohaline region which can be explained by the stress-subsidy concept where tolerant organisms thrive but non-tolerant organisms are absent.
7	Estuaries have more human-induced pressures than other systems and these include both exogenic unmanaged pressures and endogenic managed pressures. Consequently, their management has not only to accommodate the causes and consequences of pressures within the system but, more than other ecosystems, it has to be sensitive to the consequences of both external natural and anthropogenic influences.
8	Estuaries provide a wider variety of ecosystem services and an increased delivery of societal goods and benefits than many other ecosystems. Hence, estuaries are some of the most valuable aquatic ecosystems serving human needs, but for this to continue they require functional links with the adjoining terrestrial, freshwater and marine systems.

and the ability to manage these systems in relation to internal and external influences. The first management-related paradigm (#7) showed that, in effect, the anthropogenic influences, which are significant in coastal zones around the world, are superimposed onto already high natural fluctuations within these ecosystems, thus placing many estuaries under a large amount of additional stress. This results in a relatively low species richness but high abundance and biomass of biota when compared with the adjacent coastal marine environment (Whitfield et al. 2012). Estuarine management therefore needs to be aware of, and thus accommodate, the consequences of exogenic pressures, such as increasing sea level or reduced river flow due to climate change, with a willingness to engage and ameliorate endogenic managed pressures such as overfishing or pollution (Elliott 2011).

The second management-related paradigm (#8) states that estuaries are widely and intensively used by society for a wide range of activities such as human settlements, harbours, recreation and fishing. This paradigm also emphasises the importance of connectivity to estuaries, with the linkages between inland catchment areas and coastal marine environments being particularly important for the biota and ecological functioning of these systems (Atkins et al. 2011).

Illustrative case studies

Given these comments, it is valuable to indicate the history, lessons learned and the way ahead for two illustrative geographical examples (Boxes 16.1 and 16.2). From different hemispheres, these illustrate the main challenges facing estuaries and adjacent areas worldwide, the governance systems of their countries and the available management regimes.

Box 16.1 Case illustration: ecoengineering for natural and societal benefits – the St Lucia estuarine system, KwaZulu-Natal, South Africa

Past developments and history

The St Lucia estuarine system is a World Heritage Site and Ramsar Site of International Importance that has been well studied since the 1960s (e.g. Cyrus, Vivier and Jerling 2010; Vivier, Cyrus and Jerling 2010; Perissinotto, Stretch and Taylor 2013). Earlier system management decisions during the 1950s were made without scientific inputs and were triggered by massive sedimentation of the St Lucia Estuary caused by draining and canalisation of the Mfolozi Swamps that used to filter the silt-laden river water before it entered the estuary. The artificial creation of a separate mouth for the Mfolozi River resulted in Lake St Lucia losing its major freshwater supply and becoming extremely hypersaline for prolonged periods (Whitfield and Taylor 2009). When, towards the end of the last century, it was realised that the St Lucia Estuary should not be artificially managed as a permanently open system, the mouth was allowed to close naturally. Unfortunately, this coincided with a decade-long drought and, owing to a lack of connectivity with the Mfolozi River, resulted in the St Lucia system almost drying out (Whitfield et al. 2013). The iSimangaliso Wetland Park Management Authority then decided to reconnect the Mfolozi River to the St Lucia Estuary, despite the high sediment loads carried by the inflowing water. There followed a decadal scenario of river inflow but no marine connectivity owing to a closed estuary mouth, resulting in a gradual 'freshening' of the lake system and loss of marine biota previously characteristic of the system (Schutte, Vivier and Cyrus 2020). The management authority, based on scientific advice from a specially commissioned GEF (Global Environment Facility) Project, supported a natural mouth breaching policy. In retrospect, however, the absence of an adaptive management policy when more than a decade had elapsed with no mouth opening caused some estuary-associated marine species to disappear from the system (Whitfield 2021). In addition, flooding of floodplain agricultural land during high Mfolozi River flow periods led to protests by farmers whose livelihood was threatened owing to the closed estuary mouth phase. Furthermore, recreational and subsistence fishers in the St Lucia system complained that marine fish species had virtually disappeared as a result of the lack of marine connectivity between the estuary and sea.

Future possibilities and management

Fortunately, the iSimangaliso Wetland Park Management Authority has now strongly indicated a flexible approach towards managing the estuary mouth state, and will be applying a more adaptive management policy. This will have major benefits for aquatic connectivity between this 35,000 ha St Lucia system and the sea, promote estuarine complexity and biodiversity, increase ecological functioning and productivity, and address the societal needs of people in the region. These societal needs include important ecotourism businesses based at St Lucia Village, recreational and subsistence fishing concerns in the St Lucia Estuary and associated lakes, and reduced river flooding of agricultural lands on the Mfolozi floodplain.

Box 16.2   Case illustration: whole-estuary management for natural and societal benefits – the Humber Estuary, Eastern England, UK

Past developments and history

The estuary drains a fifth of England, covering the main industrial areas of the eighteenth and nineteenth centuries (coal mining, textile industries and steelmaking), supplemented by petrochemical and metal smelting industries in the twentieth century. It became a sink for persistent contaminants, with many studies initially focusing on the levels of contaminants, especially trace metals, and then on the biological effects of those contaminants; that is, pollution *per se*, with some of those contaminants being sequestered in the estuarine sediments unless disturbed by dredging (e.g. García-Alonso et al. 2011). Such studies covered all levels of biological organisation from the cell to the ecosystem. At the same time, over three centuries, the Humber wetlands were drained for agricultural and urban land, with approximately 50% of the estuarine wetlands being removed, thereby distorting the estuary shape and removing space to accommodate North Sea storm-surges. This change in shape in turn reduced resilience to sea-level rise and, because the remaining land was protected by dykes, led to coastal squeeze (Wolanski and Elliott 2015). The historical removal of wetlands and their resultant allochthonous and autochthonous organic matter could have had repercussions for the trophic functioning of the estuary except that this natural organic matter appeared to have been supplemented by anthropogenic organic matter from urban and industrial waste (Boyes and Elliott 2006). However, those inputs also created a dissolved oxygen sag in the upper estuary Turbidity Maximum Zone, which in turn created a barrier for fish migrations. The estuary was 'developed' to support four major ports, with the need for selective dredging and dredged material disposal being important. The local heavily eroding coastline and the low-lying areas dictated the need for flood and coastal protection in the national interest, mainly in urban or industrial areas.

Future possibilities and management

The above threats and pressures required a holistic management system involving the many statutory bodies, others with some control, and voluntary organisations where they could play a role (Lonsdale et al. 2018). The control of industries and urban discharges has been effected, either because old industries have been closed down, or because greater pollution control and treatment have been implemented. There is an increasing reliance on cost–benefit approaches to determining which flood and erosion protection measures are required, and there is a continued movement towards creating/re-creating wetlands (via managed realignment) to accommodate sea-level rise and storm surges (Elliott et al. 2016). Hence, this has led to the management emphasis over this coming decade on net-gain and nature-based solutions, especially to obtain the so-called triple-wins – for ecology, economy and human safety. However, there are also major developments proposed to economically regenerate the area, such as a large artificial lagoon along the Hull waterfront with the added benefits of an amenity and recreation area, a better transport route and flood protection for the city. Those societal benefits will be tempered by an irrevocable change in the estuary shape and functioning – hence giving an excellent example of the need to balance nature and societal consequences, a feature relevant to both examples given here (Fig. 16.2).

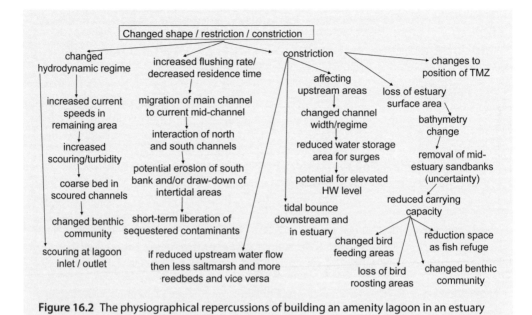

Figure 16.2 The physiographical repercussions of building an amenity lagoon in an estuary

Discussion – the future

The above text suggests that in many areas, the estuarine and coastal pollution problems have been solved, or at least we know how to solve them, and that activities and their pressures inside a management area can be controlled with the aid of strict Environmental Impact Assessments and associated plethora of legislation (e.g. Boyes and Elliott 2014). However, it is the pressures acting from outside a management area, the exogenic unmanaged pressures (Elliott 2011), which may be global, that now give the greatest challenges for management.

In recent decades, the emphasis on local threats to estuarine, coastal and marine areas has been overshadowed by global climate change (Elliott et al. 2015). For example, environmental management within an area has to tackle the consequences of climate change by increasing the resistance and resilience of areas to these changes. However, the causes actually require international action such as the 2016 Paris Agreement. Global climate impacts are diverse, ranging from sea-level rise to increased flooding and erosion, all of which are a threat to life and property. Furthermore, much environmental legislation for an area considers whether changes are within or outside the control of environmental managers, the so-called force majeure. The paradox is that the more that is known about climate change then the less a management authority can claim it is outside their control (Saul, Barnes and Elliott 2016).

The impact and interactions between climate change and anthropogenic impacts are beginning to be understood – or at least the links have now been created (Fig. 16.3). Science therefore has to respond, and is responding, to the holistic and synergistic effects of climate change (global heating, ocean acidification, hypoxia, sea-level rise, polar melting, thermal expansion), and salinity changes linked to altered run-off patterns that are related to climate oscillations (e.g. the North Atlantic Oscillation (NAO) and El Niño).

Estuarine, coastal and marine management now and in the coming decade has a lever provided by several notable international and global initiatives. These include the intergovernmental G7 Future of the Oceans and Seas Initiative, the UN Decade of Ocean

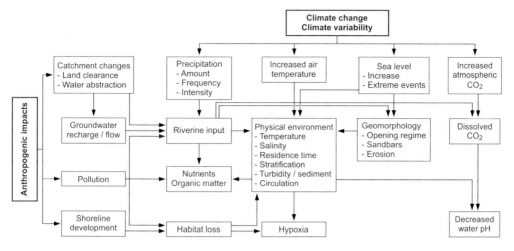

Figure 16.3 The interactions between climate change-induced variability and anthropogenic impacts on estuarine ecosystems (after Gillanders et al. 2022)

Science for Sustainability 2021–2030, the UN Decade of Ecosystem Restoration 2021–2030 (Waltham et al. 2020), responses in the Second World Oceans Assessment (United Nations 2021), and attaining the Sustainable Development Goals, especially SDG14, Life Below Water (Cormier and Elliott 2017). The above initiatives have placed these important issues in the spotlight and it is up to scientists, managers and society to respond.

Science has to respond to the quest for greater amounts of resources from estuaries, coasts and the seas. In addition, decarbonising (transition from a fossil-fuel economy) and its energy demands from the seas and estuaries via tidal and wave power projects is also increasing; for example, large-scale wind power developments such as the 1,000 turbines to be placed around the Dogger Bank (North Sea). Economics will become even more important, leading to the need for merging natural and social sciences, with Natural Capital Accounting further quantifying the ecosystem services and societal goods and benefits provided by estuarine, coastal and marine systems (Turner and Schaafsma 2015).

Other scientific hot topics include microplastic pollution, eDNA (environmental DNA) tracing, eutrophication and HABs (harmful algal blooms) – but again there is a paradox of concern in that these rely on classical skills and knowledge such as taxonomy, which is undergoing a considerable global loss of research skills. Similarly, the reduction in funding for monitoring, and a redirection of funding away from the fundamental sciences towards analysing and modelling existing data, is also of concern (Borja and Elliott 2021). As such, the overall decrease in science funding in recent decades will reduce the ability of scientists to answer important global change questions in estuaries and the adjacent coastal zone. Furthermore, in a post-COVID world, environmental research may be limited by economic constraints as funding is directed elsewhere.

At a species and habitat level, estuarine and coastal science is needed to ensure that habitat restoration projects will be successful and will continue to flourish as a means of increasing resistance and resilience to climate and other change, including the growing use of concepts such as Net Gain and biodiversity offsets. Where adverse effects of human activities cannot be mitigated, for instance the loss of habitats owing to infrastructure building, then environmental compensation will need to be used. This is based on three alternatives – to compensate the users (e.g. pay fishermen for any damage caused to fish stocks), compensate the resource (by restocking and replanting) and/or compensate natural habitats by re-creating habitats through ecoengineering (Elliott et al. 2016).

The science of habitat and species restoration also centres on the challenges of clearing up historical problems. For example, as marine oil and gas fields become exhausted, and there is lowered demand on fossil fuels through increasing environmental awareness, we need the science related to oil and gas rig decommissioning and habitat restoration (Burdon et al. 2018). There is also an important need to integrate marine pollution research and control in support of marine conservation development, planning and management. This includes studies to give a better and more defendable science and management for Cumulative Effect Assessment (Lonsdale et al. 2020).

Finally, the links between science and policy adaptive management and the creation of nature-based solutions are gaining momentum, as is the recognition that a complex ecological and societal system needs a systems analysis approach, the 'wicked problem' of social scientists (Elliott, Borja and Cormier 2020b). Manipulating systems to provide benefits to nature, while protecting society at the same time, is a much-needed approach – but scientists and managers need to be flexible and act quickly as systems change. Preconceived patterns of action from the past need to be critically examined and discarded if found wanting – new information and innovative reactions are imperative. Estuaries, coasts and the seas are highly dynamic ecosystems, and their management should also be dynamic and adaptable to changing scenarios.

References

Al-Naswari, A.K.M., Hamylton, S.M. and Jones, B.G. (2018) An assessment of anthropogenic and climatic stressors on estuaries using a spatio-temporal GIS-modelling approach for sustainability: Towamba Estuary, southeastern Australia. *Environmental Monitoring and Assessment* 190: 375. https://doi.org/10.1007/s10661-018-6720-5

Andersen, J.H., Al-Hamdani, Z., Harvey, E.T., Kallenbach, E., Murray, C. and Stock, A. (2020) Relative impacts of multiple human stressors in estuaries and coastal waters in the North Sea–Baltic Sea transition zone. *Science of The Total Environment* 704: 135316. https://doi.org/10.1016/j.scitotenv.2019.135316

Atkins, J.P., Burdon, D., Elliott, M. and Gregory, A.J. (2011) Management of the marine environment: integrating ecosystem services and societal benefits with the DPSIR framework in a systems approach. *Marine Pollution Bulletin* 62: 215–26. https://doi.org/10.1016/j.marpolbul.2010.12.012

Bayne, B.L., Clarke, K.R. and Gray, J.S. (eds) (1988) Biological effects of pollutants: results of a practical workshop. *Marine Ecology Progress Series* 46: 1–278. https://doi.org/10.3354/meps046001

Borja, Á. and Elliott, M. (2013) Marine monitoring during an economic crisis: the cure is worse than the disease. *Marine Pollution Bulletin* 68: 1–3. https://doi.org/10.1016/j.marpolbul.2013.01.041

Borja, Á. and Elliott, M. (2019) Editorial – so when will we have enough papers on microplastics and ocean litter? *Marine Pollution Bulletin* 146: 312–16. https://doi.org/10.1016/j.marpolbul.2019.05.069

Borja, Á. and Elliott, M. (2021). From an economic crisis to a pandemic crisis: the need for accurate marine monitoring data to take informed management decisions. *Advances in Marine Biology*, 89: 79–114, ISBN 978-0-12-824623-8, ISSN 0065-2881, https://doi.org/10.1016/bs.amb.2021.08.002

Borja, Á., Elliott, M., Carstensen, J., Heiskanen, A-S. and van de Bund, W. (2010) Marine management – towards an integrated implementation of the European Marine Strategy Framework and the Water Framework Directives. *Marine Pollution Bulletin* 60: 2175–86. https://doi.org/10.1016/j.marpolbul.2010.09.026

Boyes, S.J. and Elliott, M. (2006) Organic matter and nutrient inputs to the Humber Estuary, England. *Marine Pollution Bulletin* 53: 136–43. https://doi.org/10.1016/j.marpolbul.2005.09.011

Boyes, S.J. and Elliott, M. (2014) Marine Legislation – the ultimate 'horrendogram': international law, European directives and national implementation. *Marine Pollution Bulletin* 86: 39–47. https://doi.org/10.1016/j.marpolbul.2014.06.055

Burdon, D., Barnard, S., Boyes, S.J. and Elliott, M. (2018) Oil and gas infrastructure decommissioning in marine protected areas: system complexity, analysis and challenges. *Marine Pollution Bulletin* 135: 739–58. https://doi.org/10.1016/j.marpolbul.2018.07.077

Carson, R. (1962). *Silent Spring*. Boston, MA: Houghton Mifflin.

Cormier, R. and Elliott, M. (2017) SMART marine goals, targets and management – is SDG 14 operational or aspirational, is '*Life Below Water*' sinking or swimming? *Marine Pollution Bulletin* 123: 28–33. https://doi.org/10.1016/j.marpolbul.2017.07.060

Cormier, R., Elliott, M. and Rice, J. (2019) Putting on a bow-tie to sort out who does what and why in the complex arena of marine policy and management.

Science of the Total Environment 648: 293–305. https://doi.org/10.1016/j.scitotenv.2018.08.168

Cyrus, D.P., Vivier, L. and Jerling, H.L. (2010) Effect of hypersaline and low lake conditions on ecological functioning of St Lucia estuarine system, South Africa: an overview 2002–2008. *Estuarine, Coastal and Shelf Science* 86: 535–42. https://doi.org/10.1016/j.ecss.2009.11.015

Defeo, O. and Elliott, M. (2021) Editorial – the 'triple whammy' of coasts under threat – why we should be worried! *Marine Pollution Bulletin*, 163: 111832. https://doi.org/10.1016/j.marpolbul.2020.111832

Elliott, M. (2003) Biological pollutants and biological pollution – an increasing cause for concern. *Marine Pollution Bulletin* 46: 275–80. https://doi.org/10.1016/S0025-326X(02)00423-X

Elliott, M. (2011) Marine science and management means tackling exogenic unmanaged pressures and endogenic managed pressures – a numbered guide. *Marine Pollution Bulletin* 62: 651–5. https://doi.org/10.1016/j.marpolbul.2010.11.033

Elliott, M., Able, K.W., Blaber, S.J.M. and Whitfield, A.K. (2022) Chapter 13 Synthesis and future directions. In A.K. Whitfield, K.W. Able, S.J.M. Blaber and M. Elliott (eds) *Fish and Fisheries in Estuaries: A Global Perspective*. Oxford: John Wiley & Sons (in press).

Elliott, M., Borja, Á. and Cormier, R. (2020a) Activity-footprints, pressures-footprints and effects-footprints – walking the pathway to determining and managing human impacts in the sea. *Marine Pollution Bulletin* 155: 111201. https://doi.org/10.1016/j.marpolbul.2020.111201

Elliott, M., Borja, Á. and Cormier, R. (2020b) Managing marine resources sustainably: a proposed integrated systems analysis approach. *Ocean and Coastal Management* 197: 105315. https://doi.org/10.1016/j.ocecoaman.2020.105315

Elliott, M., Borja, Á., McQuatters-Gollop, A., Mazik, K., Birchenough, S., Andersen, J.H., Painting, S. and Peck, M. (2015) Force majeure: will climate change affect our ability to attain Good Environmental Status for marine biodiversity? *Marine Pollution Bulletin* 95: 7–27. https://doi.org/10.1016/j.marpolbul.2015.03.015

Elliott, M., Day, J.W., Ramachandran, R. and Wolanski, E. (2019) Chapter 1. A synthesis: what future for coasts, estuaries, deltas, and other transitional habitats in 2050 and beyond? In E. Wolanski, J.W. Day, M. Elliott and R. Ramachandran (eds) *Coasts and Estuaries: The Future*, pp. 1–28. Amsterdam: Elsevier. https://doi.org/10.1016/B978-0-12-814003-1.00001-0

Elliott, M., Franco, A., Smyth, K. and Cormier, R. (2020c). Industry and policy-driven conceptual frameworks of climate change impacts on aquaculture and fisheries. Deliverable 5.1, CERES project, report. Available at: https://www.ceresproject.eu (Accessed: 8 September 2021).

Elliott, M., Mander, L., Mazik, K., Simenstad, C., Valesini, F., Whitfield, A. and Wolanski, E. (2016) Ecoengineering with ecohydrology: successes and failures in estuarine restoration. *Estuarine, Coastal and Shelf Science* 176: 12–35. https://doi.org/10.1016/j.ecss.2016.04.003

Elliott, M. and Whitfield, A.K. (2011) Challenging paradigms in estuarine ecology and management. *Estuarine, Coastal and Shelf Science* 94: 306–14. https://doi.org/10.1016/j.ecss.2011.06.016

García-Alonso, J., Greenway, G.M., Munshi, A., Gómez, J.C., Mazik, K., Knight, A.W., Hardege, J.D. and Elliott, M. (2011) Biological responses to contaminants in the Humber Estuary: disentangling complex relationships. *Marine Environmental Research* 71: 295–303. https://doi.org/10.1016/j.marenvres.2011.02.004

Gillanders, B.M., McMillan, M.N., Reis-Santos, P., Baumgartner, L.J., Brown, L.R., Conallin, J., Feyrer, F.V., Henriques, S., James, N.C., Jaureguizar, A.J., Pessanha, A.L.M., Vasconcelos, R.P., Vu, A.V., Walther, B. and Wibowo, A. (2022) Chapter 7. Climate change and fishes in estuaries. In A.K. Whitfield, K.W. Able, S.J.M. Blaber and M. Elliott (eds) *Fish and Fisheries in Estuaries: A Global Perspective*. Oxford: John Wiley & Sons (in press).

Hewitt, J.E., Ellis, J.I. and Thrush, S.F. (2016) Multiple stressors, non-linear effects and the influence of climate change impacts on marine coastal systems. *Global Change Biology* 22: 2665–75. https://doi.org/10.1111/gcb.13176

Lonsdale, J., Nicholson, R., Weston, K., Elliott, M., Birchenough, A. and Sühring, R. (2018) A user's guide to coping with estuarine management bureaucracy: an Estuarine Planning Support System (EPSS) tool. *Marine Pollution Bulletin* 127: 463–77. https://doi.org/10.1016/j.marpolbul.2017.12.032

Lonsdale, J-A., Nicholson, R., Judd, A., Elliott, M. and Clarke, C. (2020) A novel approach for cumulative impacts assessment for marine spatial planning. *Environmental Science and Policy* 106: 125–35. https://doi.org/10.1016/j.envsci.2020.01.011

McIntyre, A.D. and Pearce, J.B. (eds) (1980) Biological effects of marine pollution and the problems of monitoring. Rapports et Proces-Verbaux des Reunions No. 179, Conseil International pour l'Exploration de la Mer, Copenhagen.

McLusky, D.S. and Elliott, M. (2004) *The Estuarine Ecosystem: Ecology, Threats and Management*, 3rd edn. Oxford: Oxford University Press. https://doi.org/10.1093/acprof:oso/9780198525080.001.0001

McLusky, D.S. and Elliott, M. (2007) Transitional Waters: a new approach, semantics or just muddying the waters? *Estuarine, Coastal and Shelf Science* 71: 359–63. https://doi.org/10.1016/j.ecss.2006.08.025

Nedwell, D.B., Dong, L.F., Sage, A. and Underwood, G.J.C. (2002) Variations of the nutrients loads to the mainland UK estuaries: correlation with catchment areas, urbanization and coastal eutrophication. *Estuarine, Coastal and Shelf Science* 54: 951–70. https://doi.org/10.1006/ecss.2001.0867

Pearson, T.H. and Rosenberg, R. (1978) Macrobenthic succession in relation to organic enrichment and pollution of the marine environment. *Oceanography and Marine Biology: An Annual Review* 6: 229–31.

Peck, M.A., Arvanitidis, C., Butenschon, M., Canu, D.M., Chatzinikolaou, E., Cucco, A., Domenici, P., Fernandes, J.A., Gasche, L., Huebert, K.B., Hufnagl, M., Jones, M.C., Kempf, A., Keyl, F., Maar, M., Mahévas, S., Marchal, P., Nicolas, D., Pinnegar, J.K., Rivot, E., Rochette, S., Sell, A.F., Sinerchia, M., Solidoro, C., Somerfield, P.J., Teal, L.R., Travers-Trolet, M. and van de Wolfshaar, K.E. (2018) Projecting changes in the distribution and productivity of living marine resources: a critical review of the suite of modelling approaches used in the large European project VECTORS. *Estuarine, Coastal and Shelf Science* 201: 40–55. https://doi.org/10.1016/j.ecss.2016.05.019

Perissinotto, R., Stretch, D.D. and Taylor, R.H. (2013) *Ecology and Conservation of Estuarine Ecosystems: Lake St Lucia as a global model.* Cambridge: Cambridge University Press. https://doi.org/10.1017/CBO9781139095723

Ribeiro, R., Pardal, M.A., Martino, F., Margalho, R., Tiritan, M.E., Rocha, E. and Rocha, M.J. (2008) Distribution of endocrine disruptors in the Mondego River estuary, Portugal. *Environmental Monitoring and Assessment* 149: 183–93. https://doi.org/10.1007/s10661-008-0192-y

Saul, R., Barnes, R. and Elliott, M. (2016). Is climate change an unforeseen, irresistible and external factor – a *force majeure* in marine environmental law? *Marine Pollution Bulletin* 113: 25–35. https://doi.org/10.1016/j.marpolbul.2016.06.074

Schutte, Q., Vivier, L. and Cyrus, D.P. (2020) Changes in the fish community of the St Lucia estuarine system (South Africa) following Cyclone Gamede, an episodic cyclonic event. *Estuarine, Coastal and Shelf Science* 243: 106855. https://doi.org/10.1016/j.ecss.2020.106855

Strong, J.A., Andonegi, E., Bizsel, K.C., Danovaro, R., Elliott, M., Franco, A., Garces, E., Little, S., Mazik, K., Moncheva, S., Papadopoulou, N., Patrício, J., Queirós, A.M., Smith, C., Stefanova, K. and Solaun, O. (2015) Marine biodiversity and ecosystem function relationships: the potential for practical monitoring applications. *Estuarine, Coastal and Shelf Science* 161: 46–64. https://doi.org/10.1016/j.ecss.2015.04.008

Thrush, S.F., Hewitt, J.E., Cummings, V.J., Ellis, J.I., Hatton, C., Lohrer, A. and Norkko, A. (2004) Muddy waters: elevating sediment input to coastal and estuarine habitats. *Frontiers in Ecology and the Environment* 2: 299–306. https://doi.org/10.1890/1540-9295(2004)002[0299:MWESIT]2.0.CO;2

Turner, R.K. and Schaafsma, M. (eds) (2015) *Coastal Zones Ecosystem Services: From Science to Values and Decision Making.* Cham: Springer Ecological Economic Series. https://doi.org/10.1007/978-3-319-17214-9

United Nations (2021) *World Oceans Assessment II* (2 Volumes). New York: United Nations. Available at: https://www.un.org/regularprocess/woa2launch (Accessed: 8 September 2021).

van Niekerk, L., Adams, J.B., Bate, G.C., Forbes, A.T., Forbes, N.T., Huizinga, P., Lamberth, S.J., MacKay, C.F., Petersen, C., Taljaard, S., Weerts, S.P., Whitfield, A.K. and Wooldridge, T.H. (2013) Country-wide assessment of estuary health: an approach for integrating pressures and ecosystem response in a data limited environment. *Estuarine, Coastal and Shelf Science* 130: 239–51. https://doi.org/10.1016/j.ecss.2013.05.006

Vivier, L., Cyrus, D.P. and Jerling, H.L. (2010) Fish community structure of the St Lucia estuarine system under prolonged drought conditions and its potential for recovery after mouth breaching. *Estuarine, Coastal and Shelf Science* 86: 568–79. https://doi.org/10.1016/j.ecss.2009.11.012

Waltham, N.J., Elliott, M., Lee, S.Y., Lovelock, C., Duarte, C.M., Buelow, C., Simenstad, C., Nagelkerken, I., Claassens, L., Wen, C.K.-C., Barletta, M., Connolly, R.M., Gillies, C., Mitsch, W.J., Ogburn, M.B., Purandare, J., Possingham, H. and Sheaves, M. (2020) UN Decade on Ecosystem Restoration 2021–2030 – What chance for success in restoring coastal ecosystems? *Frontiers in Marine Science* 7: 71. https://doi.org/10.3389/fmars.2020.00071

Whitfield, A.K. and Taylor, R. (2009) A review of the importance of freshwater inflow to the future conservation of Lake St Lucia. *Aquatic Conservation: Marine and Freshwater Ecosystems* 19: 838–48. https://doi.org/10.1002/aqc.1061

Whitfield, A.K. (2021) When the flathead mullet left St Lucia. *African Journal of Marine Science* 43 (2): 1–9. https://doi.org/10.2989/1814232X.2021.1927179

Whitfield, A.K. and Elliott, M. (2002) Fishes as indicators of environmental and ecological changes within estuaries – a review of progress and some suggestions for the future. *Journal of Fish Biology* 61 (Suppl. A): 229–50. https://doi.org/10.1111/j.1095-8649.2002.tb01773.x

Whitfield, A.K., Bate, G.C., Forbes, T. and Taylor, R.H. (2013) Relinkage of the Mfolozi River to the St Lucia estuarine system – urgent imperative for the long-term management of a Ramsar and World Heritage Site. *Aquatic Ecosystem Health and Management* 16: 104–10. https://doi.org/10.1080/14634988.2013.759081

Whitfield, A.K., Elliott, M., Basset, A., Blaber, S.J.M. and West, R.J. (2012) Paradigms in estuarine ecology – a review of the Remane diagram with a suggested revised model for estuaries. *Estuarine, Coastal and Shelf Science* 97: 78–90. https://doi.org/10.1016/j.ecss.2011.11.026

Whitfield, A.K. and Elliott, M. (2011) Chapter 1.07: Ecosystem and Biotic Classifications of Estuaries and Coasts. In E. Wolanski and D.S. McLusky (eds) *Treatise on Estuarine & Coastal Science*, Volume 1, Classification of estuarine and nearshore coastal ecosystems, ed. C. Simenstad and T. Yanagi, pp. 99–124. Amsterdam: Elsevier. https://doi.org/10.1016/B978-0-12-374711-2.00108-X

Wolanski, E. and Elliott, M. (2015) *Estuarine Ecohydrology: An Introduction*. Amsterdam: Elsevier. https://doi.org/10.1016/B978-0-444-63398-9.00001-5

Wolanski, E. and McLusky, D.S. (eds) (2011) *Treatise on Estuarine and Coastal Science* (12 volumes). Amsterdam: Elsevier.

Yoskowitz, D. and Russell, M. (2015) Human dimensions of our estuaries and coasts. *Estuaries and Coasts* 38: 1–8. https://doi.org/10.1007/s12237-014-9926-y

CONCLUSIONS

Trajectories and Challenges in Estuarine and Coastal Science

JOHN HUMPHREYS and SALLY LITTLE

Abstract

The chapters in this 50th anniversary volume collectively reflect many of today's challenges in estuarine and coastal science. Comparison of these with the inaugural volume of the Estuarine and Coastal Sciences Association (ECSA) provides a good indication of the development of the field in the last half-century. Here we reflect on this contrast, along with some wider literature, to identify significant trajectories that in turn provide insights into the challenges ahead. We discuss how the balance of study has shifted from the characterisation of coastal environments and species to a major and necessary preoccupation with anthropogenic impacts, along the way making important contributions to wider fields such as functional ecology. We consider how developments in technology have enabled new methods such as remote data collection and an explosion in the use and sophistication of mathematical modelling, the latter greatly enhancing our understanding of ecosystems through the simulation of both physical and biological processes. Whereas such developments may be seen as part of the natural maturation of a relatively new multidisciplinary subject, recognition of estuaries and coasts as an epicentre of climate change impacts, adaptation and mitigation could not have been anticipated 50 years ago. While all the chapters in this book reference climate change, none did in the inaugural volume. This wholly new trajectory has resulted in profound changes in environmental science and policy, and it is clear that climate change will continue to be a pre-eminent influence on our subject. We conclude by highlighting that despite the many challenges for estuarine and coastal science, we can be confident that the coming decades will see continuing advances in our understanding of these complex environments and of the necessary environmental management solutions to anthropogenic impacts. However, what is less certain is the extent to which governments and their environmental and coastal policies will deliver the necessary outcomes.

Correspondence: jhc@jhc.co

Trajectories

The contents of the association's 1972 inaugural volume (see Preface, Table P1), when compared with those of this volume, is indicative of considerable development in the subject over the last half-century. This clear contrast, set in the wider contexts of environmental sciences, legislative priorities and political movements, allows us to discern some major

John Humphreys and Sally Little, 'Trajectories and Challenges in Estuarine and Coastal Science' in: *Challenges in Estuarine and Coastal Science*. Pelagic Publishing (2022). © John Humphreys and Sally Little. DOI: 10.53061/ IWAO4211 changes ranging from specific preoccupations within the subject, to the scientific methods employed. For brevity, where we below make reference to a chapter in this volume, we cite only the name of the author or first two authors without date and in brackets.

Some aspects of estuarine and coastal science are common to both 1971 and 2021. The subject, then as now, relates to 'complex environments existing at the borders of the sea, rivers and land' (Barnes and Green 1972, Preface). Indeed, complexity, as a consequence of the wide range of varying and interacting influences, is paradigmatic of estuarine environments (Elliott and Whitfield), and creates many scientific challenges, some of which we touch on below.

The editors of the inaugural volume of what is now ECSA also emphasised the interdisciplinary nature of the subject, and positioned the new association as a basis for better coordination and communication between specialists of different disciplines (Barnes and Green 1972). While it is clear that ECSA has achieved these ambitions (see Foreword), 50 years on, the world is a very different place.

The most conspicuous distinction between this anniversary volume and the original inaugural equivalent is the extent to which anthropogenic climate change is influencing the subject. Whereas the consequences of climate change are referenced in all the chapters of this book, the subject did not feature at all in the 1972 volume. In examining the future implications of this, it is worth noting that in the 1970s the modern environmental movement was only beginning to achieve momentum (Bramwell 1989). Moreover, although the 1972 UN Conference on the Human Environment marked the first time the international community had come together to address global environmental issues, climate change did not figure significantly. At that time, even among scientists, many more papers weighed the relative likelihood of global cooling, as did unequivocally predict warming on the basis of greenhouse gas emissions (Weart 2008).

The first 50 years of the association can therefore be understood as a period of profound changes in terms of both environmental science and policy, whose many consequences for the multidisciplinary subject of estuarine and coastal science can be clearly seen. However, changes as a consequence of the general acceptance of an incipient, and for many across the world already manifest, 'climate crisis' have been superimposed on a young multidisciplinary subject itself developing.

In the context of a burgeoning public interest in anthropogenic environmental degradation, Krebs (1978) commented that until the 1960s, ecology was not regarded as an important subject. The inauguration of the association coincided with a period of significant progress in general ecological theory during which a more descriptive ecology of distribution, abundance and habitat assemblages was replaced by a functional emphasis. This trend, from the marine perspective, a 'second era' (Kaiser et al. 2005) commencing in the late 1960s/early 1970s, provided a more profound understanding of ecosystem functioning through systematic studies of competition, population dynamics, trophic relations and energy flows, to which our subject significantly contributed. In this respect, the Ythan represents a classic estuarine site as exemplified by Milne and Dunnet (Preface, Table 1) in the inaugural volume, which provides a first synthesis of a quantitative research programme that commenced in 1964.

Moreover, for the association, an ostensibly exclusive emphasis on the biology of estuarine and brackish waters was quicky recognised as inconsistent with its multidisciplinary purpose. Consequently, while the Prologue of the 1972 book noted the foundation of the Estuarine and Brackish-Water Biological Association, the Summarising Review acknowledged the deletion of the word 'Biological' in favour of 'Sciences' (see also Wilkinson and McLusky). This change was important since, whereas work on species, habitats and physical aspects of estuaries had been conducted long before the 1960s, recognition of the need for more focused and integrated attention was developing. This emerged from a combination of pure and applied motivations: a growing interest in an environment that was increasingly understood as presenting its inhabitants with 'peculiar and difficult conditions' (Tait 1968: 203), combined with the view that, in the context of pollution, it had become 'necessary to try to visualise an estuary as a whole, not just as a collection of plants, animals, water movements, or sediment movements, but rather as a completed jigsaw puzzle' (Perkins 1974: Preface).

Not surprisingly, a number of the founders and original members of the association were directly involved in increasing the profile of estuaries as subjects for study, not only through the creation of the association, but also through their own research and influential early books on the subject (e.g. Green 1968; McLusky 1971; Dyer 1973; Barnes 1974). These in turn inspired others to introduce their students to estuaries as particularly effective sites for learning fundamental ecological concepts (Humphreys 1981, 1985). However, in the subsequent development of the subject, the balance of effort has shifted from the inherently interesting characteristics of estuaries and their inhabitants, to environmental degradation and climate impacts as a result of human activities as the prime motivators of an increasingly applied estuarine science. This in turn has led to the concept of ecosystem goods and services, assigning monetary values to the benefits provided by the environment for humans. While perceived by some as an 'instrumental or utilitarian view of nature', it has become increasingly important to be able to identify and value the goods and services provided by the coastal zone (Paprotny et al. 2021: 2). Estuaries in particular are recognised as one of the most valuable aquatic ecosystems serving human needs, owing to the wide range of ecosystem services they provide (Elliott and Whitfield). This concentrates research and conservation attention, and also brings into sharp focus the impact of their degradation and loss for governments and politicians who have a more 'economic focus', but on whom their fate mainly depends.

A growing scientific understanding of the importance of connectivity of aquatic environments informed the development of this area but also the wider subject. Perhaps inevitably the appreciation of this has over time widened the scope of ECSA and its linked journals beyond the strictly estuarine. If we choose to define coastal waters as that part of the ocean affected by its proximity to the land (after Clark 1996), we must acknowledge that they reach from the transitional waters of upper estuaries (Little et al.) to the edge of the continental shelf, while also recognising that coastal systems are susceptible to more distant influences, from agricultural practices in extensive river catchments (Humphreys) to deep ocean and atmospheric circulation (Burningham et al.). Conversely, it is also now better understood how distant but connected places are themselves influenced not only by estuarine pollution, but also natural processes such as infaunal life histories, not least as 'nurseries' supporting juvenile fish by virtue of food availability, lower predator pressure and generally good conditions for rapid growth (Cabral).

Nevertheless, from the marine standpoint, proximity to land remains a useful rationale, both scientifically and in the policy context. In the first place, coastal waters, thus defined, tend to be areas of natural gradients and greater volatility, whether this be related to salinity, tidal streams, storm impacts or the light and temperature variations of the littoral and neritic shelf environment. Secondly, in policy terms, the UN Law of the Sea now provides sovereign states with exclusive natural resource rights up to 200 nm, these Exclusive Economic Zones often roughly coinciding to the limits of continental shelves. So, whereas the high seas are relatively stable, homogeneous, held to be in common and subject only to regional and international agreements, coastal waters in contrast are volatile and more heavily impacted. They also suffer from a plethora of legal environmental frameworks, varying not just by state but also within states, where terrestrial, freshwater and marine regulatory regimes overlap and interact in complex ways (Elliott and Whitfield; Humphreys). Such natural and anthropic features of coastal waters have provided the raison d'être for the association since its establishment.

Progress in science is often as much technically driven as it is conceptually, the use of remote data collection for habitat monitoring and management (Walker et al.) being a case in point. But undoubtedly the most significant change in this respect has been the exponential increase in computing power and availability that, with the subject now driven by environmental policy imperatives, has put modelling at the forefront of many studies. So, whereas in the late 1970s, Krebs (1978) was inclined to comment only on the potential of mathematical modelling, such approaches are today crucially important for understanding many aspects of environmental functioning, both physical (Mitchell and Uncles) and biological (Goss-Custard and Stillman).

But what do such trajectories in estuarine and coastal sciences tell us about future challenges?

In the first-place, climate change will surely continue to be a pre-eminent influence on the subject. While the general issue of global greenhouse gas emission targets will remain beyond the direct scope of our subject, the scientific evidence that change is already inevitable establishes the coastal zone as an 'epicentre' of impacts, adaptation and mitigation (IPCC 2014a; Oppenheimer et al. 2019). In the last decade, an abundance of evidence has accumulated indicating that, even in low emission scenarios, global mean temperature will continue to rise over the next half-century (IPCC 2014a). In this context, 'Coastal systems are particularly sensitive to three key drivers related to climate change: sea level, ocean temperature and ocean acidity' (IPCC 2014b: 364).

Challenges

In fact, coastal waters have long been an epicentre of anthropogenic impacts: the association's inaugural volume, for example, considers nitrates and sewage pollution as well as inshore fisheries. Continuing population and economic growth in the context of a globalised economic system would in any event have increased such risks for coastal waters. But now superimposed on such 'conventional' concerns are those associated with climate change (Little et al. 2017). This combination results in amplified established risks such as coastal erosion and non-indigenous species, and new risks ranging from industrial plastic pollution (Gallagher et al.) to habitat loss through rising sea levels. Many of today's priorities, authoritatively covered in our chapters, represent challenges that can be extrapolated well into the future of the subject.

At the geomorphological level, climate change, with its associated processes such as sea-level rise (SLR) and storm events and surges, has the potential to trigger a range of changes in future decades that will impact the morphodynamics of river mouths and associated delta shorelines (Burningham et al.). Similarly, erosion on some open coasts will accelerate (Pye and Blott). Understanding the hydro- and morphodynamics associated with coastal landforms along with monitoring and modelling erosion and deposition is crucial to deal with these future challenges and evaluate adaptation measures.

Such changes as SLR interacting with our responsive (and established) coastal civil engineering will in turn affect the existence and distribution of habitats through phenomena such as 'estuarine squeeze' (Little et al.) and 'coastal squeeze' (Pontee et al.). Squeeze can be seen as a specific case of physical change measured in terms of habitat impacts. As such, monitoring and understanding processes at local level is essential. However, settled definitions of even established terms such as coastal squeeze remain elusive, a difficulty which goes beyond semantics to the achievement of consistent approaches to monitoring (Pontee et al.). Estuarine squeeze is a newly coined term. It describes situations where increases in salinity and saline incursion (driven by SLR, climate change and other human impacts to the channel and catchment) act to squeeze out upper estuarine tidal freshwater zones against hard engineering structures (e.g. weirs and dams) (Little et al.). The loss of critical habitats (and their ecological functions) through both estuarine and coastal squeeze will have important repercussions on the ability of estuaries and coasts to continue to provide key ecosystem services and the resulting goods and benefits on which society relies (Elliott and Whitfield). In 2018, the European coastal zone provided an estimated €494 billion of ecosystem services. By 2100, 4.2–5.1% of these services could be lost owing to SLR and coastal erosion (Paprotny et al. 2021).

Advances in our general understanding of such physical and ecological phenomena provides an essential base for interpretation, but as the various case studies in our chapters show, only locally focused work can provide the information necessary for the secure determination of coastal solutions. The complexity of coastal and estuarine environments, and the consequent variability between different sites, means that generalised solutions are normally not sufficient, and therefore each site must be understood as a distinct case. In this sense, all coasts are local phenomena. Consequently, the provision of information sufficient to provide a basis for adaptation constitutes a scientific challenge of considerable proportions for coastal states, not all of which will have the capacity to deliver.

This challenge of capacity would be even more acute were it not for the growing utility of mathematical modelling in the environmental sciences and coastal science in particular. Elucidation of the serious consequences of anthropogenic impacts, ranging from basic disturbance, through civil engineering and land reclamation, to warming induced SLR and non-native species dispersal, has coincided with an explosion in the use of models particularly those that simulate physical processes. Some newer modelling approaches such as machine learning and artificial intelligence are also starting to see wider use to answer questions about coastal processes (Mitchell and Uncles).

Nevertheless, the paradigmatic complexity of coasts, and consequent challenges for modelling is amply illustrated by our chapter on the impact of SLR on estuarine tidal dynamics via changes in friction, geometry (e.g., estuary depth, width, length and entrance conditions), resonance and reflection. These in turn can influence tidal prism, tidal currents, tidal asymmetry and sediment transport dynamics (Khojasteh et al.). As our authors acknowledge, feedback loops, compounding and non-linear interactions between these processes may result in significant uncertainties regarding future risk scenarios.

The emergence of earth system modelling, driven by carbon cycle priorities, has coincided with increasing attention to ecological spheres, not just in the sense of the biogeochemical cycles. Models now predict changing patterns in global vegetation, and in the coastal context the responses and even individual behaviours of animals can now be incorporated into models to predict such things as non-native species dispersal (Herbert et al.) and estuarine bird mortality, the latter already widely applied to inform coastal management (Goss-Custard and Stillman).

Notwithstanding their utility, the fundamental purpose of models is to achieve an understanding of the complex dynamics of systems that is mathematically rigorous and empirically valid. The method, which includes expressing scientific laws in mathematical terms (such as foraging theory, Goss-Custard and Stillman), developing parameters, empirical calibration and prediction testing, also represents a step forward in the advancement of the pure science of the coastal environment.

Investment in the development of models will therefore further both scientific progress and policy and management solutions. Moreover, in the coastal context, filling research gaps and modelling have the potential to alleviate a particular problem of coastal conservation; namely, the frequent resort to the precautionary principle. In scientific terms, it should not be forgotten (but frequently is) that the almost ubiquitous use of this principle of conservation is an acknowledgement of ignorance of cause-and-effect relations in the face of coastal complexity. We suggest that a major challenge for coastal science over the next 50 years will be to largely replace resort to precaution by resort to science. This level of utility, however, will in many cases require the development of increasingly sophisticated models.

In summary, environmental modelling, although now well established, is still a comparatively new technique within the scientific method. Further development of models over the coming decades will involve both refining existing components for particular purposes, such as relating them to the calculation and use of residence times in hydrodynamic models (Birocchi et al.), along with the integration of different system elements. For instance, in order to better predict species range expansions and declines on warming coasts, bioclimatic envelope models will incorporate climatic, hydrodynamic and species/habitat components (Herbert et al.) at ever increasing levels of sophistication. While models are becoming increasingly sophisticated, it is crucial that ongoing refinement and testing of their utility against empirical field data remains a key part of model development.

As with other challenges outlined above, the issue of species range change and the arrival of non-indigenous species is a long established and 'conventional' area of conservation concern. However, it also constitutes a fundamental connection between the twin planetary level crises of climate change and biodiversity loss, through both direct (temperature, Herbert et al.) and indirect (habitat loss, Little et al. and Pontee et al.) mechanisms. From an ecological perspective, the concern relates to aspects of ecological function such as trophic complexity and ecosystem resilience.

It is, for example, on the basis of temperature projections, possible to predict consequences including the regional extinction of particular coastal species and/or communities; the corresponding 'competitive release' of others; and a climate-induced homogenisation of biota from different biogeographic realms (Herbert et al.). Moreover, species distribution and abundance will depend on a complex interaction of many anthropogenic variables. For example, ocean acidification is projected to impact highly calcified species such as molluscs and corals with greater risk in areas where eutrophication is an issue (IPCC 2014b), such as in many estuaries.

Continuing upward trends in greenhouse gas emissions ensure that approaches to coastal mitigation are also receiving increasing attention. We have seen a number of cases in which adaptation is manifest in coastal protection: 'hold the line' or 'managed retreat' (Pye and Blott). In the latter case, adaptation may result in interventions that relocate particular coastal habitats. Such adaptive interventions include the creation of new intertidal wetlands. In Great Britain by 2020, such sites covered 2,841 ha, which makes them, if optimally positioned, significant in terms of mitigation through carbon storage. But here again the complexity and local variability of the coastal environment makes estimations difficult and the determination of key variables remains a future challenge (Mossman et al.). However, even with the inclusion of managed realignment sites, it is estimated that losses of intertidal saltmarsh around the UK coastline continues to exceed gains. Long-term SLR (in addition to other drivers) will play a significant role in determining the future distribution, magnitude and vulnerability of intertidal saltmarsh owing in large part to the consequences of coastal squeeze. Yet with investment, appropriate management and changes to coastal management policy, there is potential to protect, restore and create intertidal saltmarsh and thus enhance coastal carbon storage capacity (Austin et al.).

Despite the many challenges for estuarine and coastal science, the progress made over the last half-century suggests that we can be confident that the coming decades will see continuing advances in our understanding of these complex environments, and of the necessary environmental management solutions to anthropogenic impacts. What is less certain is the extent to which governments and their environmental and coastal policies will deliver the necessary outcomes. In this volume, we have two chapters on estuarine and coastal management, each of which offer a different perspective on this question: The first (Elliott and Whitfield) we suggest is more optimistic, and the second (Humphreys) less so. In a sense, both may be right.

At the level of policy formulation, and local implementation (in the sense of aspects of site management), clear progress is often apparent: ecosystem approaches to conservation, nature-based solutions, recognition of the need for co-management and the role of the social sciences (Elliott and Whitfield). Similarly, it would be wrong to suggest that estuaries are today polluted in the same way they were 50 years ago. Conversely, it would also be wrong to ignore the persistent and continuing degradation of coastal waters. Yesterday's solved problems have been more than replaced by today's new ones. Fifty years ago, untreated sewage was a visible and therefore indefensible feature of many urban estuaries, whereas today its somewhat more diluted equivalent may be less visible but is no less problematic. Pervasive new technologies and practices such as use of inorganic fertilisers create both chronic and acute ecological problems. Plastics, now ubiquitous in coastal waters, are a highly successful range of materials that have developed at a pace far outstripping the development of any system of governance aimed at managing them (Gallagher et al.).

Conclusions

The last 50 years of estuarine and coastal science has been marked by considerable progress in the development of the field. Estuarine and coastal environments became increasingly well defined, characterised and understood. But also in this process, the subject has made important contributions to general scientific advances in areas such as functional ecology and mathematical modelling. The foundation established is now providing the basis for an application of science to current and future challenges. Some of these, such as water quality and habitat loss in the context of economic development and human population growth, remain similar to those 50 years ago, while technological and economic advances have contributed to new concerns such as manufactured inorganic fertilisers and plastics.

However, also during that period, a far-reaching realisation of the profound impacts that human activity is having at planetary and global system levels has been established: climate change, biodiversity loss and their consequences. These discoveries are now appreciably changing the trajectories of all environmental science, not least those that relate to estuaries and coasts, which are recognised as an epicentre of climate impacts.

In this Anthropocene context, technological solutions and legislative improvements struggle to keep up with continuing population and economic growth in the context of a globalised economic system, and the world's governments appear to be incapable of moderating environmental degradation or delivering effective regulation. The gap between societal aspiration and environmental conditions is widening.

Arguably, comprehensive technical reviews of coastal impacts already pose more policy and management challenges than it may be possible to address, let alone fund, even for economically developed coastal states. As a consequence, technical reports (such as Berry and Brown 2021, prepared for the UK government) read like wish-lists whose frequent references to the need for more science, more action and (by implication) more money seem to create an increasing convergence of the arguably essential and the highly unlikely. Meanwhile, in the coastal zone it is now possible to estimate with high confidence that flooding and land loss is projected to displace hundreds of millions of people (IPCC 2014b).

Should, as now seems plausible, greenhouse gas mitigation efforts fall short of those necessary to limit the magnitude of average global temperature change to +2 °C, incremental approaches to adaptation, which seek to maintain the integrity of existing systems, will be insufficient and transformational approaches will be necessary (IPCC 2014b). These will involve actions that change fundamental attributes or locations. Such approaches (already manifest in managed retreat and new habitat creation projects), are likely to become increasingly normalised in estuarine and coastal management.

But if over the next half-century maintaining the status quo becomes untenable, what are the priorities? Such questions are difficult, but not asking them may exacerbate the degradation of estuaries and coastal seas, not least through dispersing resources unstrategically across many unprioritised efforts of variable efficacy. In connected coastal seas, questions of ecological conservation become particularly challenging. To what extent do conventional approaches to coastal conservation become anachronisms? Should transformational approaches to coastal defence such as managed retreat be reconceptualised for coastal ecosystems in which regional extinction is projected as inevitable? Should a functionally more resilient ecology including non-indigenous species assets sometimes trump the dogmatic conservation of purely native and otherwise impoverished native communities (see, for example, Syvret et al. (2021))? And if so, what exactly are the purposes of coastal marine protected areas in the context of planetary level change?

The next 50 years will see an acceleration of anthropogenic impacts on estuarine and coastal environments. The scientific challenge will be to investigate, understand and predict unprecedented change, but beyond that, at the interface between science and policy, to develop innovative and increasingly transformational adaptive coastal responses. At this interface, some of the conventional paradigms of coastal management and conservation may need to make way to accommodate the already manifest realities of planetary level change; and here may lie the biggest challenge of all.

Disclaimer

The views and perspectives expressed in this concluding chapter are those of the editors only and should not be taken to represent those of ECSA or the authors of other chapters in this volume.

References

Barnes, R.S.K. (1974) *Estuarine Biology*. London: Edward Arnold.

Barnes, R.S.K. and Green, J. (1972) (eds) *The Estuarine Environment*. London: Applied Science Publishers Ltd.

Berry, P. and Brown, I. (2021) National environment and assets. In R.A. Betts, A.B. Haward and K.V. Pearson (eds) The third UK climate change risk assessment technical report, prepared for the Climate Change Committee, London.

Bramwell, A. (1989) *Ecology in the 20th Century: A History*. New Haven, CT: Yale University Press.

Clark, J. (1996) *Coastal Zone Management Handbook*. New York: Lewis Publishers.

Dyer, K.R. (1973) *Estuaries. A Physical Introduction*. Chichester: John Wiley & Sons Ltd.

Green, J. (1968) *The Biology of Estuarine Animals*. London: Sidgwick & Jackson.

Humphreys, J. (1981) Estuarine ecology as project work. *Journal of Biological Education* 15 (3): 225–32. https://doi.org/10.1080/00219266.1981.9654383

Humphreys, J. (1985) Production of *Nereis diversicolor* in an upper estuarine creek. *Journal of Biological Education* 19 (2) 141–6. https://doi.org/10.1080/00219266.1985.9654711

IPCC (2014a) *Climate Change 2013: The Physical Basis. Working Group 1 Contribution to the Fifth Assessment Report of the Intergovernmental Panel on Climate Change*. Cambridge: Cambridge University Press.

IPCC (2014b) *Climate Change 2014: Impacts, Adaptation, and Vulnerability. Part A Global and Sectoral Aspects. Working Group 2 Contribution to the Fifth Assessment Report of the Intergovernmental Panel on Climate Change*. Cambridge: Cambridge University Press.

Kaiser, M.J., Attrill, M.J., Jennings, S., Thomas, D.N., Barnes, D.K.A., Brierley, A.S., Polunin, N.V.C., Raffaelli, D.G., Williams, P.J. le B. (2005) *Marine Ecology: Processes, Systems and Impacts*. Oxford: Oxford University Press.

Krebs, C.J. (1978) *Ecology. The Experimental Analysis of Distribution and Abundance*. 2nd edition. New York: Harper & Row.

Little, S., Spencer, K.L., Schuttelaars, H.M., Millward, G.E. and Elliott, M. (2017) Unbounded boundaries and shifting baselines: estuaries and coastal seas in a rapidly changing world. *Estuarine, Coastal and*

Shelf Science 198 B: 311–19. https://doi.org/10.1016/j.ecss.2017.10.010

McLusky, D.S. (1971) *Ecology of Estuaries*. London: Heinemann.

Milne, H. and Dunnet, G.M. (1972) Standing crop. Productivity and trophic relations of the fauna of the Ythan estuary. In R.S.K. Barnes and J. Green (eds) *The Estuarine Environment*. London: Applied Science Publishers Ltd: Chapter 6, pp. 86–106.

Oppenheimer, M., Glavovic, B.C., Hinkel, J., van de Wal, R., Magnan, A.K., Abd-Elgawad, A., Cai, R., Cifuentes-Jara, M., DeConto, R.M., Ghosh, T., Hay, J., Isla, F., Marzeion, B., Meyssignac, B. and Sebesvari, Z. (2019) Sea level rise and implications for low-lying islands, coasts and communities. In: *IPCC Special Report on the Ocean and Cryosphere in a Changing Climate* (H.-O. Pörtner, D.C. Roberts, V. Masson-Delmotte, P. Zhai, M. Tignor, E. Poloczanska, K. Mintenbeck, A. Alegría, M. Nicolai, A. Okem, J. Petzold, B. Rama and N.M. Weyer (eds)). Intergovernmental Panel on Climate Change.

Paprotny, D., Terefenko, P., Giza, A., Czapliński, P. and Vousdoukas, M.I. (2021) Future losses of ecosystem services due to coastal erosion in Europe. *Science of the Total Environment* 15: 760:144310. https://doi.org/10.1016/j.scitotenv.2020.144310

Perkins, E.J. (1974) *The Biology of Estuaries and Coastal Waters*. London: Academic Press.

Syvret, M., Horsfall, S., Humphreys, J., Williams, C., Woolmer, A. and Adamson, E. (2021) *The Pacific Oyster: Why we should love them*. Report of the Shellfish Association of Great Britain, London. https://doi.org/10.13140/RG.2.2.22441.72806

Tait, R.V. (1968) *Elements of Marine Ecology*. London: Butterworth.

Weart, S.R. (2008) *The Discovery of Global Warming*. Cambridge, MA: Harvard University Press. https://doi.org/10.4159/9780674417557

Index

Page numbers to figures are in *italic*; tables are in **bold**.